Orthomodular Lattices

Mathematics and Its Applications *(East European Series)*

Orthomodular Lattices

Algebraic Approach

Ladislav Beran

Charles University, Prague

D. Reidel Publishing Company

A MEMBER OF THE KLUWER ACADEMIC PUBLISHERS GROUP

Dordrecht / Boston / Lancaster

Library of Congress Cataloging in Publication Data

CIP

Beran, Ladislav
Orthomodular lattices.

(Mathematics and its applications. East European series; v.)
Bibliography: p.
Includes indexes.
1. Orthomodular lattices. I. Title. II. Series: Mathematics and its applications (D. Reidel Publishing Company). East European series; v.
QA171.5.B4 1984 512'.7 84-6855
ISBN 90-277-1715-X

Published by D. Reidel Publishing Company,
P.O. Box 17, 3300 AA Dordrecht, Holland,
in co-edition with
Academia, Publishing House of the Czechoslovak Academy of Sciences, Prague.

Sold and distributed in the U.S.A. and Canada by
Kluwer Academic Publishers, 190 Old Derby Street, Hingham, MA 02043, U. S. A.

Sold and distributed in Albania, Bulgaria, China, Czechoslovakia, Cuba, German Democratic Republic, Hungary, Mongolia, Northern Korea, Poland, Rumania, U.S.S.R., Vietnam and Yugoslavia by
Academia, Publishing House of the Czechoslovak Academy of Sciences, Prague, Czechoslovakia.

Sold and distributed in all remaining countries by
Kluwer Academic Publishers Group,
P.O. Box 322, 3300 AH Dordrecht, Holland.

Scientific Editor: Prof. RNDr. Milan Kolibiar, DrSc.
Reviewer: Doc. RNDr. Tibor Katriňák, DrSc.

Printed in Czechoslovakia

To Jana

Contents

Editor's Preface

Growing specialization and diversification have brought a host of monographs and textbooks on increasingly specialized topics. However, the "tree" of knowledge of mathematics and related fields does not grow only by putting forth new branches. It also happens, quite often in fact, that branches which were thought to be completely disparate are suddenly seen to be related.

Further, the kind and level of sophistication of mathematics applied in various sciences has changed drastically in recent years: measure theory is used (non-trivially) in regional and theoretical economics; algebraic geometry interacts with physics; the Minkowsky lemma, coding theory and the structure of water meet one another in packing and covering theory; quantum fields, crystal defects and mathematical programming profit from homotopy theory; Lie algebras are relevant to filtering; and prediction and electrical engineering can use Stein spaces. And in addition to this there are such new emerging subdisciplines as "completely integrable systems", "chaos, synergetics and large-scale order", which are almost impossible to fit into the existing classification schemes. They draw upon widely different sections of mathematics.

This program, Mathematics and Its Applications, is devoted to such (new) interrelations as exempla gratia:

- a central concept which plays an important role in several different mathematical and/or scientific specialized areas;

- new applications of the results and ideas from one area of scientific endeavor into another;
- influences which the results, problems and concepts of one field of enquiry have and have had on the development of another.

The Mathematics and Its Applications programme tries to make available a careful selection of books which fit the philosophy outlined above. With such books, which are stimulating rather than definitive, intriguing rather than encyclopaedic, we hope to contribute something towards better communication among the practitioners in diversified fields.

Because of the welth of scholarly research being undertaken in the Soviet Union, Eastern Europe, and Japan, it was decided to devote special attention to the work emanating from these particular regions.

Thus it was decided to start three regional series under the umbrella of the main MIA programme.

Preface

The present book is designed primarily to be used as a textbook by the reader who wishes to learn about those ideas and methods of the theory of orthomodular lattices which have proved most useful for some areas of modern algebra. I have tried to make the exposition as self-contained as possible without assuming any special knowledge of lattice theory. However, for an understanding of the text a certain familiarity with the elementary concepts of the theory of ordered sets is presupposed. The basic facts about lattices used in the book are listed (with corresponding references to Szász´s book [179]) in Chapter I.

Chapter II provides an introduction to the theory of orthomodular lattices. A special effort has been made to render the exposition transparent. The results presented here are used constantly in the text and exrcises.

Chapter III introduces Boolean skew lattices and a part of it clarifies their connections to orthomodularity. This is followed by a section which shows a diagram of the free orthomodular lattice generated by two generators. The final part of chapter III, on Hilbert spaces, is one about which I have had doubts. The notions studied here are of such importance that I saw I could not avoid giving at least a brief outline of the subject. Yet much of the theory is not really a part of algebra, but, instead, is a part of functional analysis. However, I felt that I had to give the reader who is not extensively trained in functional analysis at least some idea of the methods and results in this direction.

Chapter IV centers around the notion of amalgamation and contains a fairly large number of constructions of some complicated orthomodular lattices.

Chapter V deals with the structure of generalized orthomodular lattices and constitutes the material which is the basic source for the next chapter.

The reflections and the coreflections of generalized orthomodular lattices are studied in Chapter VI and are used there to derive special results on solvability.

Chapter VII develops the results concerning commutators and finitely generated orthomodular lattices, after which follows an account of identities in orthomodular lattices and in Boolean skew lattices.

The final chapter considers various important aspects of orthomodularity. Of fundamental importance in this chapter is dimension theory and the study of orthologics.

A word about the system of bookkeeping: Each chapter is divided into sections. If we quote a result, say Theorem 10, and no chapter or section is specified, then the reader is to understand that our reference is to Theorem 10 of the section in which he is reading. If a theorem within the same chapter but not within the same section is needed, the section in which it will be found is also given. Thus Theorem 3.10 refers to Theorem 10 in the third section of that chapter. A reference to another chapter is indicated by the Roman numeral corresponding to the chapter. For example: Theorem II.3.10 refers to the tenth theorem of the third section in the second chapter. The end of a proof is denoted by the symbol //.

The text of the book arose from a collection of my lectures and seminaries given at Charles University. That students who are mathematically sufficiently mature can use the book successfully has thus been demonstrated. From this experience, and from the intention to give the reader a chance to come quickly into contact with living mathematics, stems in part the introductory character of our treatment. For the same reason we have included a large number of exercises and figures in the text. Answers to all exercises are given at the end of the book.

I would like to thank M. Kolibiar and T. Katriňák for suggesting numerous improvements in the original manuscript.

Prague, Czechoslovakia

Ladislav Beran

List of Symbols

Symbol	Meaning	Page
$a \prec b$	b covers a	3
$\sup_{(P,\leqslant)} M$	Supremum of M in $(P,\leqslant)$	2
$\inf_{(P,\leqslant)} M$	Infimum of M in $(P,\leqslant)$	2
1	Greatest element	2
0	Least element	2
0	Null operator	113
o	Zero vector	95
$\underline{2}^n$	Boolean algebra with n atoms	13
$[a,b]$	Interval determined by a,b	4
$(a]$	Principal ideal generated by a	4
$[a)$	Principal filter generated by a	4
$f \circ g$	Right composite of f and g	4
gf	Left composite of f and g	4
$b/a \searrow d/c$	(The quotient b/a is perspective down onto d/c)	8
$d/c \nearrow b/a$	(The quotient d/c is perspective up onto b/a)	8
$q/p \sim s/r$	The quotients q/p, s/r are transposed	8
$b/a \approx d/c$	b/a and d/c are projective	9
$b/a \searrow_w d/c$	b/a is weakly perspective down into d/c	200
$b/a \nearrow^w d/c$	b/a is weakly perspective up into d/c	200
$b/a \sim_w d/c$	b/a is weakly perspective into d/c	200
$q/p \approx_w s/r$	q/p is weakly projective into s/r	200

$`N_5$	Five-element nonmodular lattice (pentagon)	12
$`M_5$	Five-element modular nondistributive lattice (diamond)	13
$`M_{On}$	2(n+1)-element orthomodular lattice of length 2	35
$`L_1 \times \dots \times `L_n$	Direct product of $`L_1, \dots, `L_n$	9
$`K_1 \oplus \dots \oplus `K_n$	Direct sum of $`K_1, \dots, `K_n$	10
aCb	a commutes with b	37,306
cDd	c dually commutes with d	38,306
cMd	c,d orthogonally commute; c,d compatible	38,306
$\langle M \rangle$	Subalgebra generated by M	18
`A/D	Quotient algebra	21
`L/D	Quotient lattice	21
`L/I	Quotient lattice	24
$a \perp b$	Orthogonal elements	36,327
$(a,b) \perp (c,d)$	(The ordered pairs (a,b),(c,d) are orthogonal)	257
$\underline{\text{med}}\ (a,b,c)$	Lower median of a,b,c	66
$\overline{\text{med}}\ (a,b,c)$	Upper median of a,b,c	66
com (s,t)	(Upper) commutator of s,t	86
$\underline{\text{com}}\ (s,t)$	Lower commutator of s,t	86
$\overline{\text{com}}\ (s,t)$	Upper commutator of s,t	86
$\overline{\text{com}}\ (a_1,\dots,a_n)$	Upper commutator of $a_1,\dots,a_n$	235
$L^{\bullet}$	Boolean skew lattice	70
F_2	Free orthomodular lattice with two generators	86
$\langle x,y \rangle$	Scalar product of x,y	94
c^{*}	Complex conjugate of c	94

Chapter I

INTRODUCTION

1. A binary relation $\leqslant$ on a nonvoid set P is said to be a partial ordering of P if and only if it is reflexive, antisymmetric and transitive. The ordered pair $`P = (P, \leqslant)$ is then called a partially ordered set or a poset. By definition, $`P$ is a poset if and only if the following conditions hold for every $a,b,c \in P$:

(1) $a \leqslant a$;

(2) $(a \leqslant b \;\&\; b \leqslant a) \Rightarrow a = b$;

(3) $(a \leqslant b \;\&\; b \leqslant c) \Rightarrow a \leqslant c$.

The set P is called the base set of $`P$. For simplicity and if no ambiguity results, we sometimes speak about "the partially ordered set P".

Two elements a,b of a poset are said to be comparable if $a \leqslant b$ or $b \leqslant a$. They are called incomparable if neither $a \leqslant b$ nor $b \leqslant a$ holds.

Suppose M is a given nonvoid subset of P. An element s such that

(4) $m \leqslant s$ for all $m \in M$,

and

(5) if $m \leqslant t$ for every $m \in M$, then $s \leqslant t$,

will be called the least upper bound (supremum) of M in $`P$. If such an element exists, we write $s = \sup_{`P} M$. The greatest lower bound (infimum) of M in $`P$ is defined dually: It is an element i such that

(4′) $i \leqslant m$ for all $m \in M$,

and

(5′) if $t \leqslant m$ for every $m \in M$, then $t \leqslant i$.

In this case, we write $i = \inf_{`P} M$.

If $f = (a_i;\ i \in I)$ is a family of elements of a set A, i.e., f is an application of I into A, then the image of I under f will be denoted by $\{a_i;\ i \in I\}$. If the supremum $\sup_{`P} \{a_i;\ i \in I\}$ exists, we write $s = \sup (a_i;\ i \in I)$ and also $s = \bigvee(a_i;\ i \in I)$. A similar notation will be used for the greatest lower bound. If $a \leqslant b$, we write in the same meaning $b \geqslant a$. By $a < b$ (or, equivalently, by $b > a$) we mean that $a \leqslant b$ but $a \neq b$.

If there exists the element $\sup_{`P} P$, it is called the greatest element of $`P$ and it is denoted by 1. Dually, the element $\inf_{`P} P$ is said to be the least element of $`P$ and it is denoted by 0.

2. Lemma. Let $`P$ be a poset and let $(a_i;\ i \in I)$ be a family of elements of P. Assume that $I = J \cup K = \bigcup (I_t;\ t \in T)$ where J, K, I_t and T are nonvoid. Suppose moreover that the elements $b = \bigvee(a_i;\ i \in J)$, $c = \bigvee(a_i;\ i \in K)$ and $\bigvee(a_i;\ i \in I_t)$ exist for every $t \in T$. Then the supremum d of the set formed by b and c exists if and only if $e = \bigvee(\bigvee(a_i;\ i \in I_t);\ t \in T)$ exists and in this case $a = \bigvee(a_i;\ i \in I) = d = e$.

The proof is straightforward and will be omitted.

Let $(a_i;\ i \in \{1,2\})$ be a family of elements of P. As customary, we write $a_1 \vee a_2$ instead of $\sup_{`P} (a_i;\ i \in \{1,2\})$ if such a supremum exists. Similarly, $a_1 \wedge a_2$ stands for $\inf_{`P} (a_i;\ i \in \{1,2\})$ if such an infimum exists.

Notice the following two special cases of Lemma stated above:

A. If x, y and z are elements of P such that $x \vee y$, $y \vee z$ and $x \vee (y \vee z)$ exist, then also the supremum $(x \vee y) \vee z$ exists and it is equal to $x \vee (y \vee z)$.

B. If $n \geq 2$ and if $(a_i;\ i \in \{1,2,\dots,n\})$ is a family of elements in P such that $x = \bigvee (a_i;\ i \in \{1,2,\dots,n-1\})$ and $x \vee a_n$ exist, then $\bigvee (a_i;\ i \in \{1,2,\dots,n\})$ exists and it is equal to $x \vee a_n$.

<u>Convention</u>. The underlined letter N will always denote the set of natural numbers, i.e., $\underline{N} = \{1,2,\dots\}$, whilst $\underline{N}_0$ will denote the set $\{0,1,2,\dots\}$.

3. If $a, b \in P$, we write $a \prec b$ (or $b \succ a$) and say that b <u>covers</u> a or that a is <u>covered</u> by b if and only if $a < b$ and there is no element c such that $a < c$ and $c < b$. Suppose the poset `P has the least element 0. An element a is said to be an <u>atom</u> if and only if $0 \prec a$. If $a = a_0 \prec a_1 \prec a_2 \prec \dots$ $\dots \prec a_n = b$, then the sequence $a_0, a_1, \dots, a_n$ is called a <u>maximal chain</u> joining a to b. The number n is said to be the <u>length</u> of the chain. If there exists a number k such that the length of every maximal chain in `P does not exceed k and if every chain in `P is finite, we define `P to be a <u>poset of finite length</u>.

4. In any poset `P we can define aRb if and only if $b \leq a$. Then (P,R) is again a poset, usually denoted by $(P, \geq)$

and called the dual of `P. If we replace $\leq$ by $\geq$, we get from a valid theorem a new true theorem. This fact will be called Duality Principle.

5. For two elements $a \leq b$ of a poset `P, the interval $[a,b]$ is, by definition, the set $\{c \in P;\ a \leq c\ \&\ c \leq b\}$. A subset K of P is called convex if and only if it has the following property:

$$(a,b \in K\ \&\ a \leq b) \Rightarrow [a,b] \subset K.$$

A nonvoid subset I of P is said to be an order-ideal of `P if and only if $i \in I$ and $j \leq i$ imply that $j \in I$. A nonvoid subset $F$ of $P$ is called an order-filter of `P if $f \in F$ and $g \geq f$ always imply that $g \in F$. Given an element $a \in P$, the set $(a] = \{x \in P;\ x \leq a\}$ is said to be the principal ideal generated by a. Dually, the principal filter determined by a is the set $[a) = \{y \in P;\ a \leq y\}$.

6. A mapping $f:P \to Q$ is said to be an order-monomorphism of $(P, \leq)$ into $(Q, \leq)$ if and only if the following equivalence is true:

$$(6)\quad p_1 \leq p_2 \Leftrightarrow f(p_1) \leq f(p_2).$$

Every mapping with this property is necessarily injective. If, moreover, f maps P onto Q, then it is called an order-isomorphism.

Let $f:P \to Q$ and $g:Q \to R$ be two mappings. The composite mapping or, more specifically, the right composite mapping (cf. [30]) of f and g is the mapping $f \circ g:P \to R$ such that $f \circ g: p \mapsto g(f(p))$ for every $p \in P$. As is usual, the mapping $f \circ g$ is also denoted by gf and it is called the left composite of f and g. Thus $gf:P \to R$ and $gf: p \mapsto g(f(p))$.

7. Theorem. Let f be an order-isomorphism of the poset $(P, \leqslant)$ onto the poset $(Q, \leqslant)$. Let M be a nonvoid subset of P and let s be the supremum of M in $(P, \leqslant)$. Then $f(s)$ is the supremum of the set $f(M)$ in $(Q, \leqslant)$, i.e.,

$$f(\sup_{(P, \leqslant)} M) = \sup_{(Q, \leqslant)} f(M).$$

Proof. We shall prove that the element $t = f(s)$ is the supremum of the set $N = f(M)$. In fact, for every $x \in f(M)$ there exists $y \in M$ such that $x = f(y)$. Since $y \leqslant \sup M$, we get $x = f(y) \leqslant f(\sup M) = t$. Suppose $h \in Q$ is an element such that $n \leqslant h$ for every $n \in N$. Then $n \in N$ implies that there exists $m \in M$ satisfying $n = f(m)$. Clearly, if n ranges over N, m ranges over all the elements of M. Now f maps P onto Q. Hence there exists $k \in P$ such that $f(m) = n \leqslant h = f(k)$. It follows from (6) that $m \leqslant k$ for every $m \in M$. By the definition of $\sup M$, we then have $\sup M \leqslant k$. By (6), we can write $t = f(\sup M) \leqslant f(k) = h$. //

8. A lattice is an algebra $`L = (L, \vee, \wedge)$ satisfying for every $a,b,c \in L$ the following postulates:

(7) $a \vee b = b \vee a$; (7′) $a \wedge b = b \wedge a$;

(8) $(a \vee b) \vee c = a \vee (b \vee c)$; (8′) $(a \wedge b) \wedge c = a \wedge (b \wedge c)$;

(9) $a \wedge (a \vee b) = a$; (9′) $a \vee (a \wedge b) = a$.

The binary relation $\leqslant$ defined on L by

$$a \leqslant b \Leftrightarrow a = a \wedge b$$

is a partial ordering. Note that $a \leqslant b$ if and only if $b = a \vee b$. The relation $\leqslant$ is compatible with the operations $\vee, \wedge$, i.e., if $a \leqslant b$ and $c \leqslant d$, then $a \wedge c \leqslant b \wedge d$ and $a \vee c \leqslant b \vee d$. If $a \neq b$, then $a \vee b$ is the supremum and $a \wedge b$ the infimum of the set $\{a,b\}$ in $(L, \leqslant)$. For any lattice $(L, \vee, \wedge)$ we define its dual as the lattice $(L, \wedge, \vee)$. The Duality Principle applies to the class of all lattices.

A poset $`P$ is said to be a <u>meet-semilattice</u> if $\inf M$ there exists for any two-element subset M of P.

9. A nonvoid subset $S \subset L$ determines a <u>sublattice</u> of a lattice $`L$ if and only if $a \wedge b \in S$ and $a \vee b \in S$ for every $a, b \in S$.

Suppose $`P$ is a meet-semilattice and $\emptyset \neq T \subset P$. The set $T$ determines a <u>subsemilattice</u> of $`P$ if $\inf_{`P} \{s,t\}$ belongs to T for every $s, t \in T$.

10. Given two lattices $`L = (L, \vee, \wedge)$, $`T = (T, \vee, \wedge)$, a mapping f of L into T is called a <u>lattice-homomorphism</u> (or simply <u>homomorphism</u>) if

$$f(a \vee b) = f(a) \vee f(b) \;\&\; f(a \wedge b) = f(a) \wedge f(b)$$

for every $a, b \in L$. If, moreover, f is a bijection of L onto T, it is said to be an <u>isomorphism</u> (a <u>lattice-isomorphism</u>) of $`L$ onto $`T$. In this case, $`T$ is called the <u>isomorphic image</u> of the lattice $`L$ (in notation, $`L \cong `T$).

A lattice which is isomorphic to its dual is called <u>self-dual</u>.

11. <u>Iséki's theorem</u>. Let $`L$ and $`K$ be lattices. A mapping f of the set L onto the set K is a lattice-isomorphism if and only if f is an order-isomorphism of the poset $(L, \leq)$ onto the poset $(K, \leq)$.

<u>Proof</u>. 1. Suppose f is a lattice isomorphism and $p_1 \leq p_2$. Then $p_1 = p_1 \wedge p_2$ so that $f(p_1) = f(p_1 \wedge p_2) = f(p_1) \wedge f(p_2) \leq$ $\leq f(p_2)$. If, conversely, $f(p_1) \leq f(p_2)$, then $f(p_1) = f(p_1) \wedge$ $\wedge f(p_2) = f(p_1 \wedge p_2)$. Since f is injective, $p_1 = p_1 \wedge p_2 \leq p_2$.

2. Suppose f satisfies (6). Take $a,b \in L$. Now, f is surjective and $f(a) \wedge f(b) \in K$. Hence, there exists $c \in L$ such that $f(c) = f(a) \wedge f(b)$. It follows

$$f(c) \leqslant f(a) \ \& \ f(c) \leqslant f(b)$$

and, by (6), we see that

$$c \leqslant a \ \& \ c \leqslant b.$$

Thus $c \leqslant a \wedge b$, and so using (6) again, $f(c) \leqslant f(a \wedge b)$. Since $a \wedge b \leqslant a$, we get $f(a \wedge b) \leqslant f(a)$, by (6). In a similar manner we can show that $f(a \wedge b) \leqslant f(b)$. Consequently, $f(a \wedge b) \leqslant f(a) \wedge \wedge f(b) = f(c) \leqslant f(a \wedge b)$. That is, $f(a \wedge b) = f(c)$. Since f is injective, the latter fact yields $c = a \wedge b$ and, therefore, $f(a \wedge b) = f(a) \wedge f(b)$. The assertion $f(a \vee b) = f(a) \vee f(b)$ follows by duality. //

12. A poset $`P = (P, \leqslant)$ is called a <u>complete lattice</u> if and only if every nonvoid subset $M$ of $P$ has the supremum $\sup_{`P} M$ and the infimum $\inf_{`P} M$ in $`P$. Evidently, every complete lattice is a lattice.

13. <u>Criterion of completeness</u>. A poset $`P$ is a complete lattice if and only if it has the greatest element and every nonvoid subset $M$ of $P$ has the infimum $\inf_{`P} M$ in $`P$.

The dual theorem can be formulated in the following way: A poset $`P$ is a complete lattice if and only if it has the least element and every nonvoid subset $M$ of $P$ has the supremum $\sup_{`P} M$ in $`P$.

For the proof, see [179; Thm 17, p. 61].

14. Given a lattice $`L$, a nonvoid subset $I$ of $L$ is called an <u>ideal</u> of $`L$ if and only if the following conditions hold in $`L$:

(10) $i \vee j \in I$ for every $i, j \in I$;

(11) if $i \in I$ and $k \in L$ is such that $k \leqslant i$, then $k \in I$.

The condition (11) can be replaced by the condition

(11′) $k \wedge i \in I$ for every $k \in L$ and every $i \in I$.

Every ideal of a lattice is a sublattice of the lattice. The principal ideal $(a]$ is clearly an ideal of $`L$ for every $a \in L$. If $K$ is a nonvoid subset of $L$, then the smallest ideal of $`L$ which contains K, i.e., the <u>ideal generated by</u> K, is the set

$$\{j \in L; \exists\, n \in \underline{N}\, \exists\, k_1, k_2, \ldots, k_n \in K :: j \leqslant k_1 \vee k_2 \vee \ldots \vee k_n\}.$$

(The sign :: is read "such that".)

The proof follows from [179, Thm 78, p. 162].

15. We recall for future reference the following notions: Let $a \leqslant b$ be elements of a lattice $`L$ and let (a,b) be the corresponding ordered pair. From now on, we shall adopt another standard notation b/a in place of (a,b), calling b/a the <u>quotient</u> determined by a, b. If $b/a, d/c$ are quotients such that

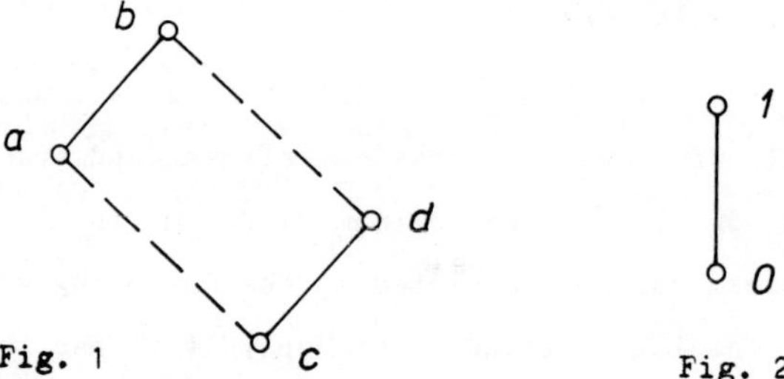

Fig. 1 Fig. 2

$c = a \wedge d$ and $b = a \vee d$ (see Figure 1), we write $b/a \searrow d/c$ (or equivalently, $d/c \nearrow b/a$). If $q/p, s/r$ are such that either $q/p \nearrow s/r$ or $s/r \nearrow q/p$, then this fact is usually written $q/p \sim s/r$ and is read "the quotients $q/p, s/r$ are <u>transposed</u>". If $q/p \sim s/r$, then $p = q$ if and only if $s = r$. Two quotients b/a

and d/c are called projective, written $b/a \approx d/c$, if there exists a sequence of quotients $b_0/a_0 = b/a, b_1/a_1, \ldots, b_n/a_n = d/c$ such that $b_i/a_i \sim b_{i+1}/a_{i+1}$ for every $i = 0,1,\ldots,n-1$.

16. Given a family $(`L_i;\ i \in I)$ of lattices, then the set of all sequences $\underline{a} = (a_i;\ i \in I)$ where $a_i \in L_i$ for every $i \in I$ represents the Cartesian product of the sets L_i. The Cartesian product becomes a lattice under the operations defined by

$$\underline{a} \vee \underline{b} = (a_i \vee b_i;\ i \in I), \quad \underline{a} \wedge \underline{b} = (a_i \wedge b_i;\ i \in I).$$

The lattice constructed in this way will be called the (complete, unrestricted) direct product of the lattices $`L_i$ and it will be denoted by $\prod(`L_i;\ i \in I)$. In the special case where we have n lattices $`L_1, `L_2, \ldots, `L_n$, an element of the corresponding Cartesian product is denoted by $(a_1, a_2, \ldots, a_n)$ and the direct product of $`L_i$ is written $`L_1 \times `L_2 \times \ldots \times `L_n$.

The direct product of n copies of the lattice $\underline{2}$ of Figure 2 will be denoted by $\underline{2}^n$. By $\underline{2}^0$ we mean the one-element lattice $\underline{1}$.

17. Lemma. Let $`L$ be a lattice with the least element 0. Then $`L$ is isomorphic to the direct product of lattices $`T, `S$ if and only if there exist sublattices $`T_1, `S_1$ of $`L$ such that

(i) $`T \cong `T_1, \quad `S \cong `S_1$;

(ii) every element of $`L$ can be uniquely written as the join of two elements of $`T_1$ and $`S_1$, respectively.

Proof. 1. Let $`L \cong `T \times `S$. Since 0 belongs to $`L$, also $`T$ and $`S$ have the least elements. Let $f: `T \times `S \to `L$ be the mentioned isomorphism. Put

$$T_1 = \{h \in L;\ \exists t \in T :: h = f((t,0))\};$$

$$S_1 = \{h \in L;\ \exists s \in S :: h = f((0,s))\}.$$

The sets T_1, S_1 are the base sets of some sublattices $`T_1, `S_1$ which are isomorphic to $`T$ and $`S$, respectively. The mapping f is an isomorphism and, therefore, for every $h \in L$ there exist elements $t \in T, s \in S$ such that

$$h = f((t,s)) = f((t,0) \vee (0,s)) = f((t,0)) \vee f((0,s)) = t_1 \vee s_1$$

where $t_1 \in T_1$ and $s_1 \in S_1$. If $h = t_2 \vee s_2$ and $t_2 = f((t_0,0)) \in$ $\in T_1$, $s_2 = f((0,s_0)) \in S_1$, then $h = f((t_0,0)) \vee f((0,s_0)) =$ $= f((t_0,s_0)) = f((t,s))$. But f is injective. Hence $(t_0,s_0) =$ $= (t,s)$ and so $t_2 = t_1, s_2 = s_1$.

2. Conversely, suppose that every element h of $`L$ can be uniquely expressed in the form $h = t_1 \vee s_1$ where $t_1 \in T_1$ and $s_1 \in S_1$. We shall prove that $`L$ is isomorphic to the direct product $`T_1 \times `S_1$. Let g be the mapping of $`T_1 \times `S_1$ into $`L$ defined by the assignment $g:(t_1,s_1) \mapsto t_1 \vee s_1$. Under the hypotheses of the theorem it is clear that $g$ is a bijection of $`T_1 \times `S_1$ onto $`L$. Next we show that the following equivalence is true:

$$(t_1,s_1) \leqslant (t_2,s_2) \Leftrightarrow g((t_1,s_1)) \leqslant g((t_2,s_2)).$$

In fact, if $(t_1,s_1) \leqslant (t_2,s_2)$, then $t_1 \leqslant t_2$ and $s_1 \leqslant s_2$. Thus

$$g((t_1,s_1)) = t_1 \vee s_1 \leqslant t_2 \vee s_2 = g((t_2,s_2)).$$

Conversely, if $g((t_1,s_1)) \leqslant g((t_2,s_2))$, then $t_1 \vee s_1 \leqslant t_2 \vee s_2$. Hence $t_2 \vee s_2 = (t_1 \vee t_2) \vee (s_1 \vee s_2)$ and, by uniqueness of such expressions, $t_2 = t_1 \vee t_2 \geqslant t_1$ and $s_2 = s_1 \vee s_2 \geqslant s_1$, i.e., $(t_1,s_1) \leqslant (t_2,s_2)$. By Iséki´s Theorem 11, the considered lattices are isomorphic. //

A lattice $`L$ is said to be the <u>direct sum of its sublattices</u> $`K_1, `K_2, \ldots, `K_n$ (notation: $`L = `K_1 \oplus `K_2 \oplus \ldots \oplus `K_n$)

if each $a \in L$ is uniquely representable in the form $a = k_1 \vee \vee k_2 \vee \ldots \vee k_n$ with $k_i \in K_i$ for every $i = 1,2,\ldots,n$.

Similarly we shall write $`P = `P_1 \oplus `P_2 \oplus \ldots \oplus `P_n$ and say that a poset $`P$ is the <u>direct sum of its subposets</u> $`P_1, `P_2, \ldots, `P_n$ if and only if every element p of P can be uniquely written in the form $p = \sup_{`P} \{p_1, p_2, \ldots, p_n\}$ where $p_i \in P_i$ for every $i = 1,2,\ldots,n$.

We shall need the following result. Its generalization for the direct sum of n sublattices is straightforward.

<u>Proposition</u>. Let $`K_1, `K_2$ be sublattices of a lattice $`L$ and let $`L = `K_1 \oplus `K_2$. Then

(i) $`L$ has the least element 0;

(ii) for any $d,e \in K_1$ and any $s,t \in K_2$, from $c \leqslant d \vee s$ and $c = e \vee t$ it follows that $e \leqslant d$ and $t \leqslant s$;

(iii) for any $p,q \in K_1$ and any $x,y \in K_2$, $(p \vee x) \wedge (q \vee y) = (p \wedge q) \vee (x \wedge y)$.

<u>Proof</u>. Ad (i). Let $x_1 \in K_1$ and $z_2 \in K_2$. By assumption, $x_1 = y_1 \vee y_2$ where $y_i \in K_i$. Since $x_1 \vee z_2 = x_1 \vee (y_2 \vee z_2)$, $z_2 = y_2 \vee z_2 \geqslant y_2$. Hence $y_2 = 0_2$ is the least element of $`K_2$. Similarly we see that $`K_1$ has also the least element, say 0_1. Clearly, $0 = 0_1 = 0_2$.

Ad (ii). We have $d \vee s = d \vee s \vee c = (d \vee e) \vee (s \vee t)$ and so

$$d = d \vee e \geqslant e, \quad s = s \vee t \geqslant t.$$

Ad (iii). Let $a = p \vee x$, $b = q \vee y$. By assumption, $a \wedge b = m \vee w$ where $m \in K_1$ and $w \in K_2$. Observe that $a \wedge b \leqslant a = p \vee x$, $a \wedge b \leqslant b = q \vee y$. Hence (ii) gives

$$m \leqslant p, \quad m \leqslant q, \quad w \leqslant x, \quad w \leqslant y$$

and, therefore, $a \wedge b = m \vee w \leqslant (p \wedge q) \vee (x \wedge y)$. However, $a \wedge b =$

$= (p \vee x) \wedge (q \vee y) \geqslant (p \wedge q) \vee (x \wedge y)$. Thus we have $a \wedge b = (p \wedge q) \vee \vee (x \wedge y)$. //

Corollary. Under the notation above,

$$`K_1 \oplus `K_2 \oplus \ldots \oplus `K_n \cong `K_1 \times `K_2 \times \ldots \times `K_n.$$

Proof. This follows from (i) of Proposition and from the lemma above. //

18. A lattice $`L$ with 0 and 1 is said to be complemented if and only if for every $a \in L$ there exists an element $b \in L$ such that $a \vee b = 1$ and $a \wedge b = 0$. Such an element b is called a complement of a.

We define a lattice $`T$ to be relatively complemented if and only if for every $a,b,c \in T$ such that $a \leqslant b \leqslant c$ there exists d satisfying $b \vee d = c$ and $b \wedge d = a$. Such an element d is called a relative complement of b in the interval $[a,c]$.

19. A lattice $`L$ is said to be distributive if $a \wedge (b \vee c) = (a \wedge b) \vee (a \wedge c)$ for every $a,b,c \in L$. It is well-known that in a distributive lattice we also have $a \vee (b \wedge c) = (a \vee b) \wedge (a \vee c)$ for every $a,b,c$ of $L$. A lattice $`L$ is modular if it satisfies the implication

$$a \leqslant c \Rightarrow a \vee (b \wedge c) = (a \vee b) \wedge c$$

for every a,b,c of L. It is immediate that every distributive lattice is modular. A lattice is modular if and only if it does not contain a sublattice isomorphic to the lattice $`N_5$ of Figure 3 (cf. [179, Thm 32, p. 90]). A lattice is distributive if and only if no its sublattice is isomorphic to the lattice $`N_5$ or to the lattice $`M_5$ shown in Figure 4 (cf. [179, Thm 33, p. 90]). It is customary to refer to the lattice $`N_5$ as the pentagon and to the lattice $`M_5$ as the <u>diamond</u>.

If $b/a \sim d/c$ and `L is modular, then the sublattices

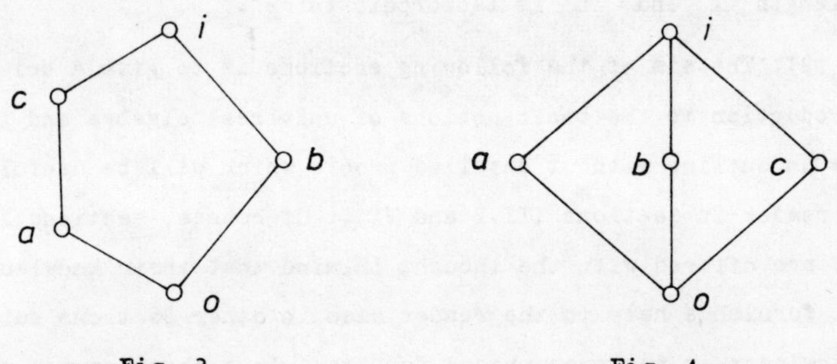

Fig. 3 Fig. 4

determined by the intervals [a,b], [c,d] are isomorphic (cf. [179, Thm 36, p. 95]). If a sublattice [a,b] of a modular lattice is a poset of finite length, then all the maximal chains between a and b have the same length (cf. [179, Thm 43, p. 107]).

20. A lattice `L which is distributive and complemented is said to be a <u>Boolean lattice</u>. Every element $a \in L$ has in this case exactly one complement which is denoted by $a'$. The mapping $f: a \mapsto a'$ defines a unary operation $'$ on L. Given a Boolean lattice `L, we may construct the algebra $(L, \vee, \wedge, ', 0, 1)$, usually called a <u>Boolean algebra</u>. However, the nullary operations 0,1 are often not considered, and by a Boolean algebra we mean, at times, the algebra $(L, \vee, \wedge, ')$. The identities $(a \vee b)' = a' \wedge b'$, $(a \wedge b)' = a' \vee b'$ hold in every Boolean algebra and are called <u>De Morgan laws</u>. Observe that any lattice $\underline{2}^n$, $n \in \underline{N}_0$, determines a Boolean algebra. In actuality, these lattices characterize finite Boolean algebras. We namely have (cf. [29, Thm 4, p. 59])

<u>Theorem</u>. Let $`L$ be a finite Boolean lattice which has exactly $k$ atoms. Then every maximal chain connecting 0 to 1 is of length $k$ and $`L$ is isomorphic to $\underline{2}^k$.

21. The aim of the following sections is to give a brief introduction to the basic notions of universal algebra and to show an outline without detailed proofs which will be useful to the reader in sections III.2 and VI.4. Of course, sections I.21 - I.27 are offered with the thought in mind that their knowledge will furnish a help to the reader also in other Sections but such a knowledge is not presupposed for them. In a short course aimed directly at the elementary theory of orthomodular lattices or at a first reading, sections I.21 - I.27 may be omitted. Readers familiar with basic ideas of universal algebra can skip at once to Chapter II.

To begin with we quote the words of one of the world's experts on the subject: "... It can be safely said that lattice theory is the best friend of universal algebra. The converse, however, is not true. But I think universal algebra has made enough contributions to lattice theory to qualify as a good friend. ..." (G. Grätzer, [1; p. 204]).

Let A be a nonvoid set and let A^n be its n-th Cartesian power. By an <u>n-ary operation</u> on A is meant any mapping f of A^n into A. The number n is called the <u>arity</u> of f. By an <u>operation</u> on A we mean an n-ary operation for an $n \in \underline{N}_0$. We say that f is a <u>binary operation</u> if $n = 2$. It is a <u>unary operation</u> if $n = 1$. In case n is 0, the operation is said to be <u>nullary</u>. Such an operation is a mapping of $A^0 = \{\emptyset\}$ into A and it is determined by a single element $f(\emptyset) \in A$. We usually say that a nullary operation selects a fixed element of A.

We have already used the notion of an algebra, e.g., in the definition of a lattice. We are now in a position to give a precise definition.

An <u>algebra</u> $`A = (A,F)$ is a pair where $A$ is a nonvoid set (called the <u>base set</u> of $`A$) and where F is a family of operations on A. To simplify the notation, we will confine ourselves to the case where F is finite in which case $(A,(f_1,f_2,\ldots,f_r))$ will be abbreviated to $(A,f_1,f_2,\ldots,f_r)$. Let n_i denote the arity of the operation f_i. The r-tuple $(n_1,n_2,\ldots,n_r)$ is called the <u>type</u> of the algebra $(A,f_1,f_2,\ldots,f_r)$.

If $`A$ and $`B$ are two algebras of the same type, it is usual to denote the corresponding operations by the same letters, i.e., we write $`A = (A,f_1,f_2,\ldots,f_r)$ and $`B = (B,f_1,f_2,\ldots,f_r)$ when there is no danger of confusion.

The following algebras will appear in the course of our arguments:

(i) Lattices (cf. I.8) where $F = \{\vee,\wedge\}$ and where $`L = (L,\vee,\wedge)$ stands for $`L = (L,F)$.

(ii) Boolean algebras (cf. I.20).

(iii) Ortholattices (cf. II.1).

(iv) Orthomodular lattices (cf. II.3).

To provide some background for a general situation, we first outline our approach by means of the well-known example of distributive lattices (cf. I.19). A distributive lattice can be regarded as an algebra $(L,\vee,\wedge)$ satisfying the "identity"

$$x \wedge (y \vee z) = (x \wedge y) \vee (x \wedge z).$$

Using $+$ and $.$ for a moment to denote the operation $\vee$ and $\wedge$, respectively, we formally get an algebra $(L,+,.)$ satisfying the "polynomial identity"

$$x.(y + z) = (x.y) + (x.z).$$

Such a terminology seems to be legitimate, since the both sides can be replaced by the polynomials $p(x,y,z) = x.(y + z)$, $q(x,y,z) = (x.y) + (x.z)$ in three indeterminates. From another point of view it is possible to speak about $p(x,y,z)$ and $q(x,y,z)$ as polynomials in the operations $+,.$. All this is exceedingly simple and known from introductory books on algebra. But what if we replace the words "polynomial in the operations $+,.$" by "polynomial in the operations $\vee,\wedge$ "? Once this question has been raised it is natural to enquire how a more general notion of a polynomial in some operations can be defined.

We now briefly indicate the corresponding idea which has been very fruitful in universal algebra theory.

Let X be a fixed set of indeterminates, $X = \{x_1, x_2, \dots\}$.

The set $P_n(F)$ of n-ary polynomials in the operations of $F \neq \emptyset$ is a set of formal expressions defined as follows:

(i) $x_i \in P_n(F)$ for every $i = 1,2,\dots,n$;

(ii) if $p_1, p_2, \dots, p_k \in P_n(F)$, then $f(p_1, p_2, \dots, p_k) \in P_n(F)$ for every $f \in F$;

(iii) n-ary polynomials in the operations of F are precisely those obtained from (i) and (ii) in a finite number of steps.

Note that in this definition, the index k depends on the arity of the chosen f. A similar simplification in notation will also be used in the sequel.

We now come to another basic definition.

Let $`A = (A,F)$ be an algebra. An n-ary polynomial function on the algebra $`A$ associated with an n-ary polynomial in the operations of F is defined in the following manner:

(i) If $p = x_i$ and $a_1, a_2, \dots, a_n \in A$, then $p(a_1, a_2, \dots, a_n) =$

$= a_i$ for every $i = 1,2,\dots,n$.

(ii) If $g = f(p_1,p_2,\dots,p_k)$, $p_i \in P_n(F)$, $f \in F$ and $p_i(a_1,a_2,\dots$ $\dots,a_n) = b_i$ $(i = 1,2,\dots,k)$, then

$$g(a_1,a_2,\dots,a_n) = f(b_1,b_2,\dots,b_k).$$

There is some variation of terminology. Instead of an n-ary polynomial function on the algebra $`A$ associated with $p$ one sometimes speaks about an n-ary polynomial on $`A$ or about an n-ary F-polynomial.

For brevity we shall replace the expression $p(a_1,a_2,\dots,a_n)$ by $p(a_1,\bullet)$ etc.

If p is an n-ary polynomial for an n, it is called simply a polynomial.

In the case of lattices, the general definition can be rephrased as follows (cf. [74]):

The set P_n of n-ary lattice polynomials is the smallest set satisfying the following conditions:

(i) $x_i \in P_n$ for every $i = 1,2,\dots,n$;

(ii) if $p_1,p_2 \in P_n$, then $(p_1 \vee p_2) \in F$ and $(p_1 \wedge p_2) \in F$.

For example, if $n \geqslant 3$, then $(x_1 \vee x_2)$, $((x_1 \vee x_2) \wedge x_3)$, $((x_1 \vee x_2) \wedge (x_1 \vee x_3))$ are n-ary lattice polynomials. If we drop the parentheses, which are not essential, we obtain more usual expressions providing examples of lattice polynomials: $x_1 \vee x_2$, $(x_1 \vee x_2) \wedge x_3$, $(x_1 \vee x_2) \wedge (x_1 \vee x_3)$ etc.

The lattice-theoretic interpretation of a polynomial function is also obvious:

Let $`L = (L,\vee,\wedge)$ be a lattice. An n-ary lattice polynomial function (or, simply, a polynomial) on $`L$ is defined as follows:

(i) If $p = x_i$ and $a_1,a_2,\dots,a_n \in L$, then $p(a_1,a_2,\dots,a_n) =$ $= a_i$ for every $i = 1,2,\dots,n$.

(ii) If $g = p_1 \vee p_2$, $h = p_1 \wedge p_2$ and $p_i(a_1,a_2,\ldots,a_n) = b_i$ $(i = 1,2)$, then $g(a_1,\bullet) = b_1 \vee b_2$ and $h(a_1,\bullet) = b_1 \wedge b_2$.

The rank $r(p)$ of a lattice polynomial p is defined recursively: The rank of the x_i's is 1. If g is equal to $p_1 \vee p_2$ or to $p_1 \wedge p_2$, then $r(g) = r(p_1) + r(p_2)$.

Another special case of a polynomial in three operations will be discussed in Section II.3.

22. Let $`A = (A,F)$ be an algebra and let $S$ be a nonvoid subset of the base set $A$. Then $`S = (S,F)$ is called a sub-algebra of $`A$ if and only if S is closed under all the operations f of F, i.e., $f(s_1,s_2,\ldots,s_k) \in S$ for every $f \in F$ and every $s_1,s_2,\ldots,s_k \in S$.

Suppose M is a nonvoid subset of A and let $`S_i$, $i \in I$, be all the subalgebras of $`A$ such that $S_i \supset M$. Then $(\bigcap(S_i;\ i \in I),F)$ is called the subalgebra generated by the set M. In this book it is denoted by $\langle M \rangle$.

If $\langle M \rangle = `A$, then $M$ is said to be a generating set of the algebra $`A$. If $M = \{m_1,m_2,\ldots,m_t\}$ and if M is a generating set of $`A$, we say that $`A$ is an algebra having t generators $m_1,m_2,\ldots,m_t$.

Let $`A = (A,f_1,f_2,\ldots,f_r)$ and $`B = (B,f_1,f_2,\ldots,f_r)$ be two algebras of the same type. Then a mapping h of A into B is called a homomorphism of $`A$ into $`B$ if and only if

$$h(f_i(a_1,a_2,\ldots,a_{k(i)})) = f_i(h(a_1),h(a_2),\ldots,h(a_{k(i)}))$$

for every $i = 1,2,\ldots,r$ and all $a_1,a_2,\ldots,a_{k(i)}$ of A.

For example, if $`A = (A,\vee,\wedge,{}',0,1)$ and $`B = (B,\vee,\wedge,{}',0,1)$ are two algebras of type $(2,2,1,0,0)$, then $h: A \to B$ is a homomorphism of $`A$ into $`B$ if and only if

(1) $h(a_1 \vee a_2) = h(a_1) \vee h(a_2)$;

(2) $h(a_1 \wedge a_2) = h(a_1) \wedge h(a_2)$;

(3) $h(a_1') = (h(a_1))'$

for every $a_1, a_2 \in A$, and if

(4) $h(0) = 0$; $h(1) = 1$.

If h is a bijective homomorphism, it is called an _isomorphism_ of `A onto `B and in this case we write `A ≅ `B.

23. In order to motivate another important notion, we start with the following very special example.

Let $\underline{2}^2 = \{(0,0),(1,0),(0,1),(1,1)\}$ be the four-element Boolean algebra. It is an algebra with one generator. Indeed, put $m_1 = (1,0)$, $M = \{m_1\}$ and consider the subalgebra `S of $\underline{2}^2$ which is generated by M. Its base set S contains

(i) m_1, by definition;

(ii) m_1', since S is closed under the operation $'$;

(iii) 0,1, since it is closed under the nullary operations.

Hence `S = $\underline{2}^2$ and so $\langle M \rangle = \underline{2}^2$.

The algebra $\underline{2}^2$ has the following property: Given a Boolean algebra `B and a mapping $g$ of $\{m_1\}$ into the base set B, there exists a homomorphism $h$ of the Boolean algebra $\underline{2}^2$ into the Boolean algebra `B such that $h(m_1) = g(m_1)$.

To prove this assertion, put

$$h((1,0)) = a, \quad h((0,1)) = a', \quad h((0,0)) = 0, \quad h((1,1)) = 1$$

where $a = g(m_1)$. Then h has the required properties.

Consequently, any Boolean algebra with one generator is a homomorphic image of $\underline{2}^2$.

Now it is easy to admit that the algebra $\underline{2}^2$ is the most general Boolean algebra having one generator in the sense we have just indicated.

The example of $\underline{2}^2$ illustrates the following definition of "the most general algebra in a given class of algebras having a given generating set".

Let $\underline{K}$ be a class of algebras $`A_i = (A_i,F)$ of the same type. Let $`A = (A,F)$ be an algebra from $\underline{K}$ and let X be a subset of A. Then the algebra $`A$ is said to be a free algebra over $\underline{K}$ freely generated by the set X if and only if it satisfies (i) and (ii):

(i) X is a generating set of $`A$.

(ii) Given any algebra $`B$ of $\underline{K}$ and any mapping $g:X\to B$, there exists a homomorphism $h$ of $`A$ into $`B$ extending the mapping $g$. If $X = \{x_1,x_2,\ldots,x_t\}$, then $`A$ is called a free algebra with t generators.

If such an algebra exists, it is unique up to isomorphism [73, Thm 1, p. 163].

Instead of speaking about a free algebra over the class of Boolean algebras, it is usual to call it a free Boolean algebra. A similar convention is used in other well-known classes of algebras.

In this connection we record that $\underline{2}^e$ $(e = 2^t)$ is the free Boolean algebra with t generators (cf. [29, Thm 7, p. 62]).

24. Let $`A = (A,F)$ be an algebra. A binary relation $D$ on $A$ is called a congruence relation of $`A$ if and only if it is an equivalence relation on A which satisfies the following property for every f (of arity k) from F:

($`A$) If $(a_i,b_i)\in D$ $(i = 1,2,\ldots,k)$, then

$$(f(a_1,a_2,\ldots,a_k),f(b_1,b_2,\ldots,b_k))\in D.$$

The condition (\`A) is somewhat clearer if we restrict consideration to lattices:

Let \`L = $(L, \vee, \wedge)$ be a lattice. A binary relation D on L is called a (lattice) congruence relation of \`L if and only if it is an equivalence relation on L which satisfies the following property:

(\`L) If $(a_i, b_i) \in D$ $(i = 1,2)$, then $(a_1 \vee a_2, b_1 \vee b_2) \in D$ and $(a_1 \wedge a_2, b_1 \wedge b_2) \in D$.

Since D is an equivalence relation, it is easily seen that the condition (\`L) can be replaced by the condition ($^+$L):

($^+$L) If $(a,b) \in D$ and $c \in L$, then $(a \vee c, b \vee c) \in D$ and $(a \wedge c, b \wedge c) \in D$.

Let D be again a congruence relation of an algebra \`A = = (A,F). The quotient algebra \`A/D is defined as an algebra of the same type as \`A having its base set equal to the quotient set A/D. The operations of the algebra \`A/D = (A/D,F) are defined as follows:

If $[a_1], [a_2], \ldots, [a_k]$ are any elements of the quotient set A/D, then we put

$$f([a_1], [a_2], \ldots, [a_k]) = [f(a_1, a_2, \ldots, a_k)].$$

Of course, here $[a_i] = \{a \in A;\ (a_i, a) \in D\}$.

Let us illustrate this general definition by its special case represented by lattices:

Given a congruence relation D of a lattice \`L = $(L, \vee, \wedge)$, we can construct the corresponding quotient algebra \`L/D = = $(L/D, \vee, \wedge)$ called a quotient lattice, the operations of which are determined by $[a_1] \vee [a_2] = [a_1 \vee a_2]$ and $[a_1] \wedge [a_2] = [a_1 \wedge a_2]$.

Let $h:L\to T$ be a lattice-homomorphism of a lattice $`L = (L,\vee,\wedge)$ into a lattice $`T = (T,\vee,\wedge)$. Define a relation D_h on L by

$$(a,b)\in D_h \Leftrightarrow h(a) = h(b).$$

Then D_h is a congruence relation of $`L$. It is called the canonical congruence associated with h. The set

$$h(L) = \{t\in T; \exists\, k\in L :: h(k) = t\}$$

is called the homomorphic image of $`L$ and it is the base set of a sublattice in $`T$.

Homomorphism Theorem. Under the notation above,

$$`L/D_h \cong (h(L),\vee,\wedge).$$

Proof. Note that $[a] = \{b\in L; (a,b)\in D_h\} = \{b\in L; h(a) = h(b)\}$ and consider the mapping $f:[a]\mapsto h(a)$. //

Lemma. Suppose $`L$ is a lattice which is the direct sum of sublattices $`K_1, `K_2$. Then

(i) K_1 and K_2 are ideals of $`L$;

(ii) the mapping $p_i:L\to K_i$ defined by $p_i:a_1\vee a_2\mapsto a_i$ ($a_1\in K_1$, $a_2\in K_2$) is a lattice-homomorphism of $`L$ onto $`K_i$ for $i = 1,2$.

Proof. Ad (i). Let $a\in K_1$ and $q\vee y = b \leqslant a$ where $q\in K_1$, $y\in K_2$. Then $q\vee y = b\leqslant a = a\vee 0$. This together with (ii) of Proposition in I.17 implies that $y = 0$. Thus $b = q\in K_1$.

Ad (ii). Using the notation of the same proposition and putting $a = p\vee x$, $b = q\vee y$, we get

$$p_1(a\vee b) = p_1(p\vee q\vee x\vee y) = p\vee q = p_1(a)\vee p_1(b);$$

$$p_1(a\wedge b) = p_1((p\wedge q)\vee(x\wedge y)) = p\wedge q = p_1(a)\wedge p_1(b). \quad //$$

By a <u>kernel</u> of a congruence relation D on a lattice $`L$ with $0$ is meant the set $\{k \in L;\ (0,k) \in D\}$. Notice that it is an ideal of $`L$.

<u>Proposition</u>. Suppose $`L$ is a lattice which is the direct sum of sublattices $`K_1, `K_2$. Let $a = a_1 \vee a_2$, $b = b_1 \vee b_2$ where $a_1, b_1 \in K_1$ and $a_2, b_2 \in K_2$. Define a relation T_1 on L so that $(a,b) \in T_1$ if and only if $a_1 = b_1$. Symmetrically, let T_2 be a relation defined on L by $(a,b) \in T_2$ if and only if $a_2 = b_2$.

Then

(i) T_1 and T_2 are congruence relations of $`L$;

(ii) K_2 is the kernel of T_1 and K_1 is the kernel of T_2;

(iii) $`L/T_1 \cong `K_1$ and $`L/T_2 \cong `K_2$.

<u>Proof</u>. Observe that T_i is the canonical congruence associated with p_i. //

Let D be a congruence relation of a relatively complemented lattice $`L$ with 0 and let I denote the kernel of D. Then $(a,b) \in D$ if and only if $(0,r) \in D$ for every relative complement of $a \wedge b$ in $[0, a \vee b]$. Indeed, if $(a,b) \in D$, then $(a, a \wedge b) = (a \wedge a, a \wedge b) \in D$ and $(a, a \vee b) = (a \vee a, a \vee b) \in D$. Hence $(a \wedge b, a \vee b) \in D$ and, therefore, $(0,r) = (r \wedge (a \wedge b), r \wedge (a \vee b)) \in D$. Conversely, if $(0,r) \in D$, then $(a \wedge b, a \vee b) = ((a \wedge b) \vee 0, (a \wedge b) \vee r) \in D$ and so $(a, a \vee b) = (a \vee (a \wedge b), a \vee (a \vee b)) \in D$. By symmetry, $(b, a \vee b) \in D$ and we find that $(a,b) \in D$.

Note that the words "for every relative complement" in this statement can be replaced by "for a relative complement".

Consequently, in the case of relatively complemented lattices with 0 we see that every element $[a]$ of $`L/D$ is completely determined by a and by the kernel I of the congruence relation D. Indeed, by definition, $[a] = \{b \in L;\ (a,b) \in D\}$. This set is equal to the set of those $b \in L$ which are such that $r \in I$ for every relative complement r of $a \wedge b$ in $[0, a \vee b]$. This remark justifies why we sometimes write `L/I instead of `L/D, provided `L is a relatively complemented lattice.

<u>Corollary</u>. Let `L be a relatively complemented lattice and let `L = `$K_1 \oplus$ `K_2. Then `L/$K_1 \cong$ `K_2 and `L/$K_2 \cong$ `K_1.

<u>Proof</u>. First observe that `L has 0 by the proposition of I.17. Moreover, $K_1$ and $K_2$ are kernels of $T_2$ and $T_1$, respectively (see the proposition above). Hence, by definition and by (iii) of the mentioned proposition, `L/K_1 = `L/$T_2 \cong$ `K_1 and `L/$K_2$ = `L/$T_1 \cong$ `K_1. //

If an algebra `A has only the trivial congruence relations, i.e., the universal relation $A \times A$ and the diagonal relation $\Delta =$ $= \{(a,a);\ a \in A\}$, it is called <u>simple</u>.

25. The construction of the direct product of lattices (cf. I.16) can be easily generalized to the construction of the direct product of algebras. Let `$A_i = (A_i, F)$, $i \in I \neq \emptyset$, be algebras of the same type. The <u>direct product</u> $\prod$(`A_i; $i \in I$) is an algebra `A = (A,F) of the same type as the algebras `A_i. Its base set A is equal to the Cartesian product $\prod(A_i;\ i \in I)$. The operations of `A are defined componentweise, i.e., if $\underline{a}_j = (a_{ji};\ i \in I) \in A$, $j = 1,2,\ldots,k$, then

$$f(\underline{a}_1, \underline{a}_2, \ldots, \underline{a}_k) = (f(a_{1i}, a_{2i}, \ldots, a_{ki});\ i \in I).$$

A subalgebra `B of the direct product $\prod$(`A_i; $i \in I$) is said to be a <u>subdirect product</u> of the algebras `A_i, $i \in I$, if and only if, given any $i \in I$ and any $a_i \in A_i$, there exists an element $\underline{b} = (b_i;\ i \in I) \in B$ such that $a_i = b_i$.

Let `B be a subdirect product of the algebras `A_i, $i \in I$.

Then the relation T_j $(j \in I)$ defined on B by

$$(\underline{a},\underline{b}) \in T_j \Leftrightarrow a_j = b_j$$

where $\underline{a} = (a_i;\ i \in I)$, $\underline{b} = (b_i;\ i \in I)$ is a congruence relation on B.

The proofs of the following classical theorems A - C due to Birkhoff are given in [73].

Theorem A. Let `B be a subdirect product of the algebras `A_i, $i \in I$. Then `$B/T_j \cong$ `A_j and $\bigcap(T_i;\ i \in I)$ is equal to the diagonal relation Δ on B.

Theorem B. Let `A be an algebra and let $E_i$, $i \in I$, be congruence relations of `A such that $\bigcap(E_i;\ i \in I) = \Delta$. Then `A is isomorphic to a subdirect product of the algebras `$A/E_i$, $i \in I$.

An algebra `A is called subdirectly irreducible when, in any family of congruence relations E_i satisfying $\bigcap(E_i;\ i \in I) = \Delta$, at least one $i \in I$ is such that $E_i = \Delta$.

Theorem C. Every algebra is isomorphic to a subdirect product of subdirectly irreducible algebras of the same type.

In the case of distributive lattices, this has the following corollary also due to Birkhoff (cf. [179, Thm 100]):

Corollary D. Every distributive lattice is a subdirect product of two-element lattices.

26. Let `A = (A,F) be an algebra and let p and q be two polynomials in the operations of F. An identity is an expression of the form p = q. Such an identity holds in the algebra `A if and only if the corresponding polynomial functions represent the same function on A, i.e., if $p(a_1,\bullet) = q(a_1,\bullet)$

for all $a_1, \bullet$ of F.

Let $\underline{K}$ be a class of algebras of the same type. The class $\underline{K}$ is said to be equational if there exists a set of identities $p_i = q_i$, $i \in I$, such that $`A \in \underline{K}$ if and only if they hold in $`A$. Such a class is also called a primitive class or a variety.

If $\underline{K}$ is an equational class of algebras, then all free algebras over $\underline{K}$ exist (cf. [73, p. 167]).

Let $\underline{K}$ be a class of algebras of the same type. Let $S\underline{K}$ denote the class of all the algebras which are isomorphic to a subalgebra of an algebra $`A \in \underline{K}$. Let $P\underline{K}$ denote the class of all the algebras which are isomorphic to a direct product of a family of algebras $`A_i \in \underline{K}$. Finally, let $H\underline{K}$ be the class which consists of all the algebras which are homomorphic images of algebras $`A_i \in \underline{K}$.

It is known (cf. [73]) that $\underline{K}$ is an equational class of algebras if and only if $\underline{K} = S\underline{K}$, $\underline{K} = P\underline{K}$ and $\underline{K} = H\underline{K}$.

27. Now we present some results on equational classes of lattices.

First of all, we shall be concerned with the smallest equational class of lattices usually denoted by $\underline{O}$. It consists of all the lattices in which the identity $x_1 = x_2$ holds. Here x_1 and x_2 denote two different indeterminates. Clearly, the class $\underline{O}$ consists of the one-element lattices.

We say that an equational class $\underline{K}_1$ of lattices is covered by an equational class $\underline{K}_2$ of lattices (or, equivalently, that $\underline{K}_2$ covers $\underline{K}_1$) if and only if $\underline{K}_1$ is a subclass of $\underline{K}_2$ and if there is no equational class $\underline{K}_3$ of lattices such that $\underline{K}_1 \subset \subset \underline{K}_3 \subset \underline{K}_2$, $\underline{K}_1 \neq \underline{K}_3$ and $\underline{K}_3 \neq \underline{K}_2$. In this case we write $\underline{K}_1 \prec \underline{K}_2$.

Our first remark will be concerned with the fact that there

is exactly one equational class of lattices which covers the class $\underline{O}$, namely the class $\underline{D}$ of distributive lattices.

Indeed, let $\underline{K}$ be an equational class of lattices which is different from the class $\underline{O}$. Since $\underline{O} \neq \underline{K}$, there is a lattice $L \in \underline{K}$ which has at least two distinct elements. From $\underline{K} = S\underline{K}$ we conclude that any two-element lattice belongs to $\underline{K}$. By $\underline{K} = S\underline{K}$ and $\underline{K} = P\underline{K}$, we find that every subdirect product of two-element lattices belongs to $\underline{K}$. By Corollary D of I.25, every distributive lattice is a subdirect product of some family of two-element lattices. Hence $\underline{O} \neq \underline{D} \subset \underline{K}$ and, therefore, $\underline{O} \prec \underline{D}$.

However, in contrast to this situation, the equational class of modular lattices does not cover the class $\underline{D}$. In [4] it is proved that there are uncountably many equational classes of modular lattices.

We add, finally, the following closely related remark, which is not used in this book, but, being of interest in itself, will be justified here.

Let $\underline{M}_5$ denote the equational class of lattices which is generated by the diamond M_5, i.e., it is the smallest equational class of lattices which contains the diamond. Similarly, let $\underline{N}_5$ be the equational class of lattices generated by the pentagon N_5. It is known that $\underline{M}_5$ and $\underline{N}_5$ are the only equational classes of lattices covering the class $\underline{D}$.

For further information on equational classes of lattices, see [46] and the periodical literature (cf. [107]).

ELEMENTARY THEORY OF ORTHOMODULAR LATTICES

Chapter II

1. ORTHOLATTICES

By an ortholattice we shall mean an algebra $`L = (L, \vee, \wedge, ', 0, 1)$ satisfying the following postulates:

(i) the algebra $(L, \vee, \wedge)$ is a lattice;

(ii) the unary operation $' : s \mapsto s'$ is such that the relations $s \vee s' = 1$, $s \wedge s' = 0$ hold for every $s \in L$;

(iii) if $s \leq t$, then $s' \geq t'$;

(iv) $(s')' = s$ for every $s \in L$.

The operation $'$ is called the orthocomplementation of the ortholattice $`L$, the element s' is called the orthocomplement of s.

Remark 1.1. (A) Since $s \leq s \vee s' = 1$, it is clear that the element 1 is the greatest element of the lattice $(L, \vee, \wedge)$ and we can similarly see that 0 is the least element of the lattice.

(B) Some authors use the notation $s^\perp$ instead of s'.

(C) By I.6 and (iv), the mapping given by the orthocomplementation is necessarily a one-to-one mapping of L onto L, i.e., it is bijective.

(D) The terminology can be illustrated by the following geometrical example: The plane a (determined by the lines x,y),

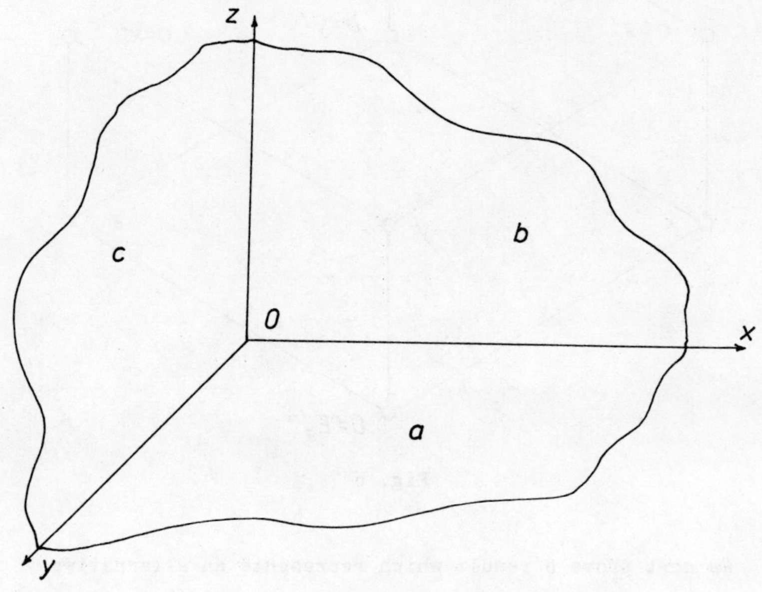

Fig. 5

the plane b (determined by x,z), the plane c (determined by y,z) and the lines x,y,z as well as the origin O and the whole space E_3 interpreted as point sets ordered by set-inclusion form an ortholattice (see Figures 5, 6).

(E) Suppose that $(L,\vee,\wedge,{}',0,1)$ is an ortholattice. If we replace the lattice $(L,\vee,\wedge)$ by the dual lattice $(L,\wedge,\vee)$, we get again an ortholattice. It follows that in proving theorems about ortholattices we can use Duality Principle. Hence, with every proved theorem also the dual one is valid.

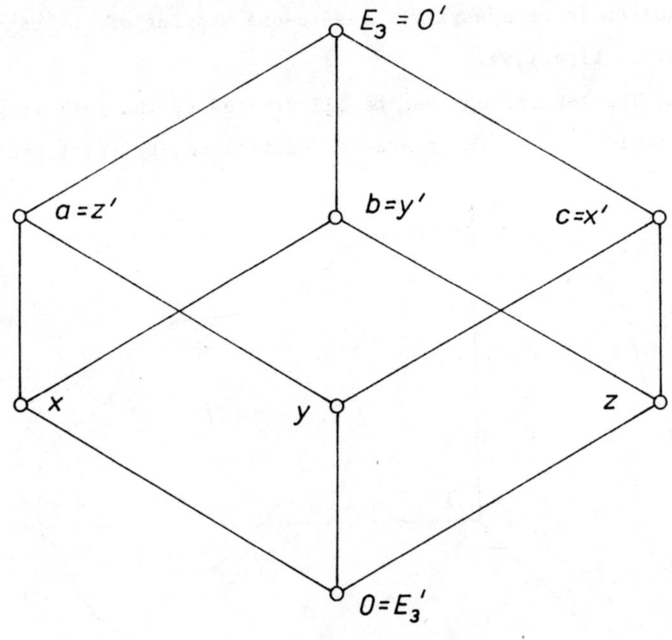

Fig. 6

We next prove a result which represents an alternative characterization of ortholattices.

Theorem 1.2 (cf. [54], [156]). Let `L be a lattice and a unary operation defined on L and satisfying on L the condition (iv) from the definition of an ortholattice. Then the condition (iii) of the definition is equivalent to De Morgan laws, i.e., to the fact that for every $s,t \in L$

$$(s \vee t)' = s' \wedge t' \quad (s \wedge t)' = s' \vee t'.$$

Proof. 1. If De Morgan laws are valid, then from $s \leq t$ it follows $s \vee t = t$ and, therefore, $t' = (s \vee t)' = s' \wedge t' \leqslant s'$.

2. Conversely, let L be a lattice satisfying (iii). Since $x \wedge y \leqslant x$, we have $(x \wedge y)' \geqslant x'$ and, similarly, $(x \wedge y)' \geqslant y'$. Therefore,

(1) $(x \wedge y)' \geqslant x' \vee y'$.

By duality,

(2) $(x \vee y)' \leqslant x' \wedge y'$.

From this and (iv) we have

(3) $(x' \wedge y')' \geqslant x'' \vee y'' = x \vee y$

and

(4) $(x' \vee y')' \leqslant x'' \wedge y'' = x \wedge y$.

Using (iv) and (iii), we see that $x' \wedge y' = (x' \wedge y')'' \leqslant (x \vee y)'$ and $x' \vee y' = (x' \vee y')'' \geqslant (x \wedge y)'$. Therefore $(x \vee y)' = x' \wedge y'$ and $(x \wedge y)' = x' \vee y'$. //

Theorem 1.3 (cf. [17]). The following three conditions are necessary and sufficient for an algebra with two binary operations $\vee, \wedge$ and a unary operation $'$ to be an ortholattice:

(b 1) $(a \vee b) \vee c = (c' \wedge b')' \vee a$ for all a,b,c;

(b 2) $a = a \wedge (a \vee b)$ for all a,b;

(b 3) $a = a \vee (b \wedge b')$ for all a,b.

Proof. 1. Evidently, the conditions (b 1) - (b 3) hold in any ortholattice.

2. Let d be an element of such an algebra defined on L and let 0 denote the element $d \wedge d'$. If $a \in L$, then

$$a \wedge a = a \wedge (a \vee 0) = a$$

by (b 3) and (b 2). Hence

(5) $a \wedge a = a$ for every $a \in L$.

We next show that

(6) $0 = 0''$.

Setting $c = b = 0$ in (b 1), using (b 3),(5) and the abbreviation $0'' = (0')'$, we get

(7) $\forall\, a \in L \quad a = 0'' \vee a$.

By (b 3) we now have $0 = 0'' \vee 0 = 0''$. Combining (6) and (7), we find

(8) $\forall\, a \in L \quad a = 0 \vee a$.

From (b 3) we conclude that

$$0 = b \wedge b' = (b \wedge b') \vee (c \wedge c') = 0 \vee (c \wedge c')$$

for every $c \in L$. Thus, by (8), $0 = c \wedge c'$ for any $c \in L$.

Setting $a = 0$ in (b 1), using (8) and (b 3), we obtain

(9) $\forall\, b,c \in L \quad b \vee c = (c' \wedge b')'$

and (with $c = b$) it follows that

(10) $\forall\, b \in L \quad b \vee b = b''$,

by (5). Choosing $c = 0$ in (9), we also have $b = (0' \wedge b')'$ for every $b \in L$. Further, by (b 1) (with $c = 0$), it is clear that

(11) $\forall\, a,b \in L \quad a \vee b = b \vee a$.

On the other hand, by (b 1), (9) and (11), $(a \vee b) \vee c = (b \vee c) \vee a = a \vee (b \vee c)$. Hence,

(12) $\forall\, a,b,c \in L \quad (a \vee b) \vee c = a \vee (b \vee c)$.

Finally, using the condition (b 2), we obtain $a' = a' \wedge (a' \vee a)$. But $a' \vee a = (a' \wedge a'')'$ by (9). Hence we get $a' = a' \wedge (a' \wedge a'')'$ and, therefore, $a'' = (a' \wedge (a' \wedge a'')')' =$

$= (a' \wedge a'') \vee a = 0 \vee a = a$, by (9) and (8). Thus

(13) $\forall a \in L \quad a = a''$.

Now we have the following consequences: First, if a runs over the elements of the set L, then by (13), a' runs over all the elements of L. Next, from (9),(13) and (11) we conclude that $c' \wedge b' = (b \vee c)' = (c \vee b)' = b' \wedge c'$. So we have De Morgan laws

(14) $(b \vee c)' = b' \wedge c', \quad (b \wedge c)' = b' \vee c'$.

The commutativity and the associativity of the operation $\wedge$ is evident from (14),(11) and (12). The second absorption law follows similarly from (b 2). Thus we see that $(L, \vee, \wedge)$ is a lattice.

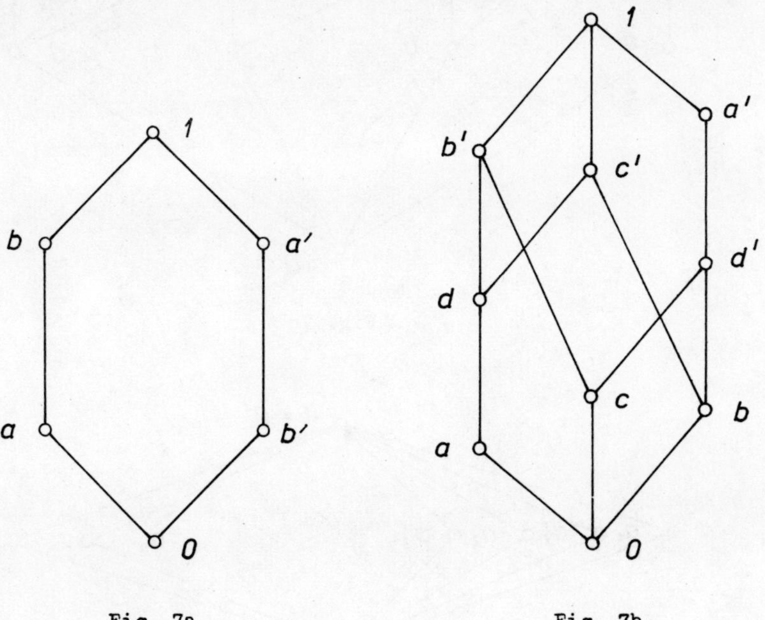

Fig. 7a Fig. 7b

Defining $0' = 1$, we get $1 = 0' = (c \wedge c')' = c' \vee c$ by (14). By Theorem 2, $(L, \vee, \wedge, ', 0, 1)$ is an ortholattice. //

Remark 1.4. (A) Every Boolean algebra is obviously an ortholattice.

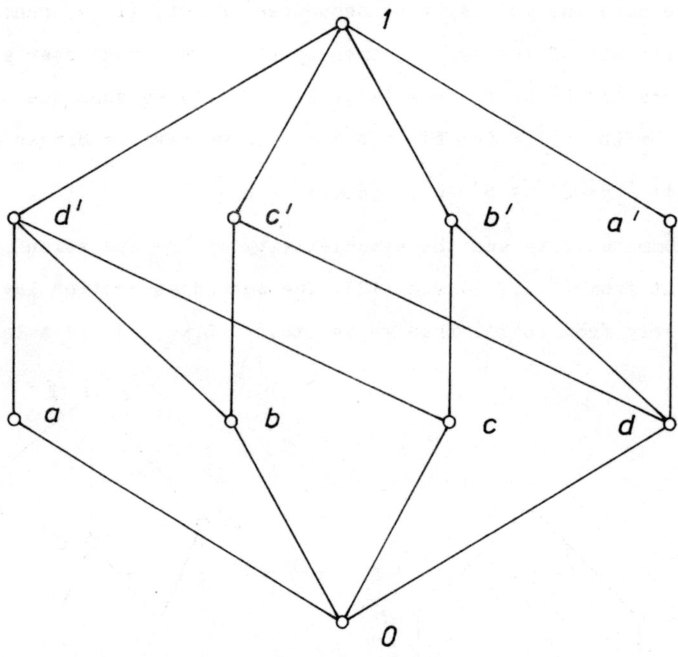

Fig. 7c

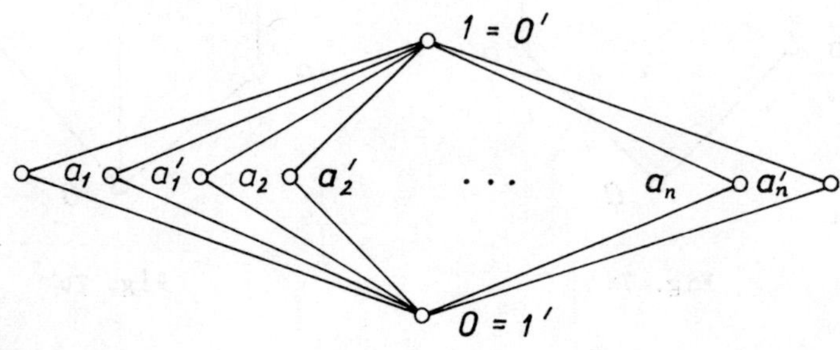

Fig. 8

(B) There exist ortholattices which are not modular. This is the case of the lattices sketched in Figures 7a-c (cf. [145]). Any ortholattice isomorphic to the ortholattice of Figure 7a will be called a benzene ring.

(C) It is easy to find an ortholattice which is modular but which is not distributive. Every lattice M_{0n}, $n > 1$, having its diagram shown in Figure 8 yields an example of such a lattice.

(D) No more than one-element finite ortholattice can have an odd number of elements. Indeed, the relation R defined on L by $(a,b) \in R$ if and only if $b = a'$ or $b = a$ is an equivalence relation on L. Since $a = a'$ implies $0 = a \wedge a' = a = a \vee a' = 1$, a contradiction, it is in the considered case clear that the equivalence classes of R are two-element sets.

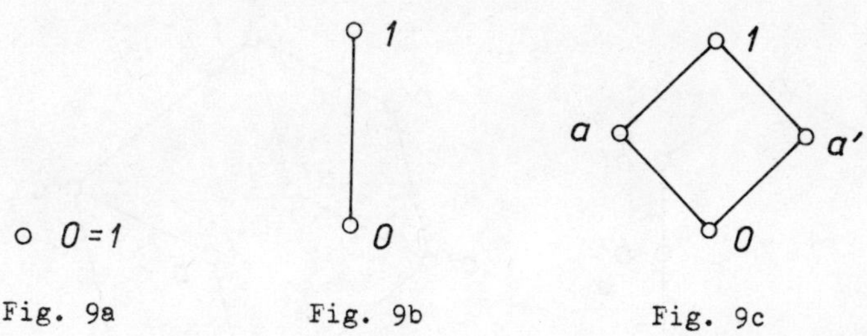

Fig. 9a Fig. 9b Fig. 9c

(E) For every ortholattice, the mapping $f: a \mapsto a'$ is an isomorphism of the lattice $(L, \vee, \wedge)$ onto the dual lattice $(L, \wedge, \vee)$. It follows that every ortholattice is self-dual. If an element a is covered in an ortholattice by n elements $b_1, b_2, \ldots, b_n$ exactly, then the element a' covers exactly n elements $b_1', b_2', \ldots, b_n'$. These two facts and the paper [129] show that between the lattices which have not more than eight elements we find (up to isomorphism) only the ortholattices depicted in Figures 9a - 9j. In the cases a,b,c,i we see the Boolean algebras $\underline{1},\underline{2},\underline{2}^2,\underline{2}^3$ and in the cases d,j the lattices $`M_{02}$ and $`M_{03}$.

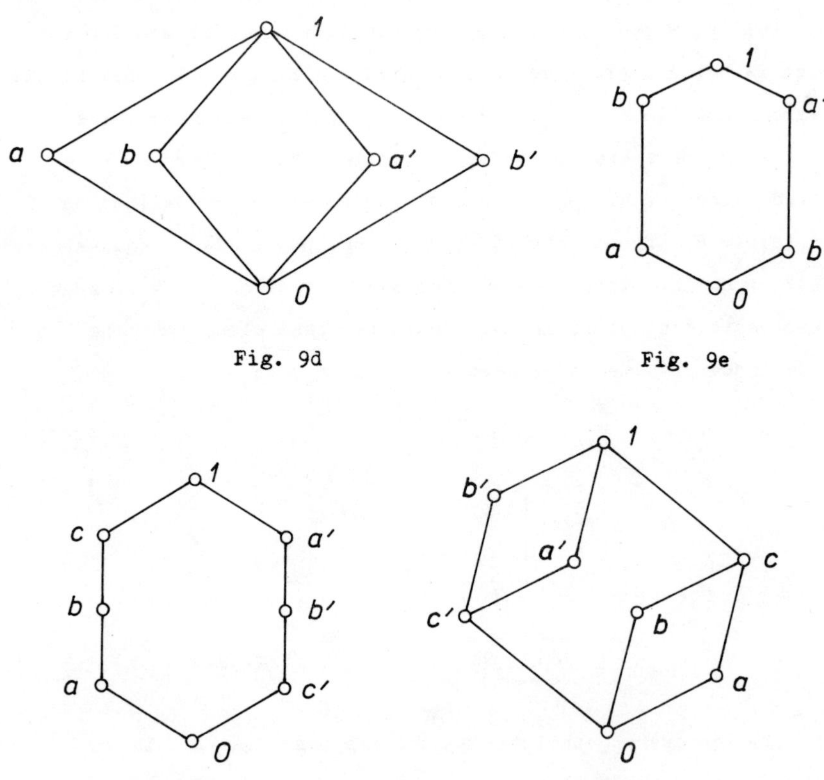

Fig. 9d

Fig. 9e

Fig. 9f

Fig. 9g

Two elements a,b of an ortholattice are said to be <u>orthogonal</u> if and only if $a \leq b'$. For such elements we write $a \perp b$.

<u>Lemma 1.5</u>. The following conditions hold in any ortholattice:

(i) $a \perp a \Leftrightarrow a = 0$; (ii) $a \perp b \Leftrightarrow b \perp a$;

(iii) $(a \perp b \;\&\; a \perp c) \Rightarrow (a \perp b \wedge c \;\&\; a \perp b \vee c)$.

Proof. Ad (i). The relation $a \perp a$ holds if and only if $a \leq a'$, i.e., if and only if $a \leq a \wedge a' = 0$.

Ad (ii). From the condition (iii) in the definition of an ortholattice we see that $a \leq b'$ if and only if $b \leq a'$.

Ad (iii). If $a \leq b'$ and $a \leq c'$, we have $a \leq b' \wedge c' = (b \vee c)'$ and also $a \leq b' \vee c' = (b \wedge c)'$, by De Morgan laws. //

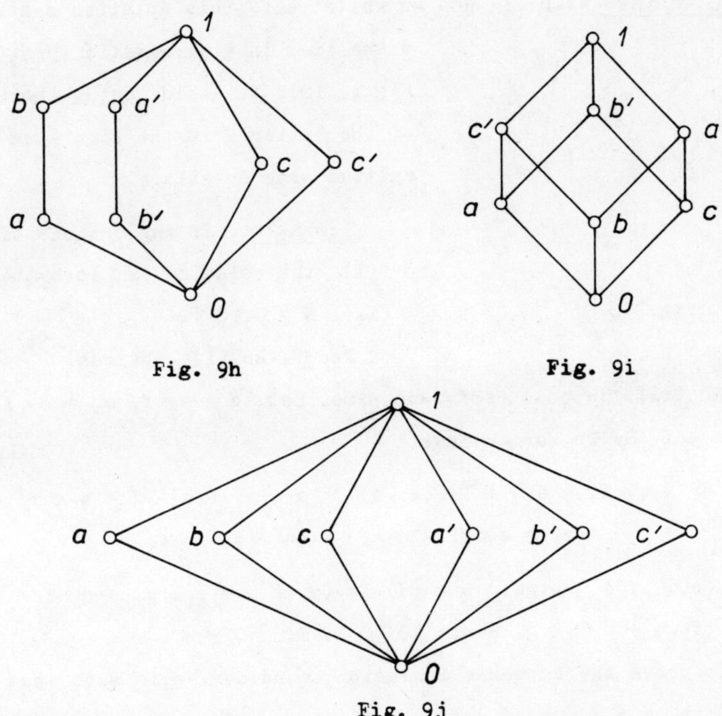

Fig. 9h

Fig. 9i

Fig. 9j

2. COMMUTATIVITY

If a,b are elements of an ortholattice L such that $a = (a \wedge b) \vee (a \wedge b')$, we write aCb and say that a commutes with b. The relation C is called the <u>commutativity relation</u> of L. If $a = (a \vee b) \wedge (a \vee b')$, we write aDb and say that a <u>dually commutes</u> with b. Finally, if there are elements $e_1, f_1, g \in L$ such that

$$e = e_1 \vee g \ \& \ f = f_1 \vee g$$

and such that $e_1 \perp g$, $f_1 \perp g$, $e_1 \perp f_1$, then we say that e <u>orthogonally commutes</u> with f and we write eMf. This notation can be visualized in diagrammatic form (see Figure 10). To avoid losing the shape of the letter M in the figure we have omitted some details.

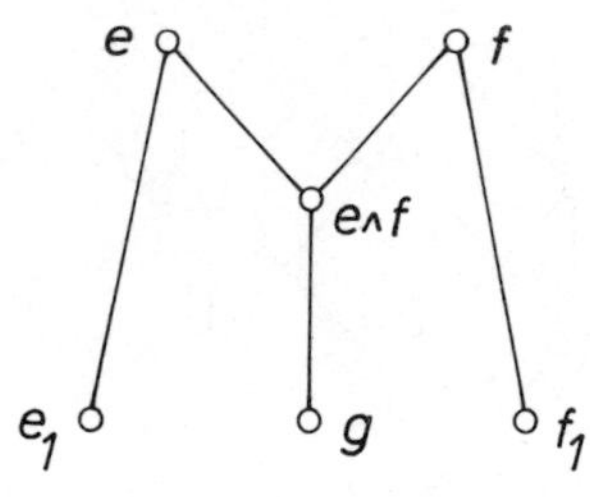

Fig. 10

<u>Lemma 2.1</u>. In any ortholattice,

(i) the relation M is symmetric;

(ii) $M = C \cap C^{-1}$.

<u>Proof</u>. Ad (i). Obvious.

Ad (ii). Suppose eCf and fCe. Let $g = e \wedge f$, $e_1 = e \wedge f'$, $f_1 = e' \wedge f$. By De Morgan laws,

$$e_1 = e \wedge f' \leqslant e' \vee f' = (e \wedge f)' = g', \ f_1 = e' \wedge f \leqslant e' \vee f' = g',$$
$$e_1 = e \wedge f' \leqslant e \vee f' = (e' \wedge f)' = f_1'.$$

Since eCf, fCe, we get $e = (e \wedge f) \vee (e \wedge f') = e_1 \vee g$ and $f = (f \wedge e) \vee (f \wedge e') = f_1 \vee g$. Hence $C \cap C^{-1} \subset M$.

To prove the converse inclusion, consider e,f such that eMf. Then $e = e_1 \vee g$, $f = f_1 \vee g$, $f_1' \geqslant e_1$ and $g' \geqslant e_1$. It follows that $f' = (f_1 \vee g)' = f_1' \wedge g' \geqslant e_1$ and, consequently, $e \wedge f' \geqslant e \wedge e_1 = (e_1 \vee g) \wedge e_1 = e_1$. From $e \wedge f = (e_1 \vee g) \wedge (f_1 \vee g) \geqslant g$ we see that $e \geqslant (e \wedge f) \vee (e \wedge f') \geqslant g \vee e_1 = e$ and so $e = (e \wedge f) \vee (e \wedge f')$. This means that eCf. Since M is symmetric, we also have fCe, i.e., $M \subset C \cap C^{-1}$. //

Remark 2.2. (A) A glance at Figure 7a shows that here aCb but b does not commute with a. Hence, by Lemma 1 (ii), it is easy to see that aMb does not hold.

Observe that the ortholattice of Figure 7a also proves that the relation C need not be symmetric.

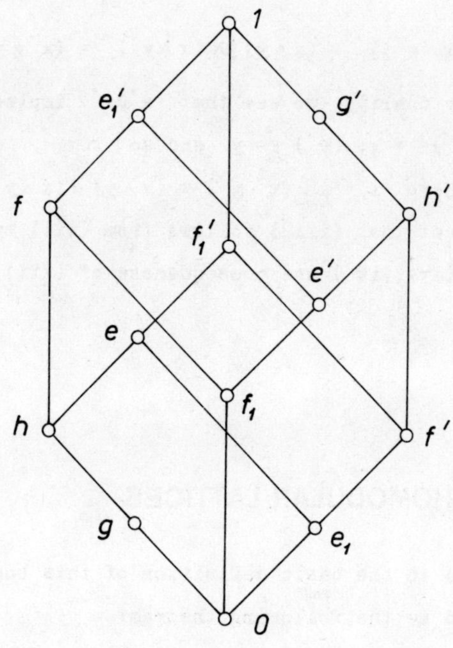

Fig. 11

(B) The lattice of Figure 11 shows an ortholattice where $e = e_1 \vee g$, $f = f_1 \vee g$ and $g \neq e \wedge f$.

(C) It is immediate that the relations C,D are dual.

Theorem 2.3. If x,y are elements of an ortholattice, then

(i) $xCy \Leftrightarrow xCy'$; (i′) $xDy \Leftrightarrow xDy'$;

(ii) $xCy \Leftrightarrow x'Dy'$;

(iii) $x \leqslant y \Rightarrow xCy$; (iii′) $x \geqslant y \Rightarrow xDy$;

(iv) $x \leqslant y' \Rightarrow xCy$; (iv′) $x \geqslant y' \Rightarrow xDy$.

Proof. The statements (i),(i′) follow by the definition of C,D.

Ad (ii). Suppose that $x = (x \wedge y) \vee (x \wedge y')$. Then, by De Morgan laws,

$$x' = [(x \wedge y) \vee (x \wedge y')]' = (x \wedge y)' \wedge (x \wedge y')' = (x' \vee y') \wedge (x' \vee y),$$

i.e., $x'Dy'$. By duality, we see that $x'Dy'$ implies $(x')'C\,C(y')'$. But $(x')' = x$, $(y')' = y$ and so xCy.

Ad (iii). Here $x = x \vee (x \wedge y') = (x \wedge y) \vee (x \wedge y')$.

It is evident that (iii′) follows from (iii) by duality. The statements (iv),(iv′) are consequences of (iii),(iii′),(i) and (i′). //

3. ORTHOMODULAR LATTICES

We now come to the basic definition of this book, which will be prepared by the following theorem:

Theorem 3.1. In any ortholattice the following conditions are equivalent:

(i) $(s \geqslant t' \;\&\; s \wedge t = 0) \Rightarrow s = t'$;

(ii) $(s \leqslant t' \;\&\; s \vee t = 1) \Rightarrow s = t'$;

(iii) $a \leqslant c \Rightarrow a \vee (a' \wedge c) = c$;

(iv) $a \geqslant c \Rightarrow a \wedge (a' \vee c) = c$.

Proof. Ad (i) $\Rightarrow$ (iii). If $a \leqslant c$, then $t' = a \vee (a' \wedge c) \leqslant \leqslant c = s$. Moreover, by De Morgan laws,

$$s \wedge t = c \wedge [a \vee (a' \wedge c)]' = c \wedge a' \wedge (a \vee c') = (a \vee c')' \wedge (a \vee c') = 0.$$

By (i), $c = s = t' = a \vee (a' \wedge c)$.

Ad (iii) $\Rightarrow$ (iv). Observe that $a \geqslant c$ implies $a' \leqslant c'$. Therefore, by (iii), $a' \vee (a'' \wedge c') = c'$. But $a'' = a$ and, by De Morgan laws,

$$c = c'' = [a' \vee (a \wedge c')]' = a \wedge (a' \vee c).$$

Ad (iv) $\Rightarrow$ (i). Suppose $s \geqslant t'$ and $s \wedge t = 0$. Let $a = s$, $c = t'$. By (iv) we then have

$$s \geqslant t' \Rightarrow s \wedge (s' \vee t') = t'.$$

By De Morgan laws it is evident that $t = s' \vee (s \wedge t) = s' \vee 0 = s'$. Thus $t' = s$.

The proved statements can be indicated in the following way:

$$\begin{array}{ccc} \text{(i)} & \Longrightarrow & \text{(iii)} \\ \nwarrow & & \swarrow \\ & \text{(iv)} & \end{array}$$

Let (n^+) denote the dual statement to (n). By duality, we also have

$$\begin{array}{ccc} (\text{i}^+) & \Longrightarrow & (\text{iii}^+) \\ \nwarrow & & \swarrow \\ & (\text{iv}^+) & \end{array}$$

But $(\text{i}^+) = (\text{ii})$, $(\text{iv}^+) = (\text{iii})$, $(\text{iii}^+) = (\text{iv})$, and so

$$\begin{array}{ccccc} \text{(ii)} & \Longrightarrow & \text{(iv)} & & \\ & \nwarrow & \swarrow & \searrow & \\ & & \text{(iii)} & \Longleftarrow & \text{(i)} \end{array}$$

Now it is obvious that the four conditions are equivalent. //

Remark 3.2. The final part in the proof of the theorem 1 shows in detail the simplification in considerations which is enabled by Duality Principle. Similar arguments will be used in what follows but in a more concise way.

An algebra $`L = (L, \vee, \wedge, ', 0, 1)$ is called an orthomodular lattice if and only if $`L$ is an ortholattice which satisfies one of the conditions in the theorem 1.

Remark 3.3 (A) Every Boolean algebra is an orthomodular lattice. Indeed, by distributivity, $a \vee (a' \wedge c) = (a \vee a') \wedge (a \vee c) = a \vee c$. Hence, $a \leqslant c$ implies that $a \vee (a' \wedge c) = c$. Clearly, any ortholattice $(L, \vee, \wedge, ', 0, 1)$ where $(L, \vee, \wedge)$ is modular determines an orthomodular lattice.

(B) The ortholattices of Figures 7a - c are not orthomodular. In the lattice of Figure 7c we for example have $s = d' \geqslant t' = a$ and $s \wedge t = d' \wedge a' = 0$ but $s \neq t'$.

(C) In Figures 9a - j only the ortholattices of 9a, 9b, 9c, 9d, 9i and 9j are orthomodular. These orthomodular lattices are either distributive or modular.

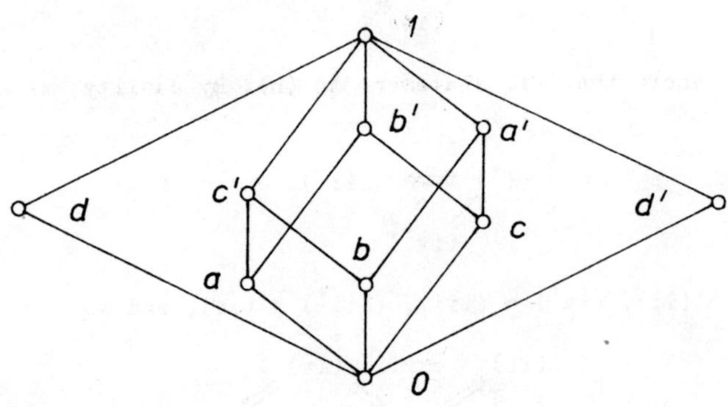

Fig. 12

We shall now exhibit an example of an orthomodular lattice which is not modular. For this reason consider the lattice whose diagram is sketched in Figure 12. The elements $0, a, c', 1, d$ form a sublattice isomorphic to $`N_5$.

(D) If $`L$ is an orthomodular lattice, then the dual lattice is also orthomodular. Hence, the Duality Principle applies to orthomodular lattices. If $`L$ is an orthomodular lattice, then the remark 1.4.D shows that a sublattice of $(L,\vee,\wedge)$ need not be even an ortholattice.

(E) If $`L_i$, $i = 1,2,\dots,n$, are orthomodular lattices, then also the direct product of these lattices is orthomodular. Recall (cf. I.16) that the direct product $`L$ is determined by the set L of all ordered n-tuples $\underline{a} = (a_1,a_2,\dots,a_n)$ where $a_i \in L_i$. If, moreover, $\underline{b} = (b_1,b_2,\dots,b_n)$, then one defines $\underline{a} \vee \underline{b}$ as $(a_1 \vee b_1, a_2 \vee b_2,\dots,a_n \vee b_n)$ and $\underline{a} \wedge \underline{b}$ as $(a_1 \wedge b_1, a_2 \wedge b_2,\dots$ $\dots,a_n \wedge b_n)$. In the lattice $`L$ we have $\underline{a} \leqslant \underline{b}$ if and only if $a_i \leqslant b_i$ for every $i = 1,2,\dots,n$. If we define $\underline{a}' = (a_1',a_2',\dots$ $\dots,a_n')$ where $a_i'$ denotes the orthocomplement of $a_i$ in the lattice $`L_i$, then $`L$ is an ortholattice. From $\underline{a} \geqslant \underline{b}'$ and $\underline{a} \wedge \underline{b} = \underline{0}$ it follows that $a_i \geqslant b_i'$ and $a_i \wedge b_i = 0$ for every $i = 1,2,\dots,n$. Since $`L_i$ are orthomodular lattices, $a_i = b_i'$ and so $\underline{a} = \underline{b}'$. Thus, $`L$ is orthomodular.

Let $K \subset L$ be a nonvoid set and let $`L = (L,\vee,\wedge,',0,1)$ be an ortholattice. If $K$ is closed under the operations $\vee,\wedge,'$, $0,1$, i.e., if for every $a,b \in K$ the elements $a \vee b$, $a \wedge b$, $a'$ belong to $K$ and $0,1 \in K$, then $(K,\vee,\wedge,',0,1)$ is called a <u>subalgebra</u> of the ortholattice $`L$ and it is denoted by $`K$.

Every subalgebra of an ortholattice determines a sublattice $(K,\vee,\wedge)$ of the lattice $(L,\vee,\wedge)$. If $`L$ is an orthomodular lattice, then every subalgebra of $`L$ is also an orthomodular lattice.

Let $X = \{x_1,x_2,\dots\}$ be a set of variables. From X one can construct intuitively polynomials in the operations $\vee,\wedge,'$ such as $x_1 \vee x_2'$, $(x_1 \wedge x_2)' \vee x_3$ etc. Clearly, this lacks the necessary precision for our purposes, so we now indicate briefly how to give a precise definition as to what a polynomial in the operations $\vee, \wedge, '$ is:

Every letter x_i with $1 \leqslant i \leqslant n$ is called an n-ary polynomial in the operations $\vee, \wedge, '$ of the rank 1. Let us suppose that we have already defined the n-ary polynomials in the operations $\vee, \wedge, '$ of the rank $k \geqslant 1$. Then the n-ary polynomials in $\vee, \wedge, '$ of the rank $k + 1$ are defined as the expressions of the form

$$(f \wedge g),\ (f \vee g),\ (f)'$$

where f and g are n-ary polynomials in $\vee, \wedge, '$ of the rank k.

For example, the expressions $(x_1 \vee (x_2)')$ and $((x_1 \wedge x_2)' \vee \vee x_3)$ both are n-ary polynomials in $\vee, \wedge, '$ for every $n \geqslant 3$. However, this notation has the unfortunate effect of making appear some superfluous parentheses. In order to avoid complicating the notation, such parentheses will be omitted.

If f is an n-ary polynomial in $\vee, \wedge, '$ for an $n \in \underline{N}$, it will be called a polynomial in $\vee, \wedge, '$.

Let `L be an ortholattice. Each n-ary polynomial in $\vee, \wedge, '$ defines a function from L^n to L. More exactly, we set the matter out in formal terms as follows:

A polynomial function (or simply, a polynomial) on the ortholattice `L associated with an n-ary polynomial f in $\vee, \wedge, '$ is defined in the following way:

(i) If $f = x_i$, then $f(h_1, h_2, \dots, h_n) = h_i$ for every $h_1, h_2, \dots, h_n \in L$ and every $i = 1, 2, \dots, n$.

(ii) If $f(h_1, h_2, \dots, h_n) = h$, $g(h_1, h_2, \dots, h_n) = k$, $f \vee g = p$, $f \wedge g = q$ and $f' = r$, then

$$p(h_1,h_2,\ldots,h_n) = h \vee k, \quad q(h_1,h_2,\ldots,h_n) = h \wedge k,$$
$$r(h_1,h_2,\ldots,h_n) = h'.$$

Note the traditional abuse of language which allows us to use the same symbol f for an n-ary polynomial and for the associated polynomial function.

If we drop the operation $'$ in the construction described above, then we are left with so-called lattice polynomials (cf. [74] and I.21).

Suppose $M \subset L$ is a subset of an ortholattice $`L$. Let $`K_i$, $i \in I$, be all the subalgebras of $`L$ such that $M \subset K_i$. Then $\bigcap(K_i;\ i \in I)$ determines a subalgebra of $`L$. This subalgebra is denoted by $\langle M \rangle$ and it is called the <u>subalgebra generated</u> by the set M. By [73], an elements $a \in L$ belongs to $\langle M \rangle$ if and only if there is a polynomial p in the considered operations $\vee, \wedge, '$ of $`L$ and if there are elements $b_1,b_2,\ldots,b_n \in M$ such that $a = p(b_1,b_2,\ldots,b_n)$.

<u>Theorem 3.4</u>. An ortholattice is orthomodular if and only if its commutativity relation is symmetric.

<u>Proof</u>. 1. Suppose that the relation C of an ortholattice $`L$ is symmetric. Let $x \leqslant y$ be elements of $`L$. By Theorem 2.3, xCy and so yCx. Thus, by definition, $y = (y \wedge x) \vee (y \wedge x') =$ $= x \vee (x' \wedge y)$ and we conclude that $`L$ is orthomodular.

2. Let $`L$ be an orthomodular lattice and let xCy. Clearly $s = y \geqslant (y \wedge x) \vee (y \wedge x') = t'$ and $s \wedge t = y \wedge (x' \vee y') \wedge (x \vee y')$. However, $x = (x \wedge y) \vee (x \wedge y')$ and, therefore, $x \vee y' = (x \wedge y) \vee$ $\vee\, y'$. By Theorem 1 (iv), $(x' \vee y') \wedge [(x \wedge y) \vee y'] = y'$. Hence $s \wedge t = y \wedge (x' \vee y') \wedge [(x \wedge y) \vee y'] = y \wedge y' = 0$. This implies that $s = t'$, i.e., yCx. //

<u>Theorem 3.5</u>. An ortholattice is orthomodular if and only if

its commutativity relation has the following property: If a, b are elements of the ortholattice and aCb, then $a'Cb$.

Proof. 1. Let $`L$ be an orthomodular lattice and let aCb. By Theorem 4, bCa and from Theorem 2.3 (i) we have bCa'. Again, by Theorem 4, $a'Cb$.

2. Conversely, if C has the stated property and if $x \leq y$, then $y' \leq x'$. As shown in Theorem 2.3 (iii), $y'Cx'$. Hence yCx' and thus $y = (y \wedge x') \vee (y \wedge x) = (y \wedge x') \vee x$ so that we can conclude that $`L$ is orthomodular. //

Theorem 3.6. An ortholattice is orthomodular if and only if $C = D$.

Proof. 1. Suppose $C = D$ and $a \leq b$. Then by Theorem 2.3 (iii′) we see that bDa, and consequently, bCa which yields $b = (b \wedge a) \vee (b \wedge a') = a \vee (a' \wedge b)$. It follows that $`L$ is orthomodular.

2. If $`L$ is orthomodular and if aCb, then by Theorem 5, $a'Cb$. Using Theorem 2.3 (i), we get $a'Cb'$. From Theorem 2.3 (ii) we now conclude that aDb and so $C \subset D$. By duality, $D \subset C$ and, consequently, $C = D$. //

Theorem 3.7. Let $`L$ be an orthomodular lattice. Then the following conditions are equivalent:

(i) aCb; (ii) $(a \vee b') \wedge b = (b \vee a') \wedge a$;

(iii) $(a \vee b') \wedge b = a \wedge b$; (iii′) $(b \vee a') \wedge a = a \wedge b$;

(iv) $(a \wedge b') \vee b = (b \wedge a') \vee a$;

(v) $(a \wedge b') \vee b = a \vee b$; (v′) $(b \wedge a') \vee a = b \vee a$.

Proof. Ad (i) ⇒ (ii). Suppose that aCb. Now by Theorem 6, aDb and thus $a = (a \vee b) \wedge (a \vee b')$. Therefore, $a \wedge b = (a \vee b') \wedge b$. Similarly, from aCb it follows that bCa by Theorem 4. Hence, $b \wedge a = (b \vee a') \wedge a$. But then $a \wedge b = (a \vee b') \wedge b = (b \vee a') \wedge a$.

Ad (ii) ⇒ (iii). If $(a \vee b') \wedge b = (b \vee a') \wedge a$, then $(a \vee b') \wedge \wedge b = [(b \vee a') \wedge a] \wedge b = a \wedge b$.

Ad (iii) ⇒ (i). If $(a \vee b') \wedge b = a \wedge b$, then $(a \wedge b) \vee (a' \wedge \wedge b) = (a' \wedge b) \vee [(a \vee b') \wedge b]$. By orthomodularity, $(a' \wedge b) \vee [(a \vee \vee b') \wedge b] = b$. Therefore, $(a \wedge b) \vee (a' \wedge b) = b$, i.e., bCa. By Theorem 4, aCb.

By duality we see that aDb is equivalent to (iv) and (v). However, by Theorem 6, aDb means the same assertion as aCb.

Since the commutativity relation is symmetric in any orthomodular lattice, the assertion concerning (iii′) and (v′) is apparent. //

Lemma 3.8. Let a, b_1, b_2 be elements of an orthomodular lattice such that aCb_1 and aCb_2. Then

$$a \wedge (b_1 \vee b_2) = (a \wedge b_1) \vee (a \wedge b_2), \quad b_1 \wedge (a \vee b_2) = (b_1 \wedge a) \vee (b_1 \wedge b_2),$$
$$b_2 \wedge (a \vee b_1) = (b_2 \wedge a) \vee (b_2 \wedge b_1).$$

Proof. In any lattice,

$$s = a \wedge (b_1 \vee b_2) \geqslant (a \wedge b_1) \vee (a \wedge b_2) = t'.$$

Consider the meet of s and t. First note that

$$s \wedge t = a \wedge (b_1 \vee b_2) \wedge (a' \vee b_1') \wedge (a' \vee b_2').$$

By assumption, aCb_1 and, by Theorems 4 and 5, $b_1'Ca$. From Theorem 7 (iii) we have $a \wedge (a' \vee b_1') = a \wedge b_1'$. Similarly, from aCb_2 it follows that $b_2'Ca$. Hence, $a \wedge (a' \vee b_2') = a \wedge b_2'$ and therefore

$$s \wedge t = (b_1 \vee b_2) \wedge (a \wedge b_1') \wedge (a \wedge b_2') = (b_1 \vee b_2) \wedge (b_1 \vee b_2)' \wedge a = 0.$$

By orthomodularity we conclude that $s = t'$.

To prove the second assertion, let

$$s_1 = b_1 \wedge (a \vee b_2) \geqslant (b_1 \wedge a) \vee (b_1 \wedge b_2) = t_1'.$$

Here $s_1 \wedge t_1 = b_1 \wedge (a \vee b_2) \wedge (b_1' \vee a') \wedge (b_1' \vee b_2')$. Since aCb_1, we have $a'Cb_1$ by Theorem 5. Therefore, by Theorem 7 (iii), $b_1 \wedge \wedge (b_1' \vee a') = a' \wedge b_1$. From Theorems 4, 2.3 (i) and aCb_2 we get b_2Ca'. Hence $a' \wedge (a \vee b_2) = a' \wedge b_2$. We summarize our result:

$$s_1 \wedge t_1 = (a \vee b_2) \wedge a' \wedge b_1 \wedge (b_1' \vee b_2') = a' \wedge b_2 \wedge b_1 \wedge (b_1' \vee b_2') =$$
$$= a' \wedge (b_1 \wedge b_2) \wedge (b_1 \wedge b_2)' = 0.$$

By orthomodularity, $s_1 = t_1'$.

The last assertion follows by symmetry. //

Remark 3.9. The validity of the identities from Theorem 8 does not, in general, imply the relations aCb_1, aCb_2. To see it, put in the orthomodular lattice of Figure 12 $b_1 = d$, $b_2 = 1$. Here we have

$$a \wedge (b_1 \vee b_2) = a \wedge (d \vee 1) = a \wedge 1 = a, \quad (a \wedge b_1) \vee (a \wedge b_2) =$$
$$= (a \wedge d) \vee (a \wedge 1) = 0 \vee a = a.$$

It is a simple matter to verify that

$$b_1 \wedge (a \vee b_2) = d \wedge (a \vee 1) = d \wedge 1 = d,$$
$$(b_1 \wedge a) \vee (b_1 \wedge b_2) = (d \wedge a) \vee (d \wedge 1) = 0 \vee d = d.$$

However, a does not commute with d.

Foulis-Holland Theorem 3.10 (cf. [63], [97]). Let a,b,c be elements of an orthomodular lattice. Suppose that at least one of them commutes with other two. Then

$$a \wedge (b \vee c) = (a \wedge b) \vee (a \wedge c), \quad a \vee (b \wedge c) = (a \vee b) \wedge (a \vee c).$$

Proof. The first identity follows by Lemma 8. From Theorem 2.3 (i), Theorem 4 and Theorem 5 we conclude that at least one of the elements a',b',c' commutes with other two. Hence, $a' \wedge \wedge (b' \vee c') = (a' \wedge b') \vee (a' \wedge c')$. By De Morgan laws,

$$a \vee (b \wedge c) = [a' \wedge (b' \vee c')]' = [(a' \wedge b') \vee (a' \wedge c')]' =$$
$$= (a \vee b) \wedge (a \vee c). \quad //$$

Remark 3.11. A generalization of Theorem 10 is shown in the paper [84].

4. PROPERTIES OF COMMUTATIVITY IN ORTHOMODULAR LATTICES

By Theorem 3.6, two elements a, b of an orthomodular lattice commute if and only if they dually commute, i.e., aCb is equivalent to aDb. A similar result can be established for the elements which orthogonally commute. Analogously to Theorem 3.6 we have

Proposition 4.1. An ortholattice is orthomodular if and only if $C = M$.

Proof. 1. If an ortholattice is orthomodular, then $C = C^{-1}$ by Theorem 3.4. In view of Lemma 2.1 (ii) we obtain $M = C \cap C^{-1} = C \cap C = C$.

2. Suppose, conversely, that $C = M$. By Lemma 2.1 (i), we see that C is symmetric. The desired assertion now follows from Theorem 3.4. //

We next come to a very important theorem which shows an interesting property of the commutativity relation C.

Theorem 4.2. Let a, b, c be elements of an orthomodular lattice. If a commutes with b and c, then a commutes also with $b \wedge c$ and $b \vee c$.

Proof. Since aCb and aCc, we may apply Theorem 3.10 to show that $(b \vee c) \wedge a = (b \wedge a) \vee (c \wedge a)$. By Theorem 3.5 we know that $a'Cb$ and $a'Cc$ so that once again $(b \vee c) \wedge a' = (b \wedge a') \vee (c \wedge a')$. From Theorem 3.4 we conclude that aCb implies bCa.

i.e., $b = (b \wedge a) \vee (b \wedge a')$. Similarly, $c = (c \wedge a) \vee (c \wedge a')$. Consequently,

$$[(b \vee c) \wedge a] \vee [(b \vee c) \wedge a'] = (b \wedge a) \vee (c \wedge a) \vee (b \wedge a') \vee (c \wedge a') = \\ = b \vee c$$

which shows that $b \vee c$ commutes with a. Thus, from Theorem 3.4, we obtain $aCb \vee c$.

By Theorem 2.3 it is clear that the relations aCb, aCc imply aCb' and aCc'. The same argument shows that $aCb' \vee c'$. Thus according to the same theorem, $aC(b' \vee c')'$, i.e., $aCb \wedge c$. //

Remark 4.3. An immediate question presents itself: Given two elements a and b of an orthomodular lattice, how can we determine whether or not they commute? Of course, we could compute it, but this procedure is not always the quickest one. Instead we try to use the comparability. To this end let us summarize our results. Clearly, for aCb it is sufficient that either (i) the elements a,b be comparable or (ii) the elements a,b' be comparable or (iii) the elements a',b be comparable or, finally, (iv) the elements a',b' be comparable. Indeed, if $a \leqslant b$, then we may use Theorem 2.3. If $a \geqslant b$, then bCa by the same theorem. Therefore, by Theorem 3.4, aCb. To prove (ii), note that we have aCb' as a consequence of (i). From Theorem 2.3 it follows aCb. If a' and b are comparable, then $a'Cb$ by (i). With the aid of Theorem 3.5 we obtain $(a')'Cb$, i.e., aCb. If a',b' are comparable, then, a fortiori, a,b are comparable and so aCb by (i).

A word of caution: Our remark does not state that if aCb then one of the conditions (i)-(iv) must of necessity occur; this is false. An example which illustrates this fact is the following: Let $M = \{x,y,z,v\}$, $A = \{x,y\}$ and $B = \{y,z\}$. In the Boolean algebra, $(P(M), \cup, \cap, ')$ of all the subsets of M we have $A' = \{z,v\}$, $B' = \{x,v\}$. Hence the conditions (i)-(iv) are not satisfied, while A and B commute.

If we know that b is a meet or a join of two elements which commute with a, then aCb by Theorem 2. Proceeding in this way, we see that aCb whenever aCc_1, aCc_2, aCc_3 and $b = (c_1 \vee c_2) \wedge c_3$.

It is easy to generalize this result.

Theorem 4.4. Let p be an n-ary lattice polynomial and let a be an element which commutes with every element c_i, $i = 1, 2, \ldots, n$. If $\{d_1, d_2, \ldots, d_n\}$ denotes a subset of the set $\{c_1, c_2, \ldots, c_n, c_1', c_2', \ldots, c_n'\}$, then $aCp(d_1, d_2, \ldots, d_n)$.

Proof (by the rank of p). 1. If the rank $r(p)$ of the polynomial p is equal to 1, then $p(x_1, x_2, \ldots, x_n) = x_i$ where $1 \leqslant i \leqslant n$. Thus, $p(d_1, d_2, \ldots, d_n) = d_i$ and either $d_i = c_j$ or $d_i = c_k'$ for suitable indices j,k. In this case the assertion $aCp(d_1, d_2, \ldots, d_n)$ follows from Theorem 2.3.

2. Suppose that the assertion of the theorem is valid for every polynomial having its rank less than $r(p)$.

2A. If $p = r \wedge s$, then, by the induction hypothesis, $aCr(d_1, \ldots, d_n)$ and also $aCs(d_1, \ldots, d_n)$. From Theorem 2 we conclude that $aCr(d_1, \ldots, d_n) \wedge s(d_1, \ldots, d_n)$, i.e., $aCp(d_1, \ldots, d_n)$.

2B. Let $p = r \vee s$. Then, by the induction hypothesis and Theorem 2, we similarly get $aCr(d_1, \ldots, d_n) \vee s(d_1, \ldots, d_n)$. Hence, $aCp(d_1, \ldots, d_n)$. //

Theorem 4.5. Let c_i, $i = 1, 2, \ldots, n$, be elements of an orthomodular lattice. Then the subalgebra generated by the set $\{c_1, c_2, \ldots, c_n\}$ is a Boolean algebra if and only if $c_i C c_j$ for every $1 \leqslant i,j \leqslant n$.

Proof. 1. Let the subalgebra be a Boolean algebra. Then, by distributivity, $(c_i \wedge c_j) \vee (c_i \wedge c_j') = c_i \wedge (c_j \vee c_j') = c_i$, i.e., $c_i C c_j$.

2. Suppose any two elements c_i, c_j commute. By Theorem 4, $c_i C p(c_1^{e_1}, c_2^{e_2}, \ldots, c_n^{e_n})$ where $e_j \in \{-1, 1\}$ and $c_j^{-1} = c_j'$, $c_j^1 = c_j$ for every j. Since C is symmetric, we have, again by Theorem 4, $\bar{p} C \bar{q}$ whenever $\bar{p} = p(c_1^{e_1}, c_2^{e_2}, \ldots, c_n^{e_n})$, $\bar{q} = q(c_1^{f_1}, c_2^{f_2}, \ldots, c_m^{f_m})$ where $f_i \in \{-1, 1\}$. By Theorem 3.10 we now can see that the distributive identities are valid for every three elements $\bar{p}, \bar{q}, \bar{r}$ of the subalgebra. Clearly, with every element the subalgebra contains also its complement. Thus, it is a Boolean algebra. //

Corollary 4.6. An orthomodular lattice is a Boolean algebra if and only if aCb for any a, b. //

5. CHARACTERISTIC PROPERTIES OF ORTHOMODULAR LATTICES

Our first concern in this section will be with a short characterization of an orthomodular lattice as an algebra satisfying three axioms.

Theorem 5.1. Let $`A$ be an ortholattice. Then $`A$ is orthomodular if and only if the condition

$$a \vee b = [(a \vee b) \wedge a] \vee [(a \vee b) \wedge a']$$

holds for every $a, b \in A$.

Proof. 1. If $`A$ is orthomodular, then, from $a \leqslant a \vee b$ and Remark 4.3, we conclude that $a \vee b C a$. By the definition of C, it means the condition of the theorem.

2. Suppose the condition is fulfilled. If $a \leqslant b$, then

$$b = a \vee b = (b \wedge a) \vee (b \wedge a') = a \vee (b \wedge a').$$

Thus the orthomodularity of $`A$ is established. //

The following theorem shows three simple axioms for the equational class of orthomodular lattices.

Theorem 5.2. An algebra $`A = (A, \vee, \wedge, ')$ determines an orthomodular lattice if and only if the following conditions are satisfied for every $a,b,c \in A$:

(OM 1) $a = a \vee (b \wedge b')$;

(OM 2) $(a \vee b) \vee c = (c' \wedge b')' \vee a$;

(OM 3) $a \vee b = [(a \vee b) \wedge (a \vee c)] \vee [(a \vee b) \wedge a']$.

Proof. 1. Clearly, the conditions (OM 1)-(OM 3) hold in any orthomodular lattice.

2. Suppose that the conditions are satisfied in $`A$. Write $a \wedge a' = 0$ and set $b = 0$ in (OM 3). Then

$$\begin{aligned} a &= a \vee (a \wedge a') && \text{by (OM 1);} \\ &= [(a \vee 0) \wedge (a \vee c)] \vee [(a \vee 0) \wedge a'] && \text{by (OM 3);} \\ &= [a \wedge (a \vee c)] \vee (a \wedge a') && \text{by (OM 1);} \\ &= a \wedge (a \vee c) && \text{by (OM 1).} \end{aligned}$$

From Theorem 1.3 it is now clear that $`A$ is an ortholattice. By Theorem 1 it is an orthomodular lattice. //

An ortholattice $`L$ is said to be <u>uniquely orthocomplemented</u> (cf. [54]) if and only if for every element of L there exists only one orthogonal complement, i.e., if for every $a \in L$ there exists exactly one element b such that $a \leqslant b'$, $a \vee b = 1$ and $a \wedge b = 0$.

Proposition 5.3. An ortholattice is uniquely orthocomplemented if and only if it is orthomodular.

Proof. 1. Suppose $`L$ is uniquely orthocomplemented and consider elements $s,t$ such that $s \geqslant t'$, $s \wedge t = 0$. Since $s \geqslant t'$ and $`L$ is an ortholattice, we have $s' \leqslant t$. Hence $s' \wedge t' \leqslant t \wedge \wedge t' = 0$, i.e., $s' \wedge t' = 0$. From $s \wedge t = 0$ we obtain $s' \vee t' = = 1$ so that t' is a complement of s'. Moreover, it is an orthogonal complement, since $s' \leqslant (t')' = t$. But s is obviously also an orthogonal complement of s'. Therefore, $s = = t'$ and, consequently, $`L$ is orthomodular.

2. If $`L$ is an orthomodular lattice and if b is a complement of a which is orthogonal to a, we have $t' = a \leqslant b' = s$. At the same time, $s \wedge t = a' \wedge b' = (a \vee b)' = 1' = 0$. By orthomodularity, $s = t'$, i.e., $b = a'$. //

Theorem 5.4. An ortholattice is orthomodular if and only if it does not contain a subalgebra isomorphic to the ortholattice of Figure 7a.

Proof. 1. Suppose $`L$ is not orthomodular. Then there are elements s,t such that $s \geqslant t'$, $s \wedge t = 0$ and $s \neq t'$. To obtain the ortholattice, it is sufficient to put $a = t'$, $b = s$, $c = s'$, $d = t$. Then, e.g., $1 \geqslant s \vee t \geqslant s \vee s' = 1$ so that $s \vee t = 1$. Suppose that $0 = t'$. Then $t = 1$ and $t' = 0 = s \wedge t = s \wedge 1 = s$, a contradiction. Thus the elements $0, t', s, s', t, 1$ determine a six-element subalgebra of the mentioned form.

2. If $`L$ contains a subalgebra with the stated properties, then $s = b \geqslant t' = a$, $s \wedge t = b \wedge a' = b \wedge b' = 0$ and $b = s \neq t' = = a$ which is impossible in an orthomodular lattice. //

Remark 5.5. Note that the word "subalgebra" is essential for the validity of Theorem 4. A typical situation reveals the orthomodular lattice of Figure 13.

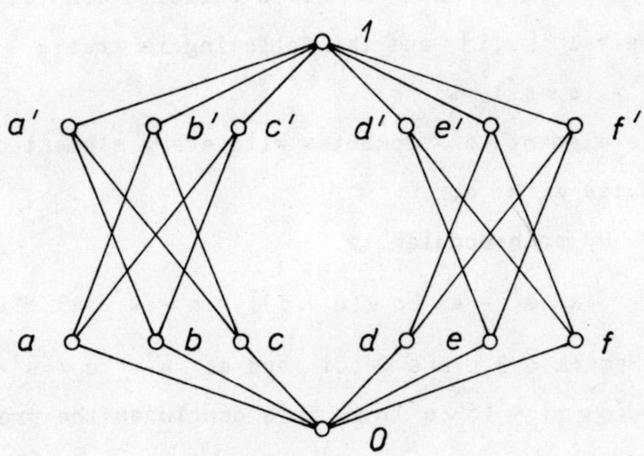

Fig. 13

Here the elements 0,c,b´,d,e´,1 determine a sublattice isomorphic to the lattice of Figure 7a. Of course, the conditions of Theorem 4 are not satisfied.

6. INTERVAL ALGEBRA

In this section we indicate how to associate an algebra with every interval of an orthomodular lattice in such a way that we get again an orthomodular lattice.

Remark 6.1. In an ortholattice an interval [o,i], $o \leqslant i$, need not be an ortholattice. A corresponding example is furnished by the lattice of Figure 9f where we choose [o,i] = [0,b´].

Lemma 6.2. Let o,i be elements of an orthomodular lattice `L such that $o \leqslant i$. If a belongs to the interval [o,i] and if $a^+ = o \vee (a' \wedge i)$, then a^+ is a relative complement of a in the interval $[o,i]$ and the following is true:

(i) $a^+ = (o \vee a') \wedge i$;

(ii) the element a^+ commutes with every element of $[o,i]$ which commutes with a.

Proof. By orthomodularity,

$$a \vee a^+ = a \vee [o \vee (a' \wedge i)] = a \vee (a' \wedge i) = i.$$

Again, by Remark 4.3, oCa', oCi and so $a^+ = o \vee (a' \wedge i) = (o \vee a') \wedge (o \vee i) = (o \vee a') \wedge i$. This concludes the proof of (i). By dual arguments,

$$a \wedge a^+ = a \wedge [i \wedge (a' \vee o)] = a \wedge (a' \vee o) = o.$$

Suppose b is an element of $[o,i]$ which commutes with a. Then bCa' and from $o \leqslant b \leqslant i$ it follows that b commutes with o and i. This, together with Theorem 4.4, implies that b commutes with $o \vee (a' \wedge i) = a^+$. //

The element a^+ will be called the relative orthocomplement of the element a in the interval $[o,i]$.

The mapping which assigns to each element $a \in [o,i]$ the element a^+ is obviously a unary operation on $[o,i]$. It will be considered in more detail in the following corollary.

Corollary 6.3. Let $o \leqslant i$ be elements of an orthomodular lattice. Then

(i) $([o,i], \vee, \wedge, {}^+, o, i)$ is an orthomodular lattice;

(ii) if $z \leqslant o \leqslant a \leqslant i \leqslant u$, $a^* = (z \vee a') \wedge u$ and $a^+ = o \vee (a' \wedge i)$, then $a^+ = (o \vee a^*) \wedge i = o \vee (a^* \wedge i)$, $a^* = z \vee (a' \wedge u)$.

Proof. Ad (i). Let $y \in [o,i]$. Since, by Remark 4.3, iCi' and $iCo' \wedge y$, we can first infer from Theorem 3.10 and from orthomodularity that

$$(y^+)^+ = \{i \wedge [(o \vee y') \wedge i]'\} \vee o = \{i \wedge [(o' \wedge y) \vee i']\} \vee o =$$
$$= (i \wedge o' \wedge y) \vee (i \wedge i') \vee o = (o' \wedge y) \vee o = y.$$

If $x, y \in [o,i]$ are such that $x \geqslant y^+$ and $x \wedge y = o$, write $s = x \geqslant y^+ = (o \vee y') \wedge i = t'$. In a similar manner we finally see that

$$s \wedge t = x \wedge [(o' \wedge y) \vee i'] = [(o' \wedge y) \wedge x] \vee (x \wedge i') \leqslant$$
$$\leqslant (o' \wedge o) \vee (i \wedge i') = 0.$$

By orthomodularity, $x = s = t' = y^+$. This together with the proof of Lemma 2 shows that $[o,i]$ determines an orthomodular lattice.

Ad (ii). Note that $w = o \vee (a^* \wedge i) = o \vee \{[(z \vee a') \wedge u] \wedge i\} = o \vee [(z \vee a') \wedge i]$. Evidently, $o \leqslant i$, $z \leqslant o$ and $a' \leqslant o'$. By Remark 4.3, oCi and $oCz \vee a'$. Hence, by Theorem 3.10, $w = (o \vee z \vee a') \wedge (o \vee i) = (o \vee a') \wedge i$. The remainder follows by (i) of Lemma 2 and by (i). //

The orthomodular lattice $([o,i], \vee, \wedge, ^+, o, i)$ is usually denoted by `L[o,i] and it is called the <u>interval algebra</u> of the orthomodular lattice `L determined by the interval $[o,i]$. The lattice `L[o,i] reproduces many properties of the whole lattice `L.

As an example of such a reproduction we now exhibit the following two theorems:

<u>Theorem 6.4</u>. Let $a, b \in [o,i]$ be elements of an orthomodular lattice `L. Then $a$ commutes with $b$ in `L if and only if a commutes with b in the interval algebra `L[o,i].

<u>Proof</u>. Lemma 2 gives $a \wedge b^+ = a \wedge (o \vee b') \wedge i = a \wedge (o \vee b')$. Since $o \leqslant a$, $o \leqslant b$, we also have aCo, oCb'. In accordance with Theorem 3.10 this implies that $a \wedge b^+ = (a \wedge o) \vee (a \wedge b') = o \vee (a \wedge b')$.

From $o \leq a$, $o \leq b$ we see that $o \leq a \wedge b$. Consequently, $(a \wedge \wedge b) \vee (a \wedge b^+) = (a \wedge b) \vee o \vee (a \wedge b') = (a \wedge b) \vee (a \wedge b')$. //

Let $`L = (L, \vee, \wedge, ', 0, 1)$ and $`K = (K, \vee, \wedge, ', 0, 1)$ be two ortholattices. If f is a lattice-homomorphism of $(L, \vee, \wedge)$ into $(K, \vee, \wedge)$ such that $f(a') = (f(a))'$ for every $a \in L$, then f is called an <u>orthohomomorphism</u> of $`L$ into $`K$. Suppose a lattice $(K, \vee, \wedge)$ is an isomorphic image of the lattice $(L, \vee, \wedge)$ under the lattice isomorphism f and $`L = (L, \vee, \wedge, ', 0, 1)$ is an ortholattice. Let $k \in K$. Then $k = f(m)$ for an element $m \in L$. Putting $k' = f(m')$, we define an orthocomplementation on $K$. A bijective orthohomomorphism of $`L$ onto $`K$ is called an <u>orthoisomorphism</u> and $`K$ is said to be an <u>orthoisomorphic image</u> of the ortholattice $`L$. It is clear that every orthoisomorphic image of an orthomodular lattice is again an orthomodular lattice.

Let $p \leq q$ be elements of a lattice $`L$ and let M be a nonvoid subset of L. The <u>lower contraction</u> of M into the interval $[p,q]$ is the mapping $q{\downarrow\uparrow}p: M \to [p,q]$ defined by $q{\downarrow\uparrow}p: : x \mapsto (q \wedge x) \vee p$. The <u>upper contraction</u> of M into $[p,q]$ is defined dually as the mapping $p{\uparrow\downarrow}q: M \to [p,q]$ which assigns to x the element $(p \vee x) \wedge q$.

Note that we always have $q{\downarrow\uparrow}p(x) \leq p{\uparrow\downarrow}q(x)$. Clearly, $q{\downarrow\uparrow}p = p{\uparrow\downarrow}q$ for every $p \leq q$ of L and every $M \subset L$ if and only if $`L$ is a modular lattice.

Observe that the relative orthocomplement a^+ of $a \in [o,i]$ as described in Lemma 2 can also be expressed in the form $a^+ = = i{\downarrow\uparrow}o(a') = o{\uparrow\downarrow}i(a')$.

<u>Theorem 6.5</u>. Let $`L$ be an orthomodular lattice and let $o \leq a \leq i$. Let g be the upper contraction of $[o,i]$ into

the interval $[i',o']$ and let f be the upper contraction of $[i',o']$ into the interval $[o,i]$, i.e.,

$$g = i'\uparrow\downarrow o' : a \mapsto (i' \vee a) \wedge o', \quad f = o\uparrow\downarrow i : x \mapsto (o \vee x) \wedge i.$$

Let b^* denote the relative orthocomplement of $b \in [i',o']$ in the interval $[i',o']$ and a^+ the relative orthocomplement of $a \in [o,i]$ in the interval $[o,i]$. Then

(i) $g \circ f = \mathrm{id}_{[o,i]}$ (identity mapping on $[o,i]$);

(ii) $f \circ g = \mathrm{id}_{[i',o']}$;

(iii) g is a bijection of $[o,i]$ onto $[i',o']$;

(iv) f is a bijection of $[i',o']$ onto $[o,i]$;

(v) for any relative complement c of $a \in [o,i]$ in the interval $[o,i]$, the element $g(c)$ is a complement of the element a;

(vi) for any two relative complements c_1, c_2 of $a \in [o,i]$ in the interval $[o,i]$, $c_1 \leq c_2$ if and only if $g(c_1) \leq g(c_2)$; in particular, $c_1 = c_2$ if and only if $g(c_1) = g(c_2)$;

(vii) g is an orthoisomorphism of the orthomodular lattice $`L^+ = ([o,i], \vee, \wedge, {}^+, o, i)$ onto the orthomodular lattice $`L^* = ([i',o'], \vee, \wedge, {}^*, i', o')$ and f is an orthoisomorphism of $`L^*$ onto $`L^+$. Further, $(g(x))^* = x' = g(x^+)$ for every $x \in [o,i]$.

<u>Proof</u>. Ad (i). If $a \in [o,i]$, then aCi', $i'Co'$ and so, by Theorem 3.10,

$$(g \circ f)(a) = f(g(a)) = f((a \vee i') \wedge o') = f((a \wedge o') \vee i').$$

Furthermore, $f((a \wedge o') \vee i') = [(a \wedge o') \vee i' \vee o] \wedge i$. Now, o commutes with a and o'. Thus $(a \wedge o') \vee o = (a \vee o) \wedge (o' \vee o) = a$. Since i commutes with a and i', $f(g(a)) = (a \vee i') \wedge i = (a \wedge i) \vee (i' \wedge i) = a$. By I.6, we have just verified that g is an injective mapping of $[o,i]$ into $[i',o']$ and that f

is a surjection onto $[o,i]$.

Ad (ii). In the proof of (i) we can write b,i',o',i and o instead of a,o,i,o' and i', respectively. Then f is replaced by g, g by f and our assertion follows.

Ad (iii) and (iv). The mappings f,g are bijective by (i) and (ii).

Ad (v). By computation it is easily verified that $a \vee g(c) = a \vee [(c \vee i') \wedge o']$. We note that $o' \geqslant i'$ and, thus, $o' C i'$. But $o \leqslant c$ implies that the element o' commutes with c. By Theorem 4.2 this means that o' commutes with $c \vee i'$. Since $o \leqslant a$, we can see that also the elements a,o' commute. On the basis of Theorem 3.10 we have $a \vee [(c \vee i') \wedge o'] = (a \vee c \vee i') \wedge \wedge (a \vee o')$. Since $o \leqslant a$, $1 = o \vee o' \leqslant a \vee o'$, i.e., $a \vee o' = 1$. By hypothesis, $a \vee c = i$ and, therefore, $a \vee c \vee i' = i \vee i' = 1$. Consequently, $a \vee g(c) = 1$. Now $a \wedge g(c) = a \wedge (c \vee i') \wedge o'$. From $a \leqslant i$, $c \leqslant i$ it follows aCi', cCi'. Similarly, from $o \leqslant i$ we get $o'Ci'$. Thus, by Theorem 4.2, $o' \wedge aCi'$. Since $i' \leqslant o'$, Theorem 3.10 implies that

$$a \wedge g(c) = (a \wedge o' \wedge c) \vee (a \wedge o' \wedge i') = 0 \vee (a \wedge i' \wedge o') = a \wedge i'.$$

Obviously, $a \wedge i' \leqslant i \wedge i' = 0$. Thus, $a \wedge g(c) = 0$.

Ad (vi). If $c_1 \leqslant c_2$, it is clear that $c_1 \vee i' \leqslant c_2 \vee i'$. We have therefore $g(c_1) = (c_1 \vee i') \wedge o' \leqslant (c_2 \vee i') \wedge o' = g(c_2)$. Similarly, from $d_1 \leqslant d_2$ where $d_1, d_2 \in [i',o']$ we get $f(d_1) \leqslant f(d_2)$. Hence, if $g(c_1) \leqslant g(c_2)$, then $c_1 = f(g(c_1)) \leqslant f(g(c_2)) = c_2$. The relation $g(c_1) = g(c_2)$ is equivalent to $g(c_1) \leqslant g(c_2)$ and $g(c_2) \leqslant g(c_1)$. Therefore $c_1 \leqslant c_2$, $c_2 \leqslant c_1$, and so $c_1 = c_2$.

Ad (vii). Suppose $x,y \in [o,i]$. Then the elements x,y commute with o,i and from Theorem 3.10 we conclude that

$$g(x \vee y) = (x \vee y \vee i') \wedge o' = (x \vee i' \vee y \vee i') \wedge o' =$$
$$= [(x \vee i') \wedge o'] \vee [(y \vee i') \wedge o'] = g(x) \vee g(y);$$
$$g(x \wedge y) = [(x \wedge y) \vee i'] \wedge o' = (x \vee i') \wedge (y \vee i') \wedge o' =$$
$$= (x \vee i') \wedge o' \wedge (y \vee i') \wedge o' = g(x) \wedge g(y).$$

Furthermore, $g(x^+) = (x^+ \vee i') \wedge o' = \{[(o \vee x') \wedge i] \vee i'\} \wedge o'$. But the elements o, x', i, i', o' commute. Hence by Theorem 3.10,

$$g(x^+) = (o \vee x' \vee i') \wedge (i \vee i') \wedge o' = (o \vee x' \vee i') \wedge o' =$$
$$= (o \wedge o') \vee [(x' \vee i') \wedge o'] = (x' \vee i') \wedge o'.$$

Clearly, $o \leqslant x \leqslant i$ implies $o' \geqslant x' \geqslant i'$ and so $g(x^+) = x'$. Using again Theorem 3.10, we see that

$$(g(x))^* = [i' \vee (g(x))'] \wedge o' = \{i' \vee [(x \vee i') \wedge o']'\} \wedge o' =$$
$$= [i' \vee (x' \wedge i) \vee o] \wedge o' = \{[(i' \vee x') \wedge (i' \vee i)] \vee o\} \wedge o' =$$
$$= (i' \vee x' \vee o) \wedge o' = x' = g(x^+).$$

To prove the assertion that (vii) is satisfied also for f, use the same argument as the one in the proof of (ii). //

<u>Theorem 6.6</u>. Let $`L$ be an orthomodular lattice and let $a \in [o,i]$. Then the number of complements of the element $a$ in $`L$ is at least so great as the number of its relative complements in the interval $[o,i]$.

<u>Proof</u>. Let $R(a)$ denote the set of all relative complements of a in the interval $[o,i]$. It is sufficient to show that there is an injective mapping of the set $R(a)$ into the set of all complements of a. To this end, consider the mapping $g: c \mapsto$ $\mapsto g(c)$ and use Theorem 5 (v) and (vi). //

<u>Remark 6.7</u>. (A) From Lemma 2 it follows that every orthomodular lattice is also relatively complemented. There exist relatively complemented lattices which have not the structure of

an orthomodular lattice. As in Remark 1.4.D we note that $`M_5$ has the properties stated. Another example of a relatively complemented lattice is the lattice of Figure 14. The lattice is clearly an ortholattice and it is not orthomodular (see the subalgebra determined by the elements 0,a,b′,a′,b,1 and use Theorem 5.4).

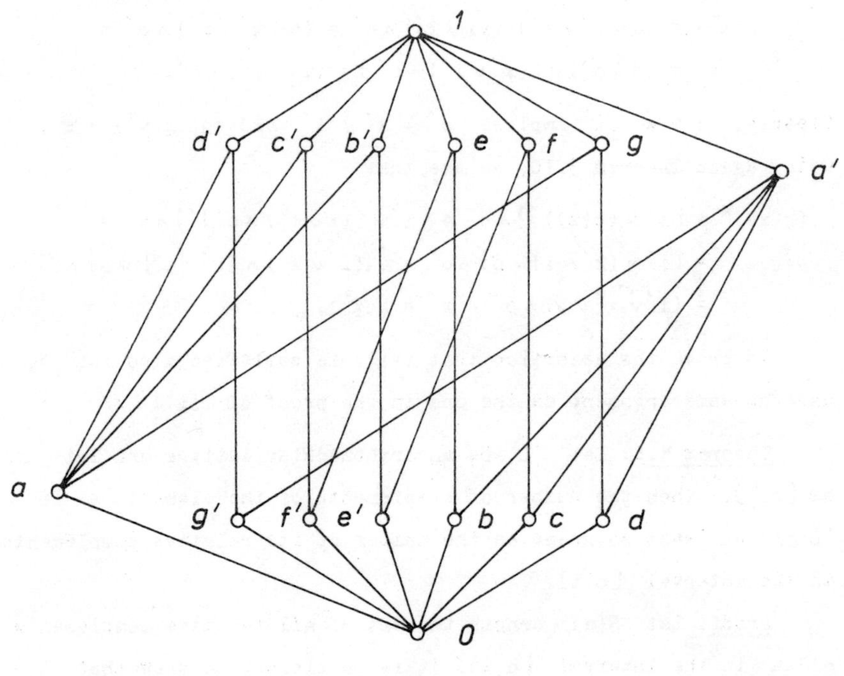

Fig. 14

(B) The element a of the orthomodular lattice of Figure 12 has three complements a′,d,d′. In the interval [0,c′] the element a has only one relative complement (namely b).

By a well-known criterion (cf. I.19) a lattice $`L$ which is not modular contains the pentagonal lattice $`N_5$ as a sublattice.

If `L is an orthomodular lattice, we may say more about the greatest and the least element of such a sublattice. Actually, one can establish the following remarkable result.

Theorem 6.8. An orthomodular lattice `L is not modular if and only if it has a sublattice which is isomorphic to the pentagon and which contains the least and the greatest element of `L.

Proof. 1. By I.19, the existence of such a sublattice implies that `L is not modular.

2. Suppose $Z < A < C < U$, B determine a sublattice isomorphic to the pentagon. By Theorem 5 (v), the elements $a = g(A) = (A \vee U') \wedge Z'$, $c = g(C) = (C \vee U') \wedge Z'$ are complements of B. Clearly, $a \leq c$. But if $a = c$, then, by Theorem 5 (i), $A = C$, a contradiction. Therefore, the elements $0, a, c, b = B, 1$ form a sublattice isomorphic to the pentagon. //

Remark 6.9. A similar assertion does not hold for sublattices isomorphic to $`M_5$. The elements $0, a, b, c, e$ in the orthomodular lattice shown by Figure 15 determine a lattice isomorphic to $`M_5$ but there is no sublattice of the lattice isomorphic to $`M_5$ and containing the element 0 and the element 1. This special situation is also typical of the general case.

Theorem 6.10. An orthomodular lattice `L is distributive if and only if it does not contain a sublattice isomorphic either to the pentagon having the least and the greatest element equal to 0 and 1 of `L, respectively, or to the lattice of Figure 16.

Note. The assertion of Theorem 10 is strengthened in Corollary III.2.12.

Proof. 1. By I.19, the existence of such a sublattice implies that the considered lattice $`L$ is not distributive.

2. Suppose the elements a,b,c,z,i satisfying $z < a,b,c < i$ form a sublattice isomorphic to the diamond $`M_5$. Let x^+ denote the relative orthocomplement of $x \in [z,1]$ in the interval $[z,1]$. By Corollary 3, the lattice $([z,1], \vee, \wedge, {}^+, z, 1)$ is orthomodular. Clearly, the mapping $f: x \mapsto x^+$ is an injection on $[z,1]$ satisfying $f(x \vee y) = f(x) \wedge f(y)$ and $f(x \wedge y) = f(x) \vee f(y)$. From this fact we infer that the elements $i^+, a^+, b^+, c^+, 1$ determine a sublattice isomorphic to $`M_5$. //

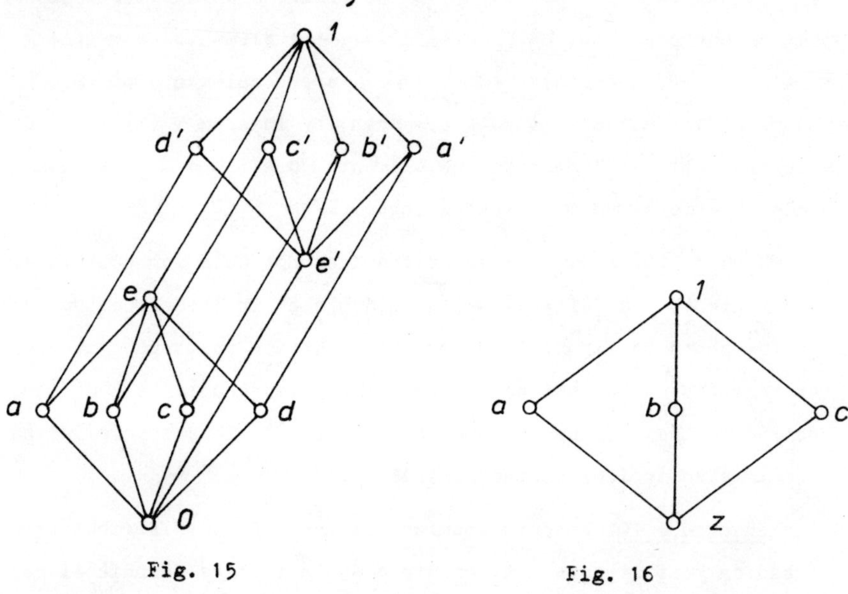

Fig. 15 Fig. 16

Recall that a lattice $`L$ having 0 and 1 is said to be <u>uniquely complemented</u> if and only if every $x \in L$ has exactly one complement. In other words, it is uniquely complemented if for every $x \in L$ there is one and only one element y such that $x \vee y = 1$ and $x \wedge y = 0$.

Theorem 6.11. Let $`L$ be an orthomodular lattice. Then the following conditions are equivalent:

(i) $`L$ is a Boolean algebra;

(ii) $`L$ is distributive;

(iii) aCb for every $a,b \in L$;

(iv) C is an equivalence relation;

(v) C is transitive;

(vi) $`L$ is uniquely complemented.

Proof. (ii) ⇔ (i): Trivial. (i) ⇔ (iii): Use Corollary 4.6. (iii) ⇔ (iv): From (iv) we conclude that $0Ca$ for every $a \in L$ and so we get (iii). (iv) ⇔ (v): Note that C is reflexive and symmetric. (i) ⇒ (vi): Trivial.

Thus our proof will be complete if we show that (vi) implies (ii). Let us suppose to the contrary that $`L$ is not distributive. Then by Theorem 10 there exists a pentagon $\{0,a,b,c,1\}$, $b < c$, as a sublattice or a diamond $\{z,a,b,c,1\}$ as a sublattice of $`L$. In the first case a has two complements b and c. In the second one, a has at least two complements by Theorem 6, which is a contradiction. //

Lemma 6.12. Every uniquely complemented ortholattice is orthomodular.

Proof. The statement follows trivially from Proposition 5.3. //

Theorem 6.13. An ortholattice is uniquely complemented if and only if it is a Boolean algebra.

Proof. 1. If $`L$ is a Boolean algebra, it is a uniquely complemented lattice by Theorem 11.

2. If $`L$ is a uniquely complemented ortholattice, it is

orthomodular by Lemma 12. In view of Theorem 11 it is a distributive lattice. //

EXERCISES

II;1. Find the couples of those elements in the lattice of Figure 7a which do not commute.

II;2. Show that the following conditions are equivalent in any ortholattice `L:

(i) `L is an orthomodular lattice;

(ii) $\forall\, s,t \in L \quad s \wedge [s' \vee (s \wedge t)] = s \wedge t$;

(iii) $\forall\, s,t \in L \quad s \vee [s' \wedge (s \vee t)] = s \vee t$;

(iv) for every s,t such that $s \leqslant t$ there exists an element r which satisfies $s \vee r = t$ and $r \leqslant s'$;

(v) for every s,t such that $s \geqslant t$ there exists an element r which satisfies $s \wedge r = t$ and $r \geqslant s'$.

II;3. Let `L be an ortholattice. Prove that the following conditions are equivalent:

(i) $(s \geqslant t' \;\&\; s \wedge t = 0) \Rightarrow s = t'$;

(ii) $s \leqslant t \leqslant u \Rightarrow s \vee (t' \wedge u) = (s \vee t') \wedge u$.

II;4. Let `L be a lattice and let $a,b,c \in L$. The <u>upper median</u> and the <u>lower median</u> are defined by

$$\overline{\text{med}}\,(a,b,c) = (a \vee b) \wedge (a \vee c) \wedge (b \vee c)$$

and

$$\underline{\text{med}}\,(a,b,c) = (a \wedge b) \vee (a \wedge c) \vee (b \wedge c),$$

respectively. Prove that the following conditions are equivalent in every ortholattice:

(i) $'L$ is an orthomodular lattice;

(ii) for every $a,b \in L$ such that $a \leq b$, $\overline{\text{med}}\,(a,b,a') = \underline{\text{med}}\,(a,b,a')$;

(iii) for every $s,t,u \in L$ such that $s \geq t$,

$$[(t' \vee u) \wedge s] \vee (t' \wedge u) = (t' \vee u) \wedge [s \vee (t' \wedge u)];$$

(iv) for every $s,t,u \in L$,

$$[(t' \vee u) \wedge (s \vee t)] \vee (t' \wedge u) = (t' \vee u) \wedge [(s \vee t) \wedge (t' \wedge u)].$$

II;5. Let $'L$ be an ortholattice. Prove that the following conditions are equivalent:

(i) $'L$ is an orthomodular lattice;

(ii) the implication

$$(b_1 C a \;\&\; a C b_2) \Rightarrow a \wedge (b_1 \vee b_2) = (a \wedge b_1) \vee (a \wedge b_2)$$

is valid for every a,b_1,b_2 of L;

(iii) the implication

$$(a C b_1 \;\&\; b_2 C a) \Rightarrow b_1 \wedge (a \vee b_2) = (b_1 \wedge a) \vee (b_1 \wedge b_2)$$

holds for every a,b_1,b_2 of L;

(iv) the implication

$$(a C b_1 \;\&\; a C b_2) \Rightarrow a \vee (b_1 \wedge b_2) = (a \vee b_1) \wedge (a \vee b_2)$$

is satisfied for every a,b_1,b_2 of L;

(v) the implication

$$(a C b_1 \;\&\; a' C b_2) \Rightarrow b_1 \vee (a \wedge b_2) = (b_1 \vee a) \wedge (b_1 \vee b_2)$$

is valid for every a,b_1,b_2 of L.

II;6. Let $'L$ be an ortholattice. Prove that the following statements are equivalent:

(i) $'L$ is orthomodular;

(ii) the implication

$$aCb \Rightarrow (a \vee b') \wedge b = a \wedge b$$

is satisfied for every a,b of L;

(iii) the implication

$$(b \vee a') \wedge a = a \wedge b \Rightarrow bCa$$

is valid for every a,b of L.

II;7. Let $`L$ be an ortholattice. Decide if the conditions (i)-(v) below are equivalent:

(i) $`L$ is orthomodular;

(ii) the implication

$$(aCb_1 \ \& \ aCb_2) \Rightarrow aCb_1 \vee b_2$$

is true for every a,b_1,b_2 of L;

(iii) in $`L$, for all $a,b_1,b_2 \in L$,

$$(aCb_1 \ \& \ aCb_2) \Rightarrow aCb_1 \wedge b_2;$$

(iv) the implication

$$(b_1Ca \ \& \ aCb_2) \Rightarrow aCb_1 \vee b_2$$

is satisfied for all $a,b_1,b_2 \in L$;

(v) the implication

$$(b_1Ca \ \& \ aCb_2) \Rightarrow aCb_1 \wedge b_2$$

holds for every a,b_1,b_2 of L.

II;8. An algebra $`A = (A, \vee, \wedge, ')$ determines a Boolean algebra if and only if the following conditions are valid for every $a,b,c \in A$:

(BA 1) $a = a \vee (b \wedge b')$;

(BA 2) $(a \vee b) \vee c = (c' \wedge b')' \vee a$;

(BA 3) $a = [a \wedge (b \vee c)] \vee (a \wedge b')$.

II;9. Prove that it is not possible to give the structure of an orthomodular lattice to the ~~ortho~~lattice of Figure 14.

II;10. Let $\mathcal{L}$ be a uniquely complemented lattice. Suppose $\mathcal{L}$ satisfies the implication

$$b \leqslant a \Rightarrow a' \leqslant b'$$

for every $a, b \in L$. Prove that $\mathcal{L}$ is a Boolean lattice.

II;11. A lattice $\mathcal{L}$ having 0 is said to be <u>sectionally complemented</u> if and only if for every $b \leqslant a$ there exists an element c which is a relative complement of b in the interval $[0,a]$.

Show that every uniquely complemented lattice $\mathcal{L}$ which is sectionally complemented has the following properties:

(i) the implication

$$b \leqslant a \Rightarrow a' \leqslant b'$$

is satisfied in $\mathcal{L}$;

(ii) $\mathcal{L}$ is a Boolean lattice.

II;12. Let $\mathcal{L}$ be a modular lattice which is complemented. Prove that the following is true:

(i) $\mathcal{L}$ is relatively complemented;

(ii) $\mathcal{L}$ is sectionally complemented;

(iii) if $\mathcal{L}$ is, moreover, uniquely complemented, then it is a Boolean lattice.

STRUCTURE OF ORTHOMODULAR LATTICES

Chapter III

1. SKEW OPERATIONS

Let $`L$ be an ortholattice and let $a,b \in L$. The element $(a \wedge b') \vee b$ will be called the <u>skew join</u> $a \dot{\vee} b$ of the elements a,b; the <u>skew meet</u> of a,b is defined dually as the element $a \dot{\wedge} b = (a \vee b') \wedge b$. The two skew operations $\dot{\vee}, \dot{\wedge}$ give rise to a new algebra $L^{\cdot} = (L, \dot{\vee}, \dot{\wedge}, ', 0, 1)$ which will be called the <u>Boolean skew lattice</u> associated with $`L$.

If $`L$ is an orthomodular lattice, then the associated Boolean skew lattice $L^{\cdot}$ can be characterized axiomatically (cf. [122]).

We now intend to examine the relation of the commutativity to the skew operations $\dot{\vee}, \dot{\wedge}$.

<u>Theorem 1.1</u>. Let a,b be elements of an orthomodular lattice. Then the following conditions are equivalent:

(i) aCb; (ii) $a \dot{\wedge} b \leqslant a$; (iii) $a \dot{\wedge} b \leqslant a \wedge b$;

(iv) $a \dot{\wedge} b = a \wedge b$; (v) $a \dot{\wedge} b' \leqslant a$; (vi) $a \dot{\wedge} b' \leqslant a \wedge b'$;

(vii) $a \dot{\vee} b = a \vee b$; (viii) $a \dot{\wedge} b = b \dot{\wedge} a$; (ix) $a \dot{\vee} b = b \dot{\vee} a$.

Proof. (i) $\Rightarrow$ (ii). If aCb, then aDb, by Theorem II.3.6, and so $a = (a \vee b) \wedge (a \vee b')$. Thus $a \geqslant a \wedge b = (a \vee b) \wedge (a \vee b') \wedge \wedge b = (a \vee b') \wedge b = a \dot{\wedge} b$.

(ii) $\Rightarrow$ (iii). By assumption, $a \dot{\wedge} b \leqslant a$. Since $a \dot{\wedge} b \leqslant b$, $a \dot{\wedge} b \leqslant a \wedge b$.

(iii) $\Rightarrow$ (iv). Since $a \wedge b \geqslant a \dot{\wedge} b = (a \vee b') \wedge b \geqslant a \wedge b$, we have $a \dot{\wedge} b = a \wedge b$.

(iv) $\Rightarrow$ (i). The implication follows by Theorem II.3.7.

(i) $\Leftrightarrow$ (v). By Theorem II.2.3 (i), aCb is equivalent to aCb′. Since we have already proved that aCb is equivalent to $a \dot{\wedge} b \leqslant a$, we see that also the following equivalences are valid:

$$aCb \Leftrightarrow aCb' \Leftrightarrow a \dot{\wedge} b' \leqslant a.$$

(i) $\Leftrightarrow$ (vi). Here we can use an entirely analogous argument as in the proof of the equivalence (i) $\Leftrightarrow$ (v).

(i) $\Leftrightarrow$ (vii). The proof is an easy consequence of Theorem II.3.7.

Clearly, if a condition is equivalent to aCb, then, by Theorem II.3.6, the dual condition is also equivalent to aCb. Hence, it remains to prove the following implications:

(i) $\Rightarrow$ (viii). We observe first that aCb is equivalent to bCa, by Theorem II.3.4. Since (i) implies (iv), we find $a \dot{\wedge} b = a \wedge b = b \wedge a = b \dot{\wedge} a$.

(viii) $\Rightarrow$ (ii). Here it is enough to note that $a \dot{\wedge} b = b \dot{\wedge} a = (b \vee a') \wedge a \leqslant a$. //

Theorem 1.2 (cf. [56]). Let a,b be elements of an orthomodular lattice and let x^+ denote the relative orthocomplement of $x \in I = [b, a \dot{\vee} b]$ in the interval I. Let y^* be the relative orthocomplement of $y \in J = [b \dot{\wedge} a, a]$ in the interval J. Let $f: I \to J$, $g: J \to I$ be defined by $f: x \mapsto x \dot{\wedge} a$, $g: y \mapsto y \dot{\vee} b$ (see Figure 17).

Then f is an orthoisomorphism of the orthomodular lattice $`L_1 = (I, \vee, \wedge, ^+, b, a \dot{\vee} b)$ onto the orthomodular lattice $`L_2 = (J, \vee, \wedge, ^*, b \dot{\wedge} a, a)$. In addition, f is equal to the upper and the lower contraction of I into J. The mapping g is an orthoisomorphism of $[b \dot{\wedge} a, a]$ onto $[b, a \dot{\vee} b]$.

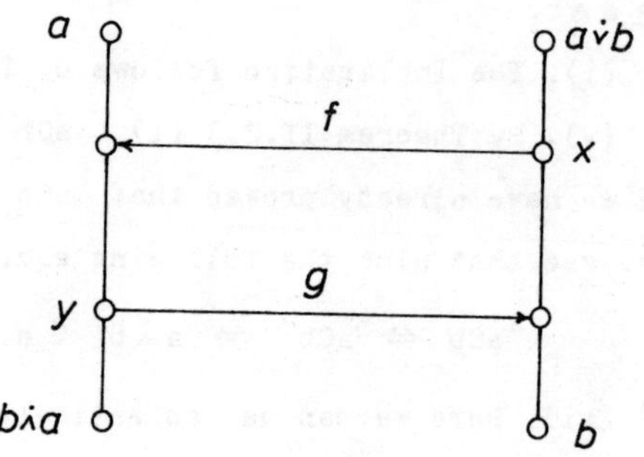

Fig. 17

Proof. 1. If $b \leq x \leq a \dot{\vee} b$, then $b \dot{\wedge} a \leq x \dot{\wedge} a \leq (a \dot{\vee} b) \dot{\wedge} a$, by Exercise III;4. In view of Exercise III;3 we have $(a \dot{\vee} b) \dot{\wedge} a \leq \leq a$. Hence f is a mapping of I into J.

2. If we interchange a and b, then, in the dual lattice, the mappings f and g are interchanged.

3. If $x \in I$, then

$$g(f(x)) = (x \dot{\wedge} a) \dot{\vee} b = [(x \vee a') \wedge a \wedge b'] \vee b.$$

Since $a \wedge b' \leq b'$, it is clear that $a \wedge b' C b$. From $x \in I$ we infer that $b \leq x$ and, a fortiori, we obtain $b \vee a' \leq x \vee a'$. Thus the element $x \vee a'$ commutes with the element $a \wedge b' = (b \vee a')'$.

By Foulis-Holland Theorem II.3.10 we therefore have

$$g(f(x)) = (x \vee a' \vee b) \wedge [(a \wedge b') \vee b] = (x \vee a' \vee b) \wedge (a \dot{\vee} b).$$

But $x \leqslant a \dot{\vee} b$ and so $xCa \dot{\vee} b$. Now, $(a' \vee b)' = a \wedge b' \leqslant (a \wedge b') \vee b = a \dot{\vee} b$ and, thus, $a' \vee bCa \dot{\vee} b$. By Theorem II.3.10,

$$g(f(x)) = \{x \wedge [(a \wedge b') \vee b]\} \vee \{(a' \vee b) \wedge [(a \wedge b') \vee b]\}.$$

But the first expression $\{\ldots\}$ is equal to x by using $x \leqslant (a \wedge b') \vee b = a \dot{\vee} b$. By orthomodularity, the expression in the second curly brackets gives b. Thus $g(f(x)) = x \vee b = x$.

4. From 2 it follows that $f(g(y)) = y$ for every $y \in J$, i.e., f and g are bijections.

5. If $x_1 \leqslant x_2$ are two elements of the interval I, then $f(x_1) \leqslant f(x_2)$, by Exercise III;4. A similar reasoning shows that $g(y_1) \leqslant g(y_2)$ whenever $y_1 \leqslant y_2$. Hence, if $f(x_1) \leqslant f(x_2)$ for some elements $x_1, x_2 \in I$, then $x_1 = g(f(x_1)) \leqslant g(f(x_2)) = x_2$, by 3. By this fact and by I.11 we see that f and g are lattice-isomorphisms.

6. First we get

$$f(x^+) = x^+ \dot{\wedge} a = (x^+ \vee a') \wedge a = \{[(b \vee x') \wedge (a \dot{\vee} b)] \vee a'\} \wedge a =$$
$$= (\{(b \vee x') \wedge [(a \wedge b') \vee b]\} \vee a') \wedge a.$$

Since $b \leqslant x$ and $a \wedge b' \leqslant b'$, bCx' and $bCa \wedge b'$. By Theorem II.3.10 we obtain $(b \vee x') \wedge [(a \wedge b') \vee b] = (x' \wedge a \wedge b') \vee b$ so that $f(x^+) = [(x' \wedge a \wedge b') \vee b \vee a'] \wedge a$. Using $b \leqslant x$, we can write $b \vee a' \leqslant x \vee a'$. Therefore, the element $b \vee a'$ commutes with $x' \wedge a = (x \vee a')'$. At the same time $b \vee a'$ commutes with b' so that

$$(x' \wedge a \wedge b') \vee b \vee a' = [(x' \wedge a) \vee b \vee a'] \wedge (b' \vee b \vee a') =$$
$$= (x' \wedge a) \vee b \vee a'.$$

Now a commutes with $x' \wedge a$ and with $b \vee a'$ and so, by Theorem II.3.10,

$$f(x^+) = (x' \wedge a) \vee [(b \vee a') \wedge a] = (x' \wedge a) \vee (b \dot{\wedge} a).$$

On the other hand,

$$(f(x))^* = (x \dot{\wedge} a)^* = ((x \vee a') \wedge a)^* = \{(b \dot{\wedge} a) \vee [(x \vee a') \wedge a]'\} \wedge a =$$
$$= [(b \dot{\wedge} a) \vee (x' \wedge a) \vee a'] \wedge a.$$

From Exercise III;3 we see that $b \dot{\wedge} a \leq a$. Since a commutes also with $x' \wedge a$ and a', we now find from Foulis-Holland Theorem II.3.10 that $(f(x))^* = (b \dot{\wedge} a) \vee (x' \wedge a) = f(x^+)$. In summary, we have shown that f is an orthoisomorphism.

7. By 2 we conclude that g is also an orthoisomorphism.

8. Let $x \in [b, a \dot{\vee} b]$. Then xCb' and $xC(a \wedge b') \vee b$. Hence, by Theorems II.4.2 and II.3.10, x commutes with $[(a \wedge b') \vee \vee b] \wedge b' = a \wedge b'$. Thus $xCa' \vee b$. Using Theorem II.3.10, we get

$$a \downarrow\uparrow b \dot{\wedge} a(x) = (a \wedge x) \vee [(b \vee a') \wedge a] = (x \vee b \vee a') \wedge a =$$
$$= (x \vee a') \wedge a = f(x).$$

On the other hand, by Theorem II.3.10 once again,

$$b \dot{\wedge} a \uparrow\downarrow a(x) = [(b \dot{\wedge} a) \vee x] \wedge a = \{[(b \vee a') \wedge a] \vee x\} \wedge a =$$
$$= (b \vee a' \vee x) \wedge (a \vee x) \wedge a = (x \vee a') \wedge a = f(x). \quad //$$

Let a be an element of an ortholattice `L. The mapping $s_a : L \to [0,a]$ defined by $s_a : x \mapsto x \dot{\wedge} a$ is called <u>Sasaki projection</u>.

<u>Example 1.3</u>. In the lattice of Figure 12 we have

x	a	a'	b	b'	c	c'	d	d'	0	1
$s_b(x)$	0	b	b	0	0	b	b	b	0	b

<u>Proposition 1.4 (cf. [62])</u>. An orthomodular lattice `L is not modular if and only if there exist elements $e, f, g \in L$ such

that $g < s_f(e')$ and $g \vee e = f \vee e$.

Proof. 1. By Theorem II.6.8, an orthomodular lattice is not modular if and only if it contains a pentagon $\{0,a,b,c,1\}$ with $0 < a < c < 1$. Now if such a sublattice exists, then choosing $g = a$, $f = c$, $e = b$, we note that on the one hand

$$g = a < s_c(b') = (b' \vee c')\wedge c = (b\wedge c)'\wedge c = 0'\wedge c = c,$$

and, on the other hand, $g \vee e = a \vee b = 1 = c \vee b = f \vee e$.

2. If there exist elements having the mentioned properties, write $A = g$, $C = s_f(e')$, $B = e$. By assumption, $A \vee B = C \vee B$. Similarly,

$$0 \leqslant A \wedge B \leqslant C \wedge B = (e' \vee f')\wedge f \wedge e = 0$$

so that $A \wedge B = C \wedge B$. Since $A < C$, it is readily seen that A, B, C generate a sublattice isomorphic to $`N_5$. //

2. FREE ORTHOMODULAR LATTICE F_2

Let $`A = (A, \vee, \wedge, ', 0, 1)$ be an orthomodular lattice and let $X = \{x_1, x_2, \ldots, x_t\} \subset A$. Then $`A$ is called a <u>free orthomodular lattice</u> (cf. I.23) <u>freely generated</u> by the set X if and only if it satisfies (i) and (ii):

(i) X is a generating set of $`A$.

(ii) Given any orthomodular lattice $`L = (L, \vee, \wedge, ', 0, 1)$ and any mapping $g: X \to L$, there exists a homomorphism $h$ of $`A$ into $`L$ extending the mapping g.

Notice that h is a homomorphism of $`A$ into $`L$ if and only if

(1) for any $a, b \in A$,

$$f(a \vee b) = f(a) \vee f(b) \;\&\; f(a \wedge b) = f(a) \wedge f(b);$$

(2) for any $a \in A$, $f(a') = (f(a))'$;

(3) $f(0) = 0 \;\&\; f(1) = 1$.

The condition (3) is a consequence of (1) and (2): $f(0) = f(a \wedge a') = f(a) \wedge f(a') = f(a) \wedge (f(a))' = 0$. Dually we get $f(1) = 1$. Hence here "a homomorphism" means the same as "an orthohomomorphism" in the sense of II.6.

In the sequel we shall be concerned with the structure of an orthomodular lattice freely generated by a two-element set. We shall usually refer to it as a <u>free orthomodular lattice with two generators</u>.

<u>Lemma 2.1</u>. Let $`L$ be an orthomodular lattice and let c be an element of L which commutes with all the elements of L. Then

(i) for any x of L, $x = (x \wedge c) \vee (x \wedge c')$;

(ii) $`L = (c] \oplus (c']$;

(iii) the mapping $f: L \to (c]$ defined by $f: x \mapsto x \wedge c$ is an orthohomomorphism of $`L$ onto the interval algebra $`L[0,c]$;

(iv) for any $p \in (c]$ and any $t \in (c']$, $(p \vee t)' = (p' \wedge c) \vee (t' \wedge c')$.

<u>Proof</u>. Ad (i). By the assumption on c, we get $x = (x \wedge c) \vee (x \wedge c')$.

Ad (ii). If $x = m \vee w$ where $m \in (c]$ and $w \in (c']$, then $m \leqslant c \leqslant w'$. Hence, by Theorem II.3.10, $x \wedge c = (m \vee w) \wedge c = (m \wedge c) \vee (w \wedge c) = m$. Analogously we find that $x \wedge c' = w$, so the result is clear by (i).

Ad (iii). It is easily seen that $f(x') = x' \wedge c = (f(x))'$ where $(f(x))'$ denotes the orthocomplement of $f(x)$ in $`L[0,c]$.

If $x,y \in L$, then $f(x \wedge y) = x \wedge y \wedge c = (x \wedge c) \wedge (y \wedge c) =$

$= f(x) \wedge f(y)$. Moreover, by Theorem II.3.10, $f(x \vee y) = (x \vee y) \wedge$ $\wedge\, c = (x \wedge c) \vee (y \wedge c) = f(x) \vee f(y)$.

Ad (iv). From (i) it follows that

$$(p \vee t)' = p' \wedge t' = (p' \wedge t' \wedge c) \vee (p' \wedge t' \wedge c').$$

By assumption, $c \leqslant t'$ and $c' \leqslant p'$. Hence

$$(p \vee t)' = (p' \wedge c) \vee (t' \wedge c'). \quad //$$

Remark 2.2. Let `T be an orthomodular lattice which is generated by two elements $s$ and $t$, i.e., `T = $\langle s,t \rangle$. Remark II.4.3 shows that $s \wedge t$ commutes with s and t. By making use of Theorem II.4.4 we find that $s \wedge t$ commutes with every element of T. The same is true for the elements $s \wedge t'$, $s' \wedge t$, $s' \wedge t'$, $s \vee t$, $s' \vee t$, $s \vee t'$, $s' \vee t'$. According to Theorem II.4.2, the elements $(s \wedge t) \vee (s \wedge t')$, $(s \wedge t) \vee (s' \wedge t)$, $\underline{c} = (s \wedge t) \vee (s \wedge t') \vee$ $\vee\, (s' \wedge t) \vee (s' \wedge t')$, $\bar{c} = (s \vee t) \wedge (s \vee t') \wedge (s' \vee t) \wedge (s' \vee t')$ also commute with every element of T.

Lemma 2.3. Let `T be an orthomodular lattice with generators $s,t$. Let `M_{O2} be the orthomodular lattice of Figure 9d with generators x_1, y_1. Then there exists an orthohomomorphism h_1 of M_{O2} onto $(\bar{c}]$ such that

(4) $h_1(x_1) = s \wedge \bar{c} = s \wedge (s' \vee t) \wedge (s' \vee t')$

and

(5) $h_1(y_1) = t \wedge \bar{c} = t \wedge (s \vee t') \wedge (s' \vee t')$.

Proof. By Lemma 1 (iii), the mapping $f_1 : T \to (\bar{c}\,]$ defined by $f_1 : x \mapsto x \wedge \bar{c}$ is an orthohomomorphism of `T onto `T$[0,\bar{c}]$. Consequently, the orthomodular lattice `T$[0,\bar{c}]$ has $f_1(s), f_1(t)$ as generators. Here

$$f_1(s) = s \wedge (s \vee t) \wedge (s \vee t') \wedge (s' \vee t) \wedge (s' \vee t') =$$

$$= s \wedge (s' \vee t) \wedge (s' \vee t')$$

and, similarly, $f_1(t) = t \wedge (s \vee t') \wedge (s' \vee t')$.

Since $\bar{c}$ commutes with every element of T, we can conclude from Theorem II.3.10 that

$$f_1(s) \vee f_1(t) = (s \wedge \bar{c}) \vee (t \wedge \bar{c}) = (s \vee t) \wedge \bar{c} = \bar{c}.$$

Analogously,

$$f_1(s') \vee f_1(t) = f_1(s) \vee f_1(t') = f_1(s') \vee f_1(t') = \bar{c}.$$

Obviously,

$$f_1(s) \wedge f_1(t) = s \wedge \bar{c} \wedge t \wedge \bar{c} = (s \wedge t) \wedge \bar{c} =$$
$$= (s \wedge t) \wedge (s \wedge t)' \wedge (s \vee t) \wedge (s \vee t') \wedge (s' \vee t) = 0.$$

An analogous argument yields $f_1(s') \wedge f_1(t) = f_1(s) \wedge f_1(t') = f_1(s') \wedge f_1(t') = 0$.

Consequently, we see that the elements $0, f_1(s), f_1(s'), f_1(t), f_1(t'), \bar{c}$ represent all the elements of the lattice $`T[0,\bar{c}]$ and that the lattice $`T[0,\bar{c}]$ is a homomorphic image of the simple lattice $`M_{O2}$. Therefore, $`T[0,\bar{c}]$ is either an isomorphic image of $`M_{O2}$ or it is a one-element lattice. Thus there exists an orthohomomorphism h_1 satisfying (4) and (5). //

<u>Corollary 2.4</u>. If $`T$ is an orthomodular lattice with generators $s,t$ and if $\bar{c} = (s \vee t) \wedge (s \vee t') \wedge (s' \vee t) \wedge (s' \vee t')$, then either $`T[0,\bar{c}] \cong `M_{O2}$ or $`T[0,\bar{c}]$ is a one-element lattice. //

<u>Lemma 2.5</u>. Let $`T$ be an orthomodular lattice which has generators $s,t$ and let $\underline{c} = (s \wedge t) \vee (s \wedge t') \vee (s' \wedge t) \vee (s' \wedge t')$. Suppose $x_2$ and $y_2$ are generators of the Boolean algebra $`B = \underline{2}^4$. Then there exists an orthohomomorphism h_2 of B onto $(\underline{c}]$ such that

(6) $h_2(x_2) = s \wedge \underline{c} = (s \wedge t) \vee (s \wedge t')$

and

(7) $h_2(y_2) = t \wedge \underline{c} = (s \wedge t) \vee (s' \wedge t)$.

Proof. In analogy with the proof of Lemma 3 we find that the mapping $f_2 : T \to (\underline{c}]$ defined by $f_2 : x \mapsto x \wedge \underline{c}$ is an orthohomomorphism of `T onto `$T[0,\underline{c}]$. Hence $\{f_2(s), f_2(t)\}$ is a generating set for `$T[0,\underline{c}]$. By Theorem II.3.10, we have

$f_2(s) = s \wedge [(s \wedge t) \vee (s \wedge t') \vee (s' \wedge t) \vee (s' \wedge t')] = (s \wedge t) \vee (s \wedge t')$.

By symmetry, $f_2(t) = (s \wedge t) \vee (s' \wedge t)$. From Remark 2 we conclude that the generators $f_2(s)$ and $f_2(t)$ commute. It follows from Theorem II.4.5 that `$T[0,\underline{c}]$ is a Boolean algebra with two generators $f_2(s), f_2(t)$. Now, referring to I.23, note that $\underline{2}^4$ is a free Boolean algebra. This shows that there exists an orthohomomorphism h_2 satisfying (6) and (7). //

Corollary 2.6. Under the notation of Lemma 5, `$T[0,\underline{c}]$ is a Boolean algebra having at most 16 elements. It is generated by $(s \wedge t) \vee (s \wedge t')$ and $(s \wedge t) \vee (s' \wedge t)$. //

Notice that $h_1(m') = (h_1(m))' \wedge \bar{c}$ for any $m \in M_{O2}$ and that $h_2(b') = (h_2(b))' \wedge \underline{c}$ for any $b \in B$. These assertions follow from the definition of an orthohomomorphism.

Lemma 2.7. Under the notations of Lemma 3 and 5, the elements $c = (x_1, x_2)$ and $d = (y_1, y_2)$ are generators of the lattice `$M_{O2} \times \underline{2}^4$.

Proof. Let `S be the subalgebra of `$M_{O2} \times \underline{2}^4$ generated by c and d. Then $c \wedge d = (0, x_2 \wedge y_2) \in S$ and $c \wedge d' = (0, x_2 \wedge y_2') \in$ $\in S$. Since $x_2 = (x_2 \wedge y_2) \vee (x_2 \wedge y_2')$ in the distributive lattice $\underline{2}^4$, $(0, x_2) \in S$. Analogously we find that $(0, x_2')$, $(0, y_2)$ and $(0, y_2')$ are elements of S. Hence $(0, b) \in S$ for all $b \in \underline{2}^4$. Dually we get that $(1, b) \in S$ for any $b \in \underline{2}^4$. Therefore, $(x_1, 0) =$

$= (x_1,x_2)\wedge(1,x_2')\in S$. Similarly, $(x_1',0)$, $(y_1,0)$ and $(y_1',0)$ belong also to S and so $(m,0)\in S$ for every m of M_{02}. Now take $m\in M_{02}$ and $b\in \underline{2}^4$. Then $(m,b) = (m,0)\vee(0,b)\in S$, that is, $`S = `M_{02}\times \underline{2}^4$. //

<u>Theorem 2.8</u>. Let $`M_{02}$ be the lattice of Figure 9d with generators $x_1,y_1$ and let $B = \underline{2}^4$ be the free Boolean algebra with generators $x_2,y_2$. Then $`F = `M_{02}\times \underline{2}^4$ is a free orthomodular lattice freely generated by the set $\{x,y\}$ where $x = (x_1,x_2)$, $y = (y_1,y_2)$.

<u>Proof</u>. Let $`L$ be an orthomodular lattice and let $g$ be a mapping of $\{x,y\}$ into L. Put $s = g(x)$ and $t = g(y)$. Let $`T = \langle s,t\rangle$ be the subalgebra of $`L$ generated by s and t. Define $h:M_{02}\times B\to T$ by $h:(u,v)\mapsto h_1(u)\vee h_2(v)$ where h_1 and h_2 are the orthohomomorphisms of Lemmas 3 and 5, $h_1:M_{02}\to(\bar{c}]$, $h_2:B\to(\underline{c}]$.

First, by the remark above,

$$h((u,v)') = h((u',v')) = h_1(u')\vee h_2(v') = [(h_1(u))'\wedge\bar{c}]\vee[(h_2(v))'\wedge\underline{c}]$$

and

$$(h((u,v)))' = (h_1(u)\vee h_2(v))' = [(h_1(u))'\wedge\bar{c}]\vee[(h_2(v))'\wedge\underline{c}],$$

by Lemma 1 (iv).

Next, given (u,v) and (z,w) of F, it can be easily shown that $h((u,v)\vee(z,w)) = h((u,v))\vee h((z,w))$.

Furthermore,

$$h((u,v)\wedge(z,w)) = h((u\wedge z,v\wedge w)) = h_1(u\wedge z)\vee h_2(v\wedge w) = [h_1(u)\wedge h_1(z)]\vee[h_2(v)\wedge h_2(w)].$$

Lastly, by (iii) in Proposition of I.17,

$$h((u,v)) \wedge h((z,w)) = [h_1(u) \vee h_2(v)] \wedge [h_1(z) \vee h_2(w)] = $$
$$= [h_1(u) \wedge h_1(z)] \vee [h_2(v) \wedge h_2(w)].$$

Finally, by Lemma 1 (i),

$$h(x) = h(x_1, x_2) = h_1(x_1) \vee h_2(x_2) = (s \wedge \bar{c}) \vee (s \wedge \underline{c}) = s$$

and, likewise, $h(y) = t$. //

There exists also another description of a free orthomodular lattice with two generators.

Let

$b_1 = (0,0,0,0)$, $b_2 = (1,0,0,0)$, $b_3 = (0,1,0,0)$, $b_4 = (0,0,1,0)$,
$b_5 = (0,0,0,1)$, $b_6 = (1,1,0,0)$, $b_7 = (1,0,1,0)$, $b_8 = (1,0,0,1)$,
$b_9 = (0,1,1,0)$, $b_{10} = (0,1,0,1)$, $b_{11} = (0,0,1,1)$, $b_{12} = (1,1,1,0)$,
$b_{13} = (1,1,0,1)$, $b_{14} = (1,0,1,1)$, $b_{15} = (0,1,1,1)$, $b_{16} = (1,1,1,1)$

denote the elements of the Boolean algebra $\underline{2}^4$. The elements of the orthomodular lattice `$M_{02}$ will be denoted by $0,a,b,a',b',1$ as in figure 9d. An element $(x,(e,f,g,h))$ of the direct product `$M_{02} \times \underline{2}^4$ will also be written in the form (x,e,f,g,h) or (x,b_j) if $b_j = (e,f,g,h)$.

Lemma 2.9. Let `L be an orthomodular lattice and let $x,y$ be two elements of L. Then the subalgebra of `L generated by x,y consists of the following elements:

$$p_1 = p_1(x,y) = 0,\ p_2 = p_2(x,y) = x \wedge y,\ \ldots$$
$$\ldots,\ p_{95} = p_{95}(x,y) = x' \vee y',\ p_{96} = p_{96}(x,y) = 1$$

(see Table 1 for the complete list and Figure 18).

Proof. The subalgebra `$S = \langle x,y \rangle$ contains all the expressions $p(x,y)$ where $p$ is a polynomial in the operations $\vee$, $\wedge$, $'$. Thus `S contains all the elements p_j.

Employing Theorem II.3.10, it is possible to prove that the

join and the meet of any two elements p_i, p_j as well as the orthocomplement of an element p_m is again an element p_k.

The reader can use Table 1 to check the corresponding results. For example, the meet $p_{20} \wedge p_{37}$ corresponds to the meet

$$(a,0,0,1,0) \wedge (b,0,0,0,1) = (0,0,0,0,0)$$

calculated in $`M_{02} \times \underline{2}^4$ (see the conventions made above). Hence $p_{20} \wedge p_{37} = p_1$. Similarly,

$$(a,0,0,1,0) \vee (b,0,0,0,1) = (1,0,0,1,1)$$

and so $p_{20} \vee p_{37} = p_{91}$ etc. //

Table 1

Fig. 18	Representation in $`L$ or in F_2	$`M_{02} \times \underline{2}^4$
01	$p_1=0$	$(0,0,0,0,0)=(0,b_1)$
02	$p_2=x\wedge y$	$(0,1,0,0,0)=(0,b_2)$
03	$p_3=x\wedge y'$	$(0,0,1,0,0)=(0,b_3)$
04	$p_4=x'\wedge y$	$(0,0,0,1,0)=(0,b_4)$
05	$p_5=x'\wedge y'$	$(0,0,0,0,1)=(0,b_5)$
06	$p_6=(x\wedge y)\vee(x\wedge y')$	$(0,1,1,0,0)=(0,b_6)$
07	$p_7=(x\wedge y)\vee(x'\wedge y)$	$(0,1,0,1,0)=(0,b_7)$
08	$p_8=(x\wedge y)\vee(x'\wedge y')$	$(0,1,0,0,1)=(0,b_8)$
09	$p_9=(x\wedge y')\vee(x'\wedge y)$	$(0,0,1,1,0)=(0,b_9)$
10	$p_{10}=(x'\wedge y')\vee(x\wedge y')$	$(0,0,1,0,1)=(0,b_{10})$
11	$p_{11}=(x'\wedge y')\vee(x'\wedge y)$	$(0,0,0,1,1)=(0,b_{11})$
12	$p_{12}=(x\wedge y)\vee(x\wedge y')\vee(x'\wedge y)$	$(0,1,1,1,0)=(0,b_{12})$
13	$p_{13}=(x\wedge y)\vee(x\wedge y')\vee(x'\wedge y')$	$(0,1,1,0,1)=(0,b_{13})$
14	$p_{14}=(x'\wedge y')\vee(x'\wedge y)\vee(x\wedge y)$	$(0,1,0,1,1)=(0,b_{14})$

Fig. 18	Representation in `L or in F_2	$`M_{02}\times \underline{2}^4$
15	$p_{15}=(x'\wedge y')\vee(x'\wedge y)\vee(x\wedge y')$	$(0,0,1,1,1)=(0,b_{15})$
16	$p_{16}=(x\wedge y)\vee(x\wedge y')\vee(x'\wedge y)\vee(x'\wedge y')$	$(0,1,1,1,1)=(0,b_{16})$
17	$p_{17}=x\wedge(x'\vee y)\wedge(x'\vee y')$	$(a,0,0,0,0)=(a,b_1)$
18	$p_{18}=x\wedge(x'\vee y)$	$(a,1,0,0,0)=(a,b_2)$
19	$p_{19}=x\wedge(x'\vee y')$	$(a,0,1,0,0)=(a,b_3)$
20	$p_{20}=(x'\wedge y)\vee[x\wedge(x'\vee y')\wedge(x'\vee y)]$	$(a,0,0,1,0)=(a,b_4)$
21	$p_{21}=(x'\wedge y')\vee[x\wedge(x'\vee y)\wedge(x'\vee y')]$	$(a,0,0,0,1)=(a,b_5)$
22	$p_{22}=x$	$(a,1,1,0,0)=(a,b_6)$
23	$p_{23}=(x'\vee y)\wedge[x\vee(x'\wedge y)]$	$(a,1,0,1,0)=(a,b_7)$
24	$p_{24}=(x'\vee y)\wedge[x\vee(x'\wedge y')]$	$(a,1,0,0,1)=(a,b_8)$
25	$p_{25}=(x'\vee y')\wedge[x\vee(x'\wedge y)]$	$(a,0,1,1,0)=(a,b_9)$
26	$p_{26}=(x'\vee y')\wedge[x\vee(x'\wedge y')]$	$(a,0,1,0,1)=(a,b_{10})$
27	$p_{27}=(x'\vee y')\wedge(x'\vee y)\wedge[x\vee(x'\wedge y')\vee(x'\wedge y)]$	$(a,0,0,1,1)=(a,b_{11})$
28	$p_{28}=x\vee(x'\wedge y)$	$(a,1,1,1,0)=(a,b_{12})$
29	$p_{29}=x\vee(x'\wedge y')$	$(a,1,1,0,1)=(a,b_{13})$
30	$p_{30}=(x'\vee y)\wedge[x\vee(x'\wedge y')\vee(x'\wedge y)]$	$(a,1,0,1,1)=(a,b_{14})$
31	$p_{31}=(x'\vee y')\wedge[x\vee(x'\wedge y)\vee(x'\wedge y')]$	$(a,0,1,1,1)=(a,b_{15})$
32	$p_{32}=x\vee(x'\wedge y)\vee(x'\wedge y')$	$(a,1,1,1,1)=(a,b_{16})$
33	$p_{33}=y\wedge(y'\vee x)\wedge(y'\vee x')$	$(b,0,0,0,0)=(b,b_1)$
34	$p_{34}=y\wedge(y'\vee x)$	$(b,1,0,0,0)=(b,b_2)$
35	$p_{35}=(x\wedge y')\vee[y\wedge(y'\vee x')\wedge(y'\vee x)]$	$(b,0,1,0,0)=(b,b_3)$
36	$p_{36}=y\wedge(y'\vee x')$	$(b,0,0,1,0)=(b,b_4)$
37	$p_{37}=(x'\wedge y')\vee[y\wedge(y'\vee x)\wedge(y'\vee x')]$	$(b,0,0,0,1)=(b,b_5)$
38	$p_{38}=(x\vee y')\wedge[y\vee(y'\wedge x)]$	$(b,1,1,0,0)=(b,b_6)$
39	$p_{39}=y$	$(b,1,0,1,0)=(b,b_7)$
40	$p_{40}=(x\vee y')\wedge[y\vee(y'\wedge x')]$	$(b,1,0,0,1)=(b,b_8)$
41	$p_{41}=(x'\vee y')\wedge[y\vee(y'\wedge x)]$	$(b,0,1,1,0)=(b,b_9)$

Fig. 18	`L or in F_2	Representation in `$M_{O2}\times \underline{2}^4$
42	$p_{42}=(x'\vee y')\wedge(x\vee y')\wedge[y\vee(x'\wedge y')\vee(x\wedge y')]$	$(b,0,1,0,1)=(b,b_{10})$
43	$p_{43}=(x'\vee y')\wedge[y\vee(y'\wedge x')]$	$(b,0,0,1,1)=(b,b_{11})$
44	$p_{44}=y\vee(y'\wedge x)$	$(b,1,1,1,0)=(b,b_{12})$
45	$p_{45}=(x\vee y')\wedge[y\vee(y'\wedge x')\vee(y'\wedge x)]$	$(b,1,1,0,1)=(b,b_{13})$
46	$p_{46}=y\vee(y'\wedge x')$	$(b,1,0,1,1)=(b,b_{14})$
47	$p_{47}=(x'\vee y')\wedge[y\vee(y'\wedge x)\vee(y'\wedge x')]$	$(b,0,1,1,1)=(b,b_{15})$
48	$p_{48}=y\vee(y'\wedge x)\vee(y'\wedge x')$	$(b,1,1,1,1)=(b,b_{16})$
49	$p_{49}=y'\wedge(y\vee x')\wedge(y\vee x)$	$(b',0,0,0,0)=(b',b_1)$
50	$p_{50}=(x\wedge y)\vee[y'\wedge(y\vee x')\wedge(y\vee x)]$	$(b',1,0,0,0)=(b',b_2)$
51	$p_{51}=y'\wedge(y\vee x)$	$(b',0,1,0,0)=(b',b_3)$
52	$p_{52}=(x'\wedge y)\vee[y'\wedge(y\vee x)\wedge(y\vee x')]$	$(b',0,0,1,0)=(b',b_4)$
53	$p_{53}=y'\wedge(y\vee x')$	$(b',0,0,0,1)=(b',b_5)$
54	$p_{54}=(x\vee y)\wedge[y'\vee(y\wedge x)]$	$(b',1,1,0,0)=(b',b_6)$
55	$p_{55}=(x\vee y)\wedge(x'\vee y)\wedge[y'\vee(x\wedge y)\vee(x'\wedge y)]$	$(b',1,0,1,0)=(b',b_7)$
56	$p_{56}=(x'\vee y)\wedge[y'\vee(y\wedge x)]$	$(b',1,0,0,1)=(b',b_8)$
57	$p_{57}=(x\vee y)\wedge[y'\vee(y\wedge x')]$	$(b',0,1,1,0)=(b',b_9)$
58	$p_{58}=y'$	$(b',0,1,0,1)=(b',b_{10})$
59	$p_{59}=(x'\vee y)\wedge[y'\vee(y\wedge x')]$	$(b',0,0,1,1)=(b',b_{11})$
60	$p_{60}=(x\vee y)\wedge[y'\vee(y\wedge x')\vee(y\wedge x)]$	$(b',1,1,1,0)=(b',b_{12})$
61	$p_{61}=y'\vee(y\wedge x)$	$(b',1,1,0,1)=(b',b_{13})$
62	$p_{62}=(x'\vee y)\wedge[y'\vee(y\wedge x')\vee(y\wedge x)]$	$(b',1,0,1,1)=(b',b_{14})$
63	$p_{63}=y'\vee(y\wedge x')$	$(b',0,1,1,1)=(b',b_{15})$
64	$p_{64}=y'\vee(y\wedge x')\vee(y\wedge x)$	$(b',1,1,1,1)=(b',b_{16})$
65	$p_{65}=x'\wedge(x\vee y')\wedge(x\vee y)$	$(a',0,0,0,0)=(a',b_1)$
66	$p_{66}=(x\wedge y)\vee[x'\wedge(x\vee y')\wedge(x\vee y)]$	$(a',1,0,0,0)=(a',b_2)$
67	$p_{67}=(x\wedge y')\vee[x'\wedge(x\vee y)\wedge(x\vee y')]$	$(a',0,1,0,0)=(a',b_3)$
68	$p_{68}=x'\wedge(x\vee y)$	$(a',0,0,1,0)=(a',b_4)$

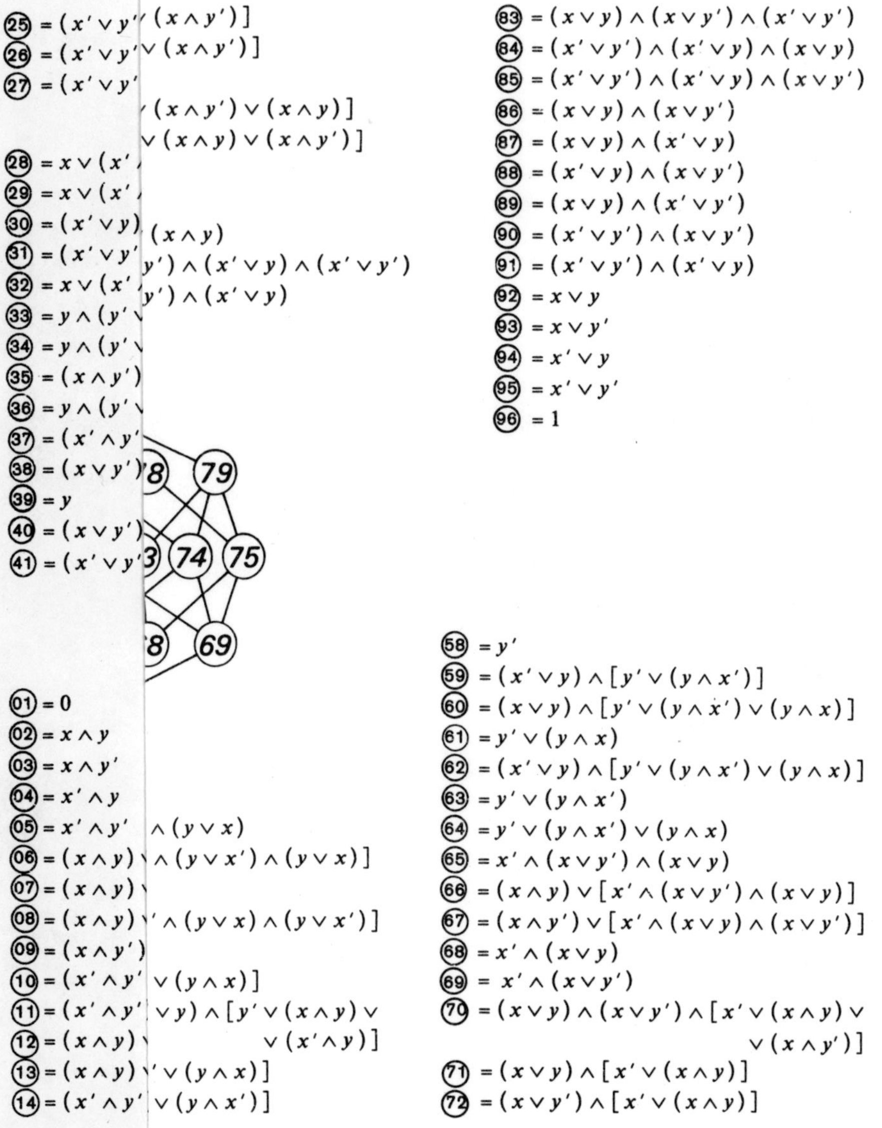

25 = (x′ ∨ y′ ∨ (x ∧ y′)]
26 = (x′ ∨ y′ ∨ (x ∧ y′)]
27 = (x′ ∨ y′
∨ (x ∧ y′) ∨ (x ∧ y)]
∨ (x ∧ y) ∨ (x ∧ y′)]
28 = x ∨ (x′
29 = x ∨ (x′
30 = (x′ ∨ y) (x ∧ y)
31 = (x′ ∨ y′ y′) ∧ (x′ ∨ y) ∧ (x′ ∨ y′)
32 = x ∨ (x′ y′) ∧ (x′ ∨ y)
33 = y ∧ (y′
34 = y ∧ (y′
35 = (x ∧ y′)
36 = y ∧ (y′
37 = (x′ ∧ y′
38 = (x ∨ y′)
39 = y
40 = (x ∨ y′)
41 = (x′ ∨ y′
78
79
73
74
75
68
69
83 = (x ∨ y) ∧ (x ∨ y′) ∧ (x′ ∨ y′)
84 = (x′ ∨ y′) ∧ (x′ ∨ y) ∧ (x ∨ y)
85 = (x′ ∨ y′) ∧ (x′ ∨ y) ∧ (x ∨ y′)
86 = (x ∨ y) ∧ (x ∨ y′)
87 = (x ∨ y) ∧ (x′ ∨ y)
88 = (x′ ∨ y) ∧ (x ∨ y′)
89 = (x ∨ y) ∧ (x′ ∨ y′)
90 = (x′ ∨ y′) ∧ (x ∨ y′)
91 = (x′ ∨ y′) ∧ (x′ ∨ y)
92 = x ∨ y
93 = x ∨ y′
94 = x′ ∨ y
95 = x′ ∨ y′
96 = 1
01 = 0
02 = x ∧ y
03 = x ∧ y′
04 = x′ ∧ y
05 = x′ ∧ y′ ∧ (y ∨ x)
06 = (x ∧ y) ∧ (y ∨ x′) ∧ (y ∨ x)]
07 = (x ∧ y)
08 = (x ∧ y) ′ ∧ (y ∨ x) ∧ (y ∨ x′)]
09 = (x ∧ y′)
10 = (x′ ∧ y′ ∨ (y ∧ x)]
11 = (x′ ∧ y′ ∨ y) ∧ [y′ ∨ (x ∧ y) ∨
12 = (x ∧ y) ∨ (x′ ∧ y)]
13 = (x ∧ y) ′ ∨ (y ∧ x)]
14 = (x′ ∧ y′ ∨ (y ∧ x′)]
58 = y′
59 = (x′ ∨ y) ∧ [y′ ∨ (y ∧ x′)]
60 = (x ∨ y) ∧ [y′ ∨ (y ∧ x′) ∨ (y ∧ x)]
61 = y′ ∨ (y ∧ x)
62 = (x′ ∨ y) ∧ [y′ ∨ (y ∧ x′) ∨ (y ∧ x)]
63 = y′ ∨ (y ∧ x′)
64 = y′ ∨ (y ∧ x′) ∨ (y ∧ x)
65 = x′ ∧ (x ∨ y′) ∧ (x ∨ y)
66 = (x ∧ y) ∨ [x′ ∧ (x ∨ y′) ∧ (x ∨ y)]
67 = (x ∧ y′) ∨ [x′ ∧ (x ∨ y) ∧ (x ∨ y′)]
68 = x′ ∧ (x ∨ y)
69 = x′ ∧ (x ∨ y′)
70 = (x ∨ y) ∧ (x ∨ y′) ∧ [x′ ∨ (x ∧ y) ∨ ∨ (x ∧ y′)]
71 = (x ∨ y) ∧ [x′ ∨ (x ∧ y)]
72 = (x ∨ y′) ∧ [x′ ∨ (x ∧ y)]

Fig. 18

Fig. 18	Representation in `L or in F_2	`$M_{02} \times \underline{2}^4$
69	$p_{69}=x'\wedge(x\vee y')$	$(a',0,0,0,1)=(a',b_5)$
70	$p_{70}=(x\vee y)\wedge(x\vee y')\wedge[x'\vee(x\wedge y)\vee(x\wedge y')]$	$(a',1,1,0,0)=(a',b_6)$
71	$p_{71}=(x\vee y)\wedge[x'\vee(x\wedge y)]$	$(a',1,0,1,0)=(a',b_7)$
72	$p_{72}=(x\vee y')\wedge[x'\vee(x\wedge y)]$	$(a',1,0,0,1)=(a',b_8)$
73	$p_{73}=(x\vee y)\wedge[x'\vee(x\wedge y')]$	$(a',0,1,1,0)=(a',b_9)$
74	$p_{74}=(x\vee y')\wedge[x'\vee(x\wedge y')]$	$(a',0,1,0,1)=(a',b_{10})$
75	$p_{75}=x'$	$(a',0,0,1,1)=(a',b_{11})$
76	$p_{76}=(x\vee y)\wedge[x'\vee(x\wedge y')\vee(x\wedge y)]$	$(a',1,1,1,0)=(a',b_{12})$
77	$p_{77}=(x\vee y')\wedge[x'\vee(x\wedge y)\vee(x\wedge y')]$	$(a',1,1,0,1)=(a',b_{13})$
78	$p_{78}=x'\vee(x\wedge y)$	$(a',1,0,1,1)=(a',b_{14})$
79	$p_{79}=x'\vee(x\wedge y')$	$(a',0,1,1,1)=(a',b_{15})$
80	$p_{80}=x'\vee(x\wedge y')\vee(x\wedge y)$	$(a',1,1,1,1)=(a',b_{16})$
81	$p_{81}=(x\vee y)\wedge(x\vee y')\wedge(x'\vee y)\wedge(x'\vee y')$	$(1,0,0,0,0)=(1,b_1)$
82	$p_{82}=(x\vee y)\wedge(x\vee y')\wedge(x'\vee y)$	$(1,1,0,0,0)=(1,b_2)$
83	$p_{83}=(x\vee y)\wedge(x\vee y')\wedge(x'\vee y')$	$(1,0,1,0,0)=(1,b_3)$
84	$p_{84}=(x'\vee y')\wedge(x'\vee y)\wedge(x\vee y)$	$(1,0,0,1,0)=(1,b_4)$
85	$p_{85}=(x'\vee y')\wedge(x'\vee y)\wedge(x\vee y')$	$(1,0,0,0,1)=(1,b_5)$
86	$p_{86}=(x\vee y)\wedge(x\vee y')$	$(1,1,1,0,0)=(1,b_6)$
87	$p_{87}=(x\vee y)\wedge(x'\vee y)$	$(1,1,0,1,0)=(1,b_7)$
88	$p_{88}=(x'\vee y)\wedge(x\vee y')$	$(1,1,0,0,1)=(1,b_8)$
89	$p_{89}=(x\vee y)\wedge(x'\vee y')$	$(1,0,1,1,0)=(1,b_9)$
90	$p_{90}=(x'\vee y')\wedge(x\vee y')$	$(1,0,1,0,1)=(1,b_{10})$
91	$p_{91}=(x'\vee y')\wedge(x'\vee y)$	$(1,0,0,1,1)=(1,b_{11})$
92	$p_{92}=x\vee y$	$(1,1,1,1,0)=(1,b_{12})$
93	$p_{93}=x\vee y'$	$(1,1,1,0,1)=(1,b_{13})$
94	$p_{94}=x'\vee y$	$(1,1,0,1,1)=(1,b_{14})$
95	$p_{95}=x'\vee y'$	$(1,0,1,1,1)=(1,b_{15})$
96	$p_{96}=1$	$(1,1,1,1,1)=(1,b_{16})$

From the 96 expressions $p_j(x,y)$ of Lemma 9 we can formally construct a poset. This construction gives a diagram of a lattice F_2 shown in Figure 18.

Using Table 1 or Figure 18 it can be shown that F_2 is orthoisomorphic to $`M_{02} \times \underline{2}^4$. Thus F_2 is also a free orthomodular lattice with two generators.

The lattice F_2 has a rich structure. We shall make free (sometimes implicit) use of its properties in many important places in next Chapters.

Suppose $`L$ is a given orthomodular lattice. The element $\overline{com}\,(s,t) = (s \vee t) \wedge (s \vee t') \wedge (s' \vee t) \wedge (s' \vee t')$ will be called the <u>upper commutator</u> of $s,t \in L$. The <u>lower commutator</u> of s,t is defined dually as the element $\underline{com}\,(s,t) = (s \wedge t) \vee (s \wedge t') \vee \vee (s' \wedge t) \vee (s' \wedge t')$. Instead of $\overline{com}\,(s,t)$ we at times write $\overline{com}_{`L}(s,t)$ or com (s,t). In particular, $\overline{com}_{[0,a]}(s,t)$ denotes the upper commutator of $s,t \in [0,a]$ in the interval algebra with the base set $[0,a]$, i.e.,

$$\overline{com}_{[0,a]}(s,t) = (s \vee t) \wedge (s \vee t^+) \wedge (s^+ \vee t) \wedge (s^+ \vee t^+)$$

where $s^+ = s' \wedge a$, $t^+ = t' \wedge a$.

In F_2 we clearly have $(\overline{com}\,(x,y)] \cong `M_{02}$ and $(\underline{com}\,(x,y)] \cong \cong \underline{2}^4$.

<u>Proposition 2.10</u>. The lattice F_2 is modular.

<u>Proof</u>. The assertion follows from the fact that F_2 is isomorphic to the direct product $`M_{02} \times \underline{2}^4$ of two modular lattices. //

<u>Theorem 2.11 (cf. [147])</u>. The following conditions are equivalent for any two elements a,b of an orthomodular lattice:

(i) aCb; (ii) $\overline{com}\,(a,b) = 0$.

Proof. 1. Observe that

$$0 = \overline{\text{com}}\,(a,b) = (a \vee b) \wedge (a \vee b') \wedge (a' \vee b) \wedge (a' \vee b') \geqslant$$
$$\geqslant a \wedge (a' \vee b) \wedge (a' \vee b').$$

Putting $s = (a' \vee b) \wedge (a' \vee b')$, $t = a$, we have $s \geqslant t'$ and $s \wedge t = 0$. By orthomodularity we can conclude that $t' = a' = (a' \vee b) \wedge (a' \vee b') = s$. Therefore, $a = (a \wedge b') \vee (a \wedge b)$, i.e., aCb.

2. If aCb, then $a' = (a' \vee b') \wedge (a' \vee b)$, by the definition of C. From Theorem II.3.6 we can see that $(a \vee b) \wedge (a \vee b') = a$. Hence $\overline{\text{com}}\,(a,b) = a \wedge a' = 0$. //

Corollary 2.12. An orthomodular lattice `L is distributive if and only if it does not contain a sublattice `M isomorphic to $`M_{O2}$ where `M has its least element equal to 0.

Proof. 1. If it contains `M, use Theorem II.6.10.

2. Suppose `L is not distributive. Then it suffices to apply Theorem 11 and Corollary 4. //

Theorem 2.13. Let c be an element of an orthomodular lattice `L. Then c commutes with every element of L if and only if $`L = (c] \oplus (c']$.

Proof. 1. Suppose $`L = (c] \oplus (c']$ and choose $x \in L$. Then $x = x_1 \vee x_2$ with $x_1 \in (c]$ and $x_2 \in (c']$. Hence $x_1 \leqslant x \wedge c$ and $x_2 \leqslant x \wedge c'$. Therefore, $x = x_1 \vee x_2 \leqslant (x \wedge c) \vee (x \wedge c') \leqslant x$. Thus xCc.

The converse assertion follows from Lemma 1 (ii). //

Remark 2.14. Note the following useful consequence of our considerations on F_2: Let `L be an orthomodular lattice and let $a, b \in L$. Suppose that $u = (a \vee b) \wedge [b' \vee (a' \wedge b)]$ and $v = (a' \vee b') \wedge [b \vee (a \wedge b')]$. Then there exists an orthohomomorphism f of $F_2 = \langle x,y \rangle$ onto the sublattice $`T = \langle a,b \rangle$ generated by a and b in `L such that $f(x) = a$ and $f(y) = b$. Let $u_1 = (x \vee y) \wedge [y' \vee (x' \wedge y)]$ and $v_1 = (x' \vee y') \wedge [y \vee (x \wedge y')]$. From Figure 18 or from Table 1 we find that $u_1 \wedge v_1 = (x \wedge y') \vee (x' \wedge y)$. Now

$$f((x \wedge y') \vee (x' \wedge y)) = \{f(x) \wedge [f(y)]'\} \vee \{[f(x)]' \wedge f(y)\} = \\ = (a \wedge b') \vee (a' \wedge b)$$

and $f(u_1 \wedge v_1) = f(u_1) \wedge f(v_1) = \ldots = u \wedge v$ and so $u \wedge v = (a \wedge b') \vee (a' \wedge b)$.

A similar argument yields the following general rule how to find $p(a,b) \wedge q(a,b)$, $p(a,b) \vee q(a,b)$ and $[p(a,b)]'$ for any polynomials p,q in $\vee, \wedge, '$: If

$$p(x,y) \wedge q(x,y) = r(x,y), \quad p(x,y) \vee q(x,y) = t(x,y)$$

and $[p(x,y)]' = s(x,y)$ is true in F_2, then

$$p(a,b) \wedge q(a,b) = r(a,b), \quad p(a,b) \vee q(a,b) = t(a,b)$$

and $[p(a,b)]' = s(a,b)$ is valid in any orthomodular lattice `L.

By Lemma 9, any $p(x,y)$ can be identified with an expression $p_j(x,y)$, $1 \leq j \leq 96$. To find the corresponding $p_j(x,y)$ for a given p, the modularity of F_2 is sometimes needed.

If a is an element of an orthomodular lattice `L, then there exists an orthoisomorphism of $\underline{2}^2$ onto the 4-element subalgebra $\{0,a,a',1\}$ generated by a in `L. The fact that $\underline{2}^2$ regarded as an orthomodular lattice has one-element set of generators, say $\{x\}$, is a straightforward exercise left to the interested reader. Again we see that any injection of $\{x\}$ into an orthomodular lattice `L can be extended to an orthohomomorphism of $\underline{2}^2$ into `L. From this point of view it is usual to call $\underline{2}^2$ the free orthomodular lattice with one generator and one can write $`F_1$ for $\underline{2}^2$. Hence, in the case $n = 1$, the structure of the free orthomodular lattice $`F_n$ is banal.

Can the methods used above be pushed further? Unfortunately, the answer is no. Example 15 below exhibits a counterexample which shows that we cannot expect finiteness of free orthomodular lattices with more than two generators.

Example 2.15. Let 0 be a fixed point of a three-dimensional Euclidean space E_3. Let L be the set consisting of all the lines $q \subset E_3$ which passe through the point 0, further of all the planes $p \subset E_3$ in which lies the point 0 and, finally, L contains also 0 and E_3. As in Remark II.1.1.D, all the elements of L are regarded as point sets and they are ordered by inclusion. The poset $(L, \subset)$ determines an ortholattice $(L, \vee, \cap, ', 0, E_3)$ with respect to the usual definition of perpendicularity. Clearly, the lattice $(L, \vee, \cap)$ does not contain a sublattice isomorphic to the pentagon and so it is modular. Therefore $`L = (L, \vee, \cap, ', 0, E_3)$ is an orthomodular lattice.

We shall show that there is a subalgebra in $`L$, which is infinite and which has three generators. In order to prove this, we shall start with three lines a, b, c of L and we shall investigate the set of all lines and planes which belong to the subalgebra generated by a, b, c. Evidently, if two different lines s, t belong to the subalgebra, then also the line $u = (s \vee t)'$ which contains 0 and is perpendicular to s and t is an element of the subalgebra. The same assertion is true for the plane $s \vee t$ which is determined by s and t.

To focus our attention, let us choose a Cartesian coordinate system with the origin 0. We shall restrict ourselves to the lines of L located in the interior of the first octant.

More exactly, let F denote the set such that a line $q \in L$ belongs to F if and only if there exists a vector $\underline{s}_q = (s_{q1}, s_{q2}, s_{q3})$ which is parallel to q and which satisfies $s_{qi} > 0$ for all $i = 1,2,3$. Hence a line q of F is the set formed by the points $[ts_{q1}, ts_{q2}, ts_{q3}]$ where t runs over all real numbers. Such a vector $\underline{s}_q$ will be called a <u>vector localized in the first octant</u>. Let $\bar{q} = \{[ts_{q1}, ts_{q2}, ts_{q3}];\ t > 0\}$.

Some terminology will help. If a and b are two distinct lines of F, then the set

$$\begin{aligned} D(a,b) = \{[c_1,c_2,c_3] \in E_3; \exists\, r_1, r_2 \in \underline{R},\ r_i > 0 \ (i = 1,2) ::\\ [c_1,c_2,c_3] = r_1[s_{a1},s_{a2},s_{a3}] + r_2[s_{b1},s_{b2},s_{b3}] =\\ = [r_1 s_{a1} + r_2 s_{b1}, \bullet]\} \end{aligned}$$

where $\underline{s}_a = (s_{a1}, s_{a2}, s_{a3})$, $\underline{s}_b = (s_{b1}, s_{b2}, s_{b3})$ are vectors localized in the first octant and parallel to a and b, respectively, is called a <u>D-complex</u>.

If a,b and c are three different lines of F, we say that the set

$$\begin{aligned} T(a,b,c) = \{[d_1,d_2,d_3] \in E_3; \exists\, r_1,r_2,r_3 \in \underline{R},\ r_i \geqslant 0 \ (i = 1,2,3) ::\\ [d_1,d_2,d_3] = r_1[s_{a1},s_{a2},s_{a3}] + r_2[s_{b1},s_{b2},s_{b3}] +\\ + r_3[s_{c1},s_{c2},s_{c3}] = [r_1 s_{a1} + r_2 s_{b1} + r_3 s_{c1}, \bullet]\} \end{aligned}$$

is a <u>T-complex</u>. Similarly as above, $\underline{s}_c$ denotes a vector localized in the first octant and parallel to the line c.

Clearly, $D(a,b) = D(b,a) \subset T(a,b,c) = T(b,c,a)$ etc.

If $T(a,b,c)$ is a T-complex, then we shall call the D-complexes $D(a,b)$, $D(b,c)$, $D(a,c)$ the <u>faces</u> of $T(a,b,c)$ (see Fig. 19).

Let a and b be two different lines of L. The line which is perpendicular to a and b and which belongs to L will be denoted by $v(a,b)$. Thus $v(a,b) = (a \vee b)'$. In referring to the plane which is determined by the lines a,b, we adopt the notation $p(a,b)$, i.e., $p(a,b) = a \vee b$.

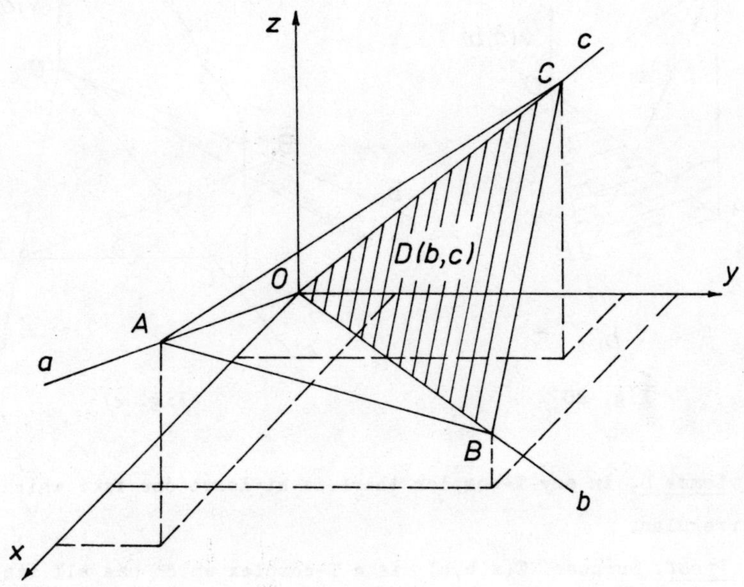

Fig. 19

Lemma A. If $T(a,b,c)$ is a T-complex, then c is not orthogonal to the plane $p(a,b)$.

Proof. Suppose the contrary. Then c is orthogonal to a. This implies that the scalar product $\langle \underline{s}_c, \underline{s}_a \rangle = s_{c1}s_{a1} + s_{c2}s_{a2} + s_{c3}s_{a3}$ is equal to 0. However, $s_{ci} > 0$ and $s_{ai} > 0$ for every $i = 1,2,3$, so $\langle \underline{s}_c, \underline{s}_a \rangle > 0$, a contradiction. //

The face $D(a,b)$ of $T(a,b,c)$ is said to be regular if and only if either $c \subset p(a,v(a,b))$ or $c \subset p(b,v(a,b))$ (see Fig. 20).

If this condition is not fulfilled, $D(a,b)$ is called irregular.

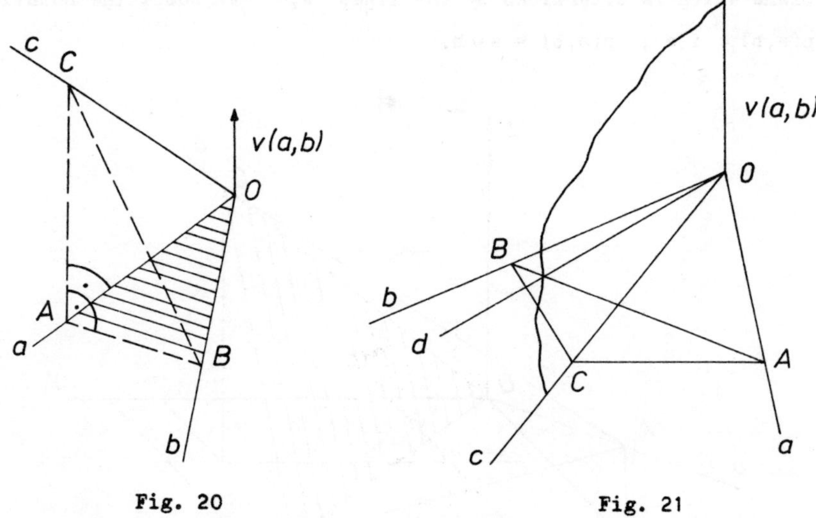

Fig. 20 Fig. 21

Lemma B. In any T-complex there is at least one face which is irregular.

Proof. Suppose $T(a,b,c)$ is a T-complex which has all its faces regular. We shall establish a contradiction. Without loss of generality we may assume that $c \subset p(a,v(a,b))$ (cf. Fig. 20), i.e., $v(a,c) \perp v(a,b)$. Since $D(b,c)$ is regular, we have either $v(a,c) \perp v(b,c)$ or $v(a,b) \perp v(b,c)$. In the first case we see that $b = p(a,b) \cap p(b,c) \supset v(a,c)$, contrary to Lemma A. In the second one, $c = p(a,c) \cap p(b,c) \supset v(a,b)$, which is again a contradiction. //

If $T(a,b,c)$ and $T(a_1,b_1,c_1)$ are two distinct T-complexes such that $T(a,b,c) \supset T(a_1,b_1,c_1)$, we write $T(a,b,c) > T(a_1,b_1,c_1)$.

We shall say that a T-complex $T(a,b,c)$ is well hewn with respect to the face $D(a,b)$ if and only if $\bar{d} \subset D(a,b)$ where d is the line $p(a,b) \cap p(v(a,b),c)$ (see Figure 21).

Note that in this case we have $T(a,b,c) > T(b,c,d)$ and that $d = (a \vee b) \cap [(a \vee b)' \vee c]$.

Since the structure of all the lines of F is sufficiently nontrivial, it is reasonable to conjecture that any three lines a,b,c of F consitute a system of generators for an infinite subalgebra $`S = \langle a,b,c \rangle$ of the orthomodular lattice `L.

We now proceed toward a more complete study of the lines which belong to the subalgebra `S determined by a given triple of lines $a,b,c \in F$. To show that `S is infinite, it suffices to prove that there exists an infinite decreasing sequence

(*) $T(a,b,c) > T(a_1,b_1,c_1) > T(a_2,b_2,c_2) > \ldots$

where $a_i,b_i,c_i \in S$ for every $i = 1,2,\ldots$.

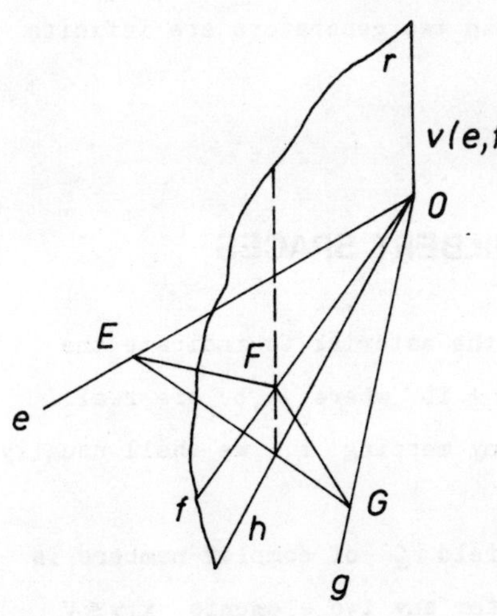

Fig. 22

By Lemma B and by the remark above it is sufficient to investigate the case of a T-complex which is not well hewn with respect to an irregular face.

Lemma C. Let $e,f,g \in S$ be distinct lines. If the T-complex $T(e,f,g)$ is not well hewn with respect to the irregular face $D(e,f)$, then there exists a T-complex $T(e_1,f_1,g_1)$ such that $T(e,f,g) > T(e_1,f_1,g_1)$ with $e_1,f_1,g_1 \in S$.

Proof. Let $r = p(v(e,f),f)$ and $s = p(v(e,f),e)$. Since $T(e,f,g)$ is not well hewn with respect to $D(e,f)$, either $r \cap D(e,g) \neq \emptyset$ or $s \cap D(f,g) \neq \emptyset$. If $r \cap D(e,g) \neq \emptyset$ (see Figure 22), then the lines $h = r \cap p(e,g)$ and g are distinct and $T(e,f,g) > T(e_1,f_1,g_1)$ where $e_1 = e$, $f_1 = f$, $g_1 = h$. Moreover, we have $r = (e \vee f)' \vee f \in S$ and $e \vee g \in S$. Thus $h = r \cap (e \vee g) \in S$.

If the case $s \cap D(f,g) \neq \emptyset$ occurs, we can use a similar argument. //

Taken together, our lemmas A - C prove the existence of a decreasing sequence described in (*). Thus the subalgebra `S is in fact infinite. The subalgebra `S provides an example of an infinite orthomodular lattice with three generators which is of finite length.

Note that from our considerations it follows that the free orthomodular lattices with more than two generators are infinite (cf. [83]).

3. INTRODUCTION TO HILBERT SPACES

In what follows we shall use the asterisk to indicate the complex conjugate, i.e., if $c = a + ib$ where a,b are real numbers, then $c^* = a - ib$. For any mapping F, we shall usually write Fx instead of $F(x)$.

A vector space V over the field $\underline{C}$ of complex numbers is said to be a pre-Hilbert space if for any two elements $x,y \in V$ there is defined a number $\langle x,y \rangle$, called their scalar product, such that the following conditions are satisfied for every x,y, $z \in V$ and all $c \in \underline{C}$:

(S 1) $\langle y,x\rangle = \langle x,y\rangle^*$;

(S 2) $\langle x + y,z\rangle = \langle x,z\rangle + \langle y,z\rangle$;

(S 3) $\langle cx,y\rangle = c\langle x,y\rangle$;

(S 4) if x is not the zero vector o of V, then $\langle x,x\rangle > 0$.

Recall that a nonvoid subset W of a vector space V over $\underline{C}$ is said to be a subspace of V if and only if (i) $w_1 + w_2 \in W$ for every $w_1, w_2 \in W$ and (ii) $cw \in W$ for each c of $\underline{C}$ and all $w \in W$.

Lemma 3.1. In any pre-Hilbert space, for all $x,y,z \in V$ and $c \in \underline{C}$,

(S 5) $\langle x,cy\rangle = c^*\langle x,y\rangle$;

(S 6) $\langle x,x\rangle = 0 \Leftrightarrow x = o$;

(S 7) $\langle x,y + z\rangle = \langle x,y\rangle + \langle x,z\rangle$.

Proof. Ad (S 5). Clearly, $\langle x,cy\rangle = (c\langle y,x\rangle)^* = c^*\langle y,x\rangle^* = = c^*\langle x,y\rangle$.

Ad (S 6). Note that $\langle o,o\rangle = \langle 0o,o\rangle = 0\langle o,o\rangle = 0$ and use (S 4).

Ad (S 7). An argument analogous to that used in the proof of (S 5) shows that

$$\langle x,y + z\rangle = \langle y + z,x\rangle^* = (\langle y,x\rangle + \langle z,x\rangle)^* = \langle y,x\rangle^* + \langle z,x\rangle^* = = \langle x,y\rangle + \langle x,z\rangle. \quad //$$

Lemma 3.2. Let V and W be two pre-Hilbert spaces. Let F,G be mappings of V into W. Then the following is true:

(S 8) If $\langle Fx,y\rangle = 0$ for every $x \in V$ and every $y \in W$, then $Fx = o$ for all $x \in V$.

(S 9) If $\langle Fx,y\rangle = \langle Gx,y\rangle$ for every $x \in V$ and every $y \in W$, then $F = G$.

(S 10) If $\langle y,Fx\rangle = 0$ for every $x \in V$ and every $y \in W$, then $Fx = o$.

(S 11) If $\langle y,Fx\rangle = \langle y,Gx\rangle$ for every $x \in V$ and every $y \in W$, then $F = G$.

Proof. Ad (S 8). By hypothesis, $\langle Fx,Fx\rangle = 0$ and so $Fx = o$, by (S 6).

Ad (S 9). Putting $Hx = Fx - Gx$, we get

$$\begin{aligned}\langle Hx,y\rangle &= \langle Fx - Gx,y\rangle = \langle Fx + (-1)Gx,y\rangle = \\ &= \langle Fx,y\rangle + (-1)\langle Gx,y\rangle = 0.\end{aligned}$$

Therefore, $o = Hx = Fx - Gx$ for every $x \in V$.

Ad (S 10), (S 11). These assertions are clear from (S 9) and (S 8). //

Recall the following usual definitions:

By a norm on a vector space U over the field $\underline{C}$ of complex numbers is meant a real-valued function on U, defined by the assignment $x \mapsto \|x\|$, which satisfies

(i) $\|x\| = 0 \Leftrightarrow x = o$;

(ii) $\|x + y\| \leqslant \|x\| + \|y\|$ for every $x,y \in U$;

(iii) $\|cx\| = |c|\,\|x\|$ for every $c \in \underline{C}$ and every $x \in U$.

A vector space with a norm is called a normed space.

A sequence $\{x_n\}$ of vectors is said to be a Cauchy sequence if for every $e > 0$ there exists $n_0 \in \underline{N}$ such that $\|x_m - x_n\| < e$ for every $m,n \geqslant n_0$. A sequence $\{x_n\}$ is said to converge to $x \in U$ if the following is true: Given $e > 0$, there exists $n_0 \in \underline{N}$ such that $\|x_n - x\| < e$ for every $n \geqslant n_0$. In this case we write $x_n \to x$ and say that the sequence $\{x_n\}$ is convergent. The vector x is then uniquely determined by the sequence $\{x_n\}$.

A subspace W of a normed space U is called complete if every Cauchy sequence of vectors of W is convergent in W.

The norm $\|x\|$ of an element x of a pre-Hilbert space is defined as the real number $\sqrt{\langle x,x\rangle}$. This is justified as follows:

Theorem 3.3. The real-valued function defined on a pre-Hilbert space by $x \mapsto \sqrt{\langle x,x\rangle}$ has the properties of a norm.

Proof. Ad (i). Clearly, $\|x\| = 0$ is equivalent to $\langle x,x\rangle = 0$. By (S 6), this is equivalent to $x = o$.

Ad (ii). To prove the assertion, we use the following argument: Let $x,y \in V$ and let s be a complex number of unit modulus, i.e., $|s| = 1$. Let t range over all the real numbers. By (S 4) and (S 6),

$$0 \leqslant \langle x - sty, x - sty\rangle = \langle x,x\rangle - st\langle y,x\rangle - s^{*}t\langle x,y\rangle + t^{2}\langle y,y\rangle = = At^{2} + Bt + C$$

where $A = \langle y,y\rangle$, $B = -(s^{*}\langle x,y\rangle + s\langle y,x\rangle)$, $C = \langle x,x\rangle$. From (S 1) we infer that

$$B^{*} = -[(s^{*}\langle x,y\rangle)^{*} + (s\langle y,x\rangle)^{*}] = B$$

and, therefore, the three numbers A,B,C are real. Note that $p(t) = At^{2} + Bt + C \geqslant 0$ for every $t \in \underline{R}$. Hence we have $B^{2} \leqslant 4AC$. Consequently, $|B| \leqslant 2\sqrt{A}\sqrt{C}$ so that

$$s^{*}\langle x,y\rangle + s\langle y,x\rangle \leqslant |s^{*}\langle x,y\rangle + s\langle y,x\rangle| \leqslant 2\sqrt{\langle y,y\rangle}\sqrt{\langle x,x\rangle}.$$

In particular, if $s = 1$,

$$\langle x + y, x + y\rangle = \langle x,x\rangle + \langle x,y\rangle + \langle y,x\rangle + \langle y,y\rangle \leqslant \leqslant \langle x,x\rangle + 2\sqrt{\langle y,y\rangle}\sqrt{\langle x,x\rangle} + \langle y,y\rangle$$

and we obtain

$$\langle x + y, x + y\rangle \leqslant (\sqrt{\langle x,x\rangle} + \sqrt{\langle y,y\rangle})^{2}.$$

This shows that

$$\|x + y\| = \sqrt{\langle x + y, x + y\rangle} \leqslant \sqrt{\langle x,x\rangle} + \sqrt{\langle y,y\rangle} = \|x\| + \|y\|.$$

Ad (iii). By (S 3) and (S 5),

$$\|cx\| = \sqrt{\langle cx, cx\rangle} = \sqrt{cc^{*}}\sqrt{\langle x,x\rangle} = |c|\,\|x\|. \quad //$$

Remark 3.4. (A) The inequality

$$|\langle x,y\rangle| \leqslant \|x\|\|y\|$$

is true for every x,y of a pre-Hilbert space and it is called the Bunjakovskij-Cauchy-Schwarz inequality.

To prove the inequality it is enough to consider the case $a = \langle x,y\rangle \neq 0$. Then we can put $s = a.|a|^{-1}$ in the proof of Theorem 3. Thus $2\sqrt{\langle y,y\rangle}\sqrt{\langle x,x\rangle} \geqslant s^*a + sa^* = 2a^*a|a|^{-1} = 2|a|$. Now it is easy to see that $\sqrt{\langle y,y\rangle\langle x,x\rangle} \geqslant |\langle x,y\rangle|$.

(B) Choosing $s = 1$, $t = 1$ and then $s = 1$, $t = -1$, we similarly have

$$\langle x - y,x - y\rangle = \langle x,x\rangle - \langle y,x\rangle - \langle x,y\rangle + \langle y,y\rangle,$$
$$\langle x + y,x + y\rangle = \langle x,x\rangle + \langle y,x\rangle + \langle x,y\rangle + \langle y,y\rangle.$$

Adding these relations, we get

$$\|x - y\|^2 + \|x + y\|^2 = 2\|x\|^2 + 2\|y\|^2.$$

(C) If x,y are such that $\langle x,y\rangle = 0$, then $\langle x \pm y,x \pm y\rangle = \langle x,x\rangle + \langle y,y\rangle$. Hence, analogously to Pythagorean Theorem we have

$$\|x \pm y\|^2 = \|x\|^2 + \|y\|^2.$$

In particular, $\|x + y\| \geqslant \|x\|$ for such $x,y \in V$.

A pre-Hilbert space V is called a Hilbert space if and only if it is complete with respect to the norm $\|x\| = \sqrt{\langle x,x\rangle}$.

Recall that a mapping F of a pre-Hilbert space V into a pre-Hilbert space W is called continuous at a point $x_0 \in V$ if it satisfies the following condition: Given any $e > 0$, there exists $d > 0$ such that

$$\|x - x_0\| < d \Rightarrow \|Fx - Fx_0\| < e.$$

We shall say that F is <u>continuous</u> (or <u>continuous on</u> V) if it is continuous at every point of V.

A mapping $F:V\to W$ is called a <u>continuous linear mapping</u> of V into W if F is continuous on V and if, moreover, it is <u>linear</u>, i.e., if $F(x + y) = Fx + Fy$ and $F(cx) = cFx$ for every $x,y \in V$ and every $c \in \underline{C}$.

If $V = W$ and F is a continuous linear mapping of V into V, then F is also called an <u>operator</u> on V.

Observe that $Fo = o$ for every linear mapping of V into W. In fact, $Fo = F(0o) = 0Fo = o$.

Let $F:V\to W$ be a continuous linear mapping. We can consider the least upper bound of all the real numbers $\|Fx\|$ which we get if x ranges over all the vectors x of the closed unit ball in V, i.e., over all the vectors x such that $\|x\| \leqslant 1$.

<u>Theorem 3.5</u>. (i) If F is a continuous linear mapping, then there exists

$$\sup_{(\underline{R}, \leqslant)} \{\|Fx\|;\ \|x\| \leqslant 1\}.$$

(ii) Let $\|F\|$ denote the supremum $\sup_{(\underline{R}, \leqslant)}\{\|Fx\|;\ \|x\| \leqslant 1\}$ of (i). Under the hypotheses of (i), $\|Fx\| \leqslant \|F\|\|x\|$ for every $x \in V$.

(iii) A linear mapping $F:V\to W$ is continuous if and only if there exists a real number r such that $\|Fx\| \leqslant r\|x\|$ for every $x \in V$.

<u>Proof</u>. Ad (i). Choose $e = 0{,}5$. Since F is continuous at the point $x = o$, there exists $d > 0$ such that $\|x\| < d$ implies $0{,}5 > \|Fx - Fo\| = \|Fx\|$. Suppose the supremum does not exist. Then, for every $n \in \underline{N}$, there exists x_n such that $\|Fx_n\| > n$ and $\|x_n\| \leqslant 1$. Denote $y_n = n^{-1}x_n$. If n_0 is the smallest natural

number satisfying $n_0^{-1} < d$, then $\|y_n\| < d$ for every $n \geqslant n_0$. For any such n we have $Fy_n = F(n^{-1}x_n) = n^{-1}Fx_n$. Thus $\|Fy_n\| = n^{-1}\|Fx_n\| > n^{-1}.n = 1$ and this yields $0{,}5 > \|Fy_n\| > 1$, a contradiction.

Ad (ii). Let $r = \|F\|$. If $o \neq y \in V$, we put $x = \|y\|^{-1}y$ and note that $\|x\| = 1$. Therefore, $\|Fx\| \leqslant r$ and from this we may conclude that $\|y\|^{-1}\|Fy\| \leqslant r$, i.e., $\|Fy\| \leqslant r\|y\|$.

Ad (iii). By (ii) it remains to prove that the existence of a number r with the indicated property implies the continuity of F. This is clear in the case $r = 0$. So let $r > 0$ and let $e > 0$. Denote $d = e.r^{-1}$. As usual, then

$$\|x - x_0\| < d \Rightarrow \|Fx - Fx_0\| = \|F(x - x_0)\| \leqslant r\|x - x_0\| < rd = e;$$

hence the desired result. //

The supremum $\|F\|$ of Theorem 5 is called the <u>norm</u> of the continuous linear mapping F.

<u>Lemma 3.6</u>. Let W be a complete subspace of a pre-Hilbert space V and let $x_0 \in V$. Then there exists a vector $y_0 \in W$ such that $\|x_0 - y_0\| \leqslant \|x_0 - y\|$ for every $y \in W$.

<u>Proof</u>. Denote $i_0 = \inf\{\|x_0 - y\|;\ y \in W\}$. Let $\{y_n\}$ be a sequence of W such that $i_0 - n^{-1} < \|x_0 - y_n\| < i_0 + n^{-1}$. Choose $e > 0$. Let n_0 be a natural number satisfying the inequality $n_0 > 4(2i_0 + 1).e^{-1}$. Let $m,n \geqslant n_0$. Using Remark 4.B where we put $x = y_m + y_n - 2x_0$, $y = y_m - y_n$, we obtain

$$4\|y_n - x_0\|^2 + 4\|y_m - x_0\|^2 = 2\|y_m + y_n - 2x_0\|^2 + 2\|y_m - y_n\|^2.$$

Consequently,

$$\|y_m - y_n\|^2 = 2\|y_n - x_0\|^2 + 2\|y_m - x_0\|^2 - \|y_m + y_n - 2x_0\|^2 \leqslant$$
$$\leqslant 2(i_0 + n^{-1})^2 + 2(i_0 + m^{-1})^2 - 4i_0^2.$$

The last inequality holds since

$$\|y_m + y_n - 2x_0\|^2 = 4\|(y_m + y_n).2^{-1} - x_0\|^2 \geq 4i_0^2,$$

by the definition of i_0. In view of the previous statement we arrive at the following conclusion:

$$\|y_m - y_n\|^2 \leq 4i_0(\tfrac{1}{n} + \tfrac{1}{m}) + 2(\tfrac{1}{n^2} + \tfrac{1}{m^2}) \leq$$

$$\leq 4i_0 \frac{2}{n_0} + \frac{4}{n_0} = \frac{4(2i_0 + 1)}{n_0} < e.$$

Hence, $\{y_n\}$ is a Cauchy sequence. By completeness of W, there exists $y_0 \in W$ such that $y_n \to y_0$. Thus, there is n_1 such that for every $n \geq n_1$ we have $\|y_n - y_0\| < e$. From Theorem 3,

$$\|y_n - x_0\| \leq \|y_n - y_0\| + \|y_0 - x_0\|,$$

$$\|y_0 - x_0\| \leq \|y_0 - y_n\| + \|y_n - x_0\|,$$

so that

$$\|y_n - x_0\| - \|y_0 - x_0\| \leq \|y_n - y_0\|,$$

$$\|y_0 - x_0\| - \|y_n - x_0\| \leq \|y_n - y_0\|.$$

Therefore,

$$\big|\|y_n - x_0\| - \|y_0 - x_0\|\big| \leq \|y_n - y_0\| < e.$$

From this we see that the sequence $\{\|y_n - x_0\|\}$ of real numbers converges to the number $\|y_0 - x_0\|$. By the definition of i_0 one sees that $\|y_0 - x_0\| = i_0$. //

Given a nonvoid subset M of V, we put $M^{\perp} = \{v \in V;\ \langle v,w\rangle = 0$ for every $w \in M\}$. We shall write $M^{\perp\perp}$ for $(M^{\perp})^{\perp}$. If $M \subset N$, then clearly $N^{\perp} \subset M^{\perp}$.

<u>Theorem 3.7</u>. Let W be a complete subspace of a pre-Hilbert space V. Then

(i) every element $p \in V$ can be uniquely written in the form $p = q + w$ where $q \in W$ and $w \in W^{\perp}$;

(ii) $W^{\perp\perp} = W$.

<u>Proof</u>. Ad (i). Choose arbitrarily $p_0 \in V$ (see Figure 23).

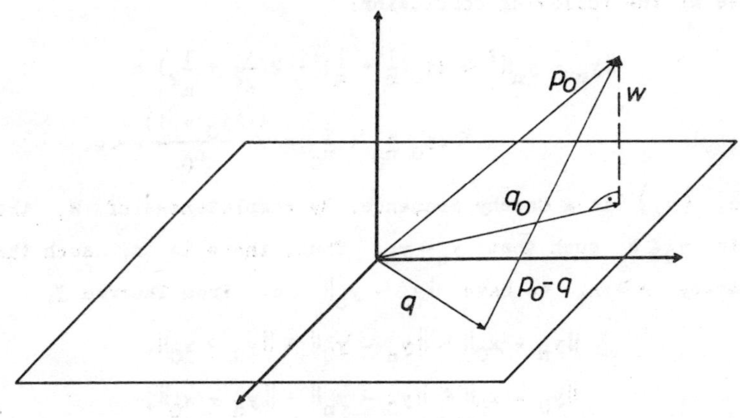

Fig. 23

Let $q_0 \in W$ be a vector obtained by Lemma 6, i.e., such that $\|p_0 - q_0\| \leq \|p_0 - q\|$ for every $q \in W$. Denote $w = p_0 - q_0$. We shall first show that $w \in W^{\perp}$. If $t \in \underline{R}$, then

$$\|w - tq\| = \|p_0 - (q_0 + tq)\| \geq \|p_0 - q_0\| = \|w\|$$

by the choice of q_0. Let $q \neq o$ be a vector of W. Put $b = ((\langle q,q\rangle))^{-1} > 0$ and $t_0 = \langle w,q\rangle.b$. Then

$$\langle w - t_0 q, w - t_0 q\rangle = \langle w,w\rangle - t_0^{*}\langle w,q\rangle - t_0\langle q,w\rangle + t_0 t_0^{*}\langle q,q\rangle =$$

$$= \langle w,w\rangle - b\langle w,q\rangle^{*}\langle w,q\rangle - b\langle w,q\rangle\langle w,q\rangle^{*} + b\langle w,q\rangle^{*}\langle w,q\rangle =$$

$$= \langle w,w\rangle - b\,|\langle w,q\rangle|^2,$$

$$\|w\|^2 = \|p_0 - q_0\|^2 \leq \|w - t_0 q\|^2 = \|w\|^2 - b\,|\langle w,q\rangle|^2 \leq \|w\|^2.$$

Thus $\|w\|^2 = \|w\|^2 - b|\langle w,q\rangle|^2$, that is, $\langle w,q\rangle = 0$.

Since $p_0 = q_0 + p_0 - q_0 = q_0 + w$, we have proved that p_0 can be written in the form mentioned in our assertion.

If $p_0 = q' + w'$ where $q' \in W$ and $w' \in W^{\perp}$, then $q_0 - q' = w' - w \in W \cap W^{\perp} = \{o\}$ and so $q_0 = q'$, $w = w'$.

Ad (ii). By the definition of $W^{\perp\perp}$, $W \subset W^{\perp\perp}$. To prove the converse inclusion, consider $p \in W^{\perp\perp}$. Using (i), write $p = q + w$ where $q \in W$ and $w \in W^{\perp}$. Then, since $q \in W \subset W^{\perp\perp}$, $w = p - q$ is a linear combination of two vectors of $W^{\perp\perp}$ and, consequently, $w \in W^{\perp\perp} \cap W^{\perp} = \{o\}$. Thus $w = o$ and so $p = q \in W$. //

The set $\underline{C}$ of complex numbers is a well-known example of a Hilbert space where $\langle x,y \rangle = xy^*$ for $x,y \in \underline{C}$. A linear mapping of a Hilbert space H into the space $\underline{C}$ is usually called a <u>linear form</u> (simply a <u>form</u>) on H. If, moreover, such a linear form $f: H \to \underline{C}$ is continuous, it is called a <u>continuous linear form</u>. The set $K = \{h \in H;\ f(h) = 0\}$ is said to be the <u>null space</u> of the form f.

<u>Remark 3.8</u>. If H is a Hilbert space and f a continuous linear form on H, then its null space K is a complete subspace of H. Indeed, if $\{h_n\}$ is a Cauchy sequence of K, then the completeness of H implies that there exists an element $h \in H$ such that $h_n \to h$. Our goal consists of showing that $h \in K$, i.e., that $f(h) = 0$.

Since f is continuous, it is clear that for every $e > 0$ there exists $d > 0$ such that

$$\|v - h\| < d \Rightarrow |f(v) - f(h)| < e.$$

From $h_n \to h$ we conclude that there is n_0 such that $\|h_n - h\| < d$ for every $n \geq n_0$. Consequently, $|f(h_n) - f(h)| < e$. Now $h_n \in K$ and so $f(h_n) = 0$. Therefore, $|f(h)| < e$ for every $e > 0$. Thus $f(h) = 0$.

Theorem 3.9. Let H be a Hilbert space and let f be a continuous linear form on H. Then there exists exactly one vector $w \in H$ such that $f(v) = \langle v,w\rangle$ for every $v \in H$.

Proof. 1. Existence. Let K be the null set of the form f. No generality is lost in assuming $K^{\perp} \neq \{o\}$. In fact, if $K^{\perp} = \{o\}$, then $K^{\perp\perp} = \{o\}^{\perp} = H$ and, by Theorem 7 (ii), $K = H$. But then we can put $w = o$.

Now let u be a non-zero vector of $K^{\perp}$. By (S 4), $\langle u,u\rangle \neq 0$. A simple calculation, employing the linearity of f shows that

$$f(f(u)v - f(v)u) = f(f(u)v) - f(f(v)u) = f(u)f(v) - f(v)f(u) = 0$$

and it follows $f(u)v - f(v)u \in K$. Since $u \in K^{\perp}$,

$$0 = \langle f(u)v - f(v)u,u\rangle = \langle f(u)v,u\rangle - \langle f(v)u,u\rangle = f(u)\langle v,u\rangle - f(v)\langle u,u\rangle.$$

From this it is evident that $f(v) = f(u).(\langle u,u\rangle)^{-1}.\langle v,u\rangle$. Denote $z = f(u)(\langle u,u\rangle)^{-1}$, $w = z^{*}u$. Then $f(v) = \langle v,w\rangle$.

2. Uniqueness. Suppose $f(v) = \langle v,w'\rangle$ for every $v \in H$. Then $0 = \langle v,w - w'\rangle$, and, in particular, $\langle w - w',w - w'\rangle = 0$, i.e., $w - w' = o$. //

Theorem 3.10. Let $F:H \to K$ be a continuous linear mapping of a Hilbert space H into a Hilbert space K. Then there exists exactly one continuous mapping $F^{*}:K \to H$ such that $\langle Fx,y\rangle = \langle x,F^{*}y\rangle$ for every $x \in H$ and every $y \in K$.

Proof. 1. Uniqueness. Suppose F^{+} is a mapping such that $\langle Fx,y\rangle = \langle x,F^{+}y\rangle$ for every $x \in H$ and $y \in K$. By (S 11) of Theorem 2, $F^{*} = F^{+}$.

2. We shall prove the existence of F^{*}. Fix $y \in K$. The

mapping $g(x) = \langle Fx,y \rangle$ is a continuous linear form on H. Indeed, we observe first that

$$g(x_1 + x_2) = \langle F(x_1 + x_2),y \rangle = \langle Fx_1 + Fx_2,y \rangle = \\ = \langle Fx_1,y \rangle + \langle Fx_2,y \rangle .$$

Similarly, $g(cx) = cg(x)$ for any $c \in \underline{C}$. Now

$$|g(x)| = |\langle Fx,y \rangle| \leqslant \|Fx\|\|y\| \leqslant \|F\|.\|y\|.\|x\|,$$

by the inequality of Remark 4.A and by Theorem 5 (ii). From Theorem 5 (iii) it follows that $g:H \rightarrow \underline{C}$ is a continuous linear form.

By Theorem 9 there exists one and only one vector $w \in H$ such that $\langle Fv,y \rangle = g(v) = \langle v,w \rangle$ for every $v \in H$. Define the mapping F^* by $F^*y = w$. It is easy to show that F^* is a linear mapping: If $c_1,c_2 \in \underline{C}$ and $y_1,y_2 \in K$, consider the elements w_1,w_2 such that $\langle Fv,y_i \rangle = \langle v,w_i \rangle$, $i = 1,2$. Then

$$\langle Fv,c_1y_1 + c_2y_2 \rangle = \langle Fv,c_1y_1 \rangle + \langle Fv,c_2y_2 \rangle = \\ = c_1^*\langle Fv,y_1 \rangle + c_2^*\langle Fv,y_2 \rangle = c_1^*\langle v,w_1 \rangle + c_2^*\langle v,w_2 \rangle = \\ = \langle v,c_1w_1 \rangle + \langle v,c_2w_2 \rangle = \langle v,c_1w_1 + c_2w_2 \rangle .$$

This shows that $F^*(c_1y_1 + c_2y_2) = c_1F^*y_1 + c_2F^*y_2$.

The mapping F^* is continuous. Indeed,

$$\|w\|^2 = |\langle w,w \rangle| = |\langle w,F^*y \rangle| = |\langle Fw,y \rangle| \leqslant \\ \leqslant \|Fw\|.\|y\| \leqslant \|F\|.\|w\|.\|y\|,$$

by Remark 4.A and Theorem 5 (ii). Since $w = F^*y$, $\|F^*y\| \leqslant \|F\|.\|y\|$. From Theorem 5 (iii) we now see the continuity of F^*. //

If $c \in \underline{C}$ and if F,G are mappings of H into K, then the mappings cF, F + G of H into K are defined by $(cF)(x) = c(Fx)$ and $(F + G)(x) = Fx + Gx$.

Theorem 3.11. Let F,G be continuous linear mappings of a Hilbert space H into a Hilbert space K. Then

(i) $F = 0 \Leftrightarrow F^* = 0$; (ii) $(cF)^* = c^*F^*$ for every $c \in \underline{C}$;
(iii) $(F + G)^* = F^* + G^*$; (iv) $F^{**} = F$; (v) $\|F\| = \|F^*\|$;
(vi) $\|FF^*\| = \|F^*F\| = \|F\|^2$; (vii) $FF^* = 0 \Leftrightarrow F = 0$.
(viii) If F_i, $i = 1,2$, are continuous linear mappings of H into H, then $(F_1F_2)^* = F_2^*F_1^*$.

Proof. Ad (i). A mapping F is the null mapping 0 if and only if $\langle Fv,w\rangle = 0$ for every $v \in H$ and every $w \in K$. This is equivalent to the statement that $\langle v,F^*w\rangle = 0$ for every $v \in H$ and $w \in K$. However, the latter fact is equivalent to $F^* = 0$.

Ad (ii) and (iii). We have

$$\langle cFv,w\rangle = c\langle Fv,w\rangle = c\langle v,F^*w\rangle = \langle v,c^*F^*w\rangle .$$

At the same time $\langle cFv,w\rangle = \langle v,(cF)^*w\rangle$. From (S 11) of Lemma 2 we infer that $(cF)^* = c^*F^*$.

The proof of (iii) is analogous.

Ad (iv). We write

$$\langle Fv,w\rangle = \langle v,F^*w\rangle = \langle F^*w,v\rangle^* = \langle w,F^{**}v\rangle^* = \langle F^{**}v,w\rangle$$

which, by the help of (S 9), gives $F = F^{**}$.

Ad (v). We know from the proof of Theorem 10 that $\|F^*y\| \leq \|F\|.\|y\|$. If $\|y\| \leq 1$, then $\|F^*y\| \leq \|F\|$ and so $\|F^*\| \leq \|F\|$. Hence $\|F\| = \|F^{**}\| \leq \|F^*\|$, i.e., $\|F^*\| = \|F\|$.

Ad (vi). We remark that $\langle F^*Fv,w\rangle = \langle Fv,F^{**}w\rangle = \langle Fv,Fw\rangle$ and, in particular, $\langle F^*Fv,v\rangle = \|Fv\|^2$. Suppose that $v \in H$ is such that $\|v\| \leq 1$. Then, by Remark 4.A and Theorem 5 (ii),

$$\|Fv\|^2 = |\langle F^*Fv,v\rangle| \leq \|F^*Fv\|\|v\| \leq \|F^*Fv\| \leq \|F^*F\|\|v\| \leq \|F^*F\|$$

and it is immediate that $\|F\|^2 \leq \|F^*F\|$. On the other hand,

$$\|(F^*F)v\| = \|F^*(Fv)\| \leq \|F^*\|\|F\|\|v\|.$$

If $\|v\| \leq 1$, then $\|(F^*F)v\| \leq \|F^*\|\|F\| = \|F\|^2$ and so $\|F^*F\| \leq \|F\|^2$.

Ad (vii). A continuous mapping is null if and only if its norm is equal to 0. Hence the statement (vii) follows from (vi).

Ad (viii). First note that

$$\langle F_1F_2v,w\rangle = \langle F_2v,F_1^*w\rangle = \langle v,F_2^*F_1^*w\rangle.$$

Further, $\langle F_1F_2v,w\rangle = \langle v,(F_1F_2)^*w\rangle$. Hence (S 11) gives $(F_1F_2)^* = F_2^*F_1^*$. //

We recall that a <u>closed subspace</u> of a Hilbert space H is every subspace H_1 of H which has the following property: If $\{x_n\}$ is a sequence of vectors $x_n \in H_1$ which is convergent in H, then its limit belongs to H_1.

<u>Remark 3.12</u>. (A) A subspace K of a Hilbert space H is closed if and only if $K = K^{\perp\perp}$. In fact, suppose first that K is a closed subspace. If $\{x_n\}$ is a Cauchy sequence of vectors belonging to K, then there exists its limit x in H. Since K is closed, $x \in K$. This means that K is a complete subspace. Thus, any closed subspace of a Hilbert space is complete. Consequently, using Theorem 7 (ii), we get $K = K^{\perp\perp}$. Finally we must show that if $K = K^{\perp\perp}$, then K is closed. It is obvious that it is sufficient to prove that $M^{\perp}$ is a closed subspace of H whenever M is a subspace of H. It is clear that $M^{\perp}$ is a subspace of H. Let $\{x_n\}$ be a sequence of vectors in $M^{\perp}$ such that $x_n \to x$. Then $\langle x_n,s\rangle = 0$ for every $s \in M$. Without loss of generality, we may assume that $s \neq 0$. Let $e > 0$ be arbitrary. Then there exists a number n_0 such that $\|x_n - x\| < e.\|s\|^{-1}$ for every $n \geqslant n_0$. Therefore, by Remark 4.A,

$$|\langle x,s\rangle| = |\langle x_n,s\rangle - \langle x,s\rangle| = |\langle x_n - x,s\rangle| \leqslant \|x_n - x\|\|s\| < e$$

and so $\langle x,s\rangle = 0$. Thus $x \in M$.

(B) If K is a subspace of a Hilbert space, then $K^{\perp\perp}$ is the smallest closed subspace which contains K. To see this, we note that $K \subset K^{\perp\perp}$ and that $K^{\perp\perp}$ is a closed subspace. If K_1 is a closed subspace such that $K \subset K_1$, then $K^{\perp\perp} \subset K_1^{\perp\perp}$. By (A), $K_1^{\perp\perp} = K_1$.

(C) In the part (A) we have seen that every closed subspace of a Hilbert space is complete. Recall that the converse is also true. It is an immediate consequence of (A) and Theorem 7 (ii).

As is customary, we shall use $\overline{K}$ to denote $K^{\perp\perp}$. The subset $\overline{K}$ will be called the <u>closure</u> of K.

<u>Theorem 3.13</u>. The closed subspaces of a Hilbert space H form a complete lattice with respect to the inclusion. If K_1, K_2 are two closed subspaces, then the lattice operations $\vee, \wedge$ are described as follows:

$$K_1 \vee K_2 = \overline{K_1 + K_2}, \quad K_1 \wedge K_2 = K_1 \cap K_2.$$

Here $\overline{K_1 + K_2}$ denotes the closure of the subspace

$$\{h \in H; \exists k_1 \in K_1 \ \exists k_2 \in K_2 :: h = k_1 + k_2\}.$$

<u>Proof</u>. We shall verify that $\overline{K_1 + K_2}$ has the properties of the supremum of the set $\{K_1, K_2\}$. Clearly, $\overline{K_1 + K_2} \supset \overline{K_1 + \{o\}} = \overline{K_1} = K_1$ and, symmetrically, $\overline{K_1 + K_2} \supset K_2$. The subspace $\overline{K_1 + K_2}$ is closed and it is also the least upper bound of $\{K_1, K_2\}$. In fact, if F is a closed subspace of the Hilbert space H containing K_1 and K_2, then also $F \supset K_1 + K_2$. Consequently, $F = \overline{F} \supset \overline{K_1 + K_2}$.

It is not hard to verify that if K_i, $i \in I \neq \emptyset$, are closed subspaces, then the intersection $\bigcap \{K_i;\ i \in I\}$ is also a closed subspace which has the property of the infimum of the set $\{K_i;\ i \in I\}$. By I.13, the lattice of all closed subspaces of H is complete. //

4. PROJECTION LATTICE OF A HILBERT SPACE

By a projection of a Hilbert space H we mean a continuous linear mapping P of H into H (i.e., an operator on H) which has the following properties:

$$\text{(i)}\quad P = P^*; \qquad \text{(ii)}\quad P^2 = P.$$

Theorem 4.1. Let P be a projection of a Hilbert space H. Then the image $P(H)$ of H under P is a closed subspace of H. If F is a closed subspace of H, then there exists one and only one projection Q of H such that $F = Q(H)$.

Proof. 1. Evidently, $P(H)$ is a subspace of H. Let $y_n = Px_n$, suppose $y_n \to y \in H$ and let $e > 0$. Then there exists, by continuity of P, a number $d > 0$ such that $\|Px - Py\| < e$ whenever $\|x - y\| < d$. Since $y_n \to y$, there exists n_0 such that $\|y_n - y\| < d$ for every $n \geqslant n_0$. Hence $\|Py_n - Py\| < e$ and so $Py_n \to Py$. But we have $Py_n = PPx_n = P^2x_n = Px_n = y_n$, by the second property (ii) of a projection. Thus $y_n \to Py$. At the same time $y_n \to y$. Therefore, $y = Py \in P(H)$ and this proves that $P(H)$ is closed.

2. Conversely, let F be a closed subspace. By Remark 3.12.C and Theorem 3.7 (i), given $p \in H$, there exists only one possibility how to express p in the form $p = q + w$ where $q \in F$ and $w \in F^{\perp}$. The mapping $Q: H \to H$ defined by $Q: p \mapsto q$ is clearly well defined and it is a mapping of H onto F, i.e., $F = Q(H)$.

Since $q = q + o$, we have $Q^2p = Qq = q$. Thus $Q^2 = Q$.

An easy argument shows that Q is linear.

By Remark 3.4.C we have $\|Qp\| = \|q\| \leqslant \|q + w\| = \|p\|$. From Theorem 3.5 (iii) we conclude immediately that the mapping Q is continuous.

To prove that $Q = Q^*$, write $x = q_1 + w_1$, $y = q_2 + w_2$, $q_i \in F$, $w_i \in F^\perp$. Observe that

$$\langle Qx,y\rangle = \langle q_1,q_2 + w_2\rangle = \langle q_1,q_2\rangle + \langle q_1,w_2\rangle = \langle q_1,q_2\rangle =$$
$$= \langle q_1,q_2\rangle + \langle w_1,q_2\rangle = \langle q_1 + w_1,q_2\rangle = \langle x,q_2\rangle = \langle x,Qy\rangle.$$

Since $\langle Qx,y\rangle = \langle x,Q^*y\rangle$, we have $\langle x,Qy\rangle = \langle x,Q^*y\rangle$ for every $x,y \in H$. It follows from Lemma 3.2 (S 11) that $Q = Q^*$.

To see that a projection G on H is uniquely determined by its image $F = G(H)$, write as above $H \ni p = q + w$, $q \in F$, $w \in F^\perp$. Because of $G = G^*$, we have $\langle Gw,v\rangle = \langle w,Gv\rangle$ for every $v \in H$. But $w \in F^\perp$, $Gv \in G(H) = F$ and so $\langle w,Gv\rangle = 0$. Thus $Gw \in$ $\in F^\perp$. Clearly $Gw \in G(H) = F$. Hence $Gw = o$ and $Gp = G(q + w) =$ $= Gq + Gw = Gq$. Now, $q \in F = G(H)$ implies that $q = Gg$ for a convenient vector $g \in H$. Consequently, $Gq = G^2g = Gg = q$ and we find that $Gp = q$. //

Given two projections P,Q on a Hilbert space H, write $P \leqslant Q$ if and only if $P(H) \subset Q(H)$.

<u>Theorem 4.2</u>. The relation $\leqslant$ is a partial ordering of the set $\underline{P}$ of all projections on the Hilbert space H.

<u>Proof</u>. The relation $\leqslant$ is evidently reflexive and transitive. Suppose that $P \leqslant Q$ and $Q \leqslant P$. Then $P(H) = Q(H)$ and from Theorem 1 it follows that $P = Q$. //

Let f denote the mapping of the set $\underline{F}$ of all closed subspaces of H into the set $\underline{P}$ of all projections on the Hilbert space H which is defined as follows: If F is a closed subspace, then $fF = Q$ where Q is the projection such that $F = Q(H)$. By Theorem 1, f is a surjective mapping. We shall show that, in addition,

$$F_1 \subset F_2 \;\Leftrightarrow\; fF_1 \leqslant fF_2.$$

In fact, the relation $P_1 = fF_1 \leq fF_2 = P_2$ is equivalent to $F_1 = P_1(H) \subset P_2(H) = F_2$.

<u>Theorem 4.3</u>. The set $\underline{P}$ forms a complete lattice under the partial ordering of Theorem 2 and the lattice $(\underline{P}, \leq)$ is isomorphic to the lattice $(\underline{F}, \subset)$ of all closed subspaces of the Hilbert space H.

<u>Proof</u>. The required assertion follows by Iséki´s Theorem I.11 and by Theorem 3.13. //

If P, Q are continuous operators on H, then $P_1 = P + Q$ and $P_2 = P - Q$ are defined in accordance with the remark preceding Theorem 3.11 by $P_1x = Px + Qx$ and by $P_2x = Px - Qx$ for every $x \in H$. The identity operator is denoted by I, the null operator is denoted by O, so that $Ix = x$ and $Ox = o$ for every $x \in H$.

If P is a projection on a Hilbert space, then $(I - P)^2 = I - P - P + P^2 = I - P - P + P = I - P$ and $(I - P)^* = I^* - P^* = I - P$. Thus $I - P$ is also a projection.

<u>Lemma 4.4</u>. Let P be a projection on a Hilbert space H and let $M = P(H)$. Then

(i) $M^\perp = (I - P)(H)$.

(ii) The relation $Px = x$ holds if and only if $x \in M$.

(iii) The relation $Px = o$ is satisfied if and only if $x \in M^\perp$.

<u>Proof</u>. Ad (i) and (iii). By Theorem 1, Remark 3.12.A and Theorem 3.7 (i), every element $p \in H$ can be uniquely expressed in the form $p = q + w$ where $q \in M$ and $w \in M^\perp$. From the second part of the proof of Theorem 1 we infer that $Pp = q$. Hence p belongs to $M^\perp$ if and only if $Pp = o$. If $p \in M^\perp$, then $p = w = p - q = p - Pp = (I - P)p$. Thus $M^\perp \subset (I - P)(H)$. Conversely, if $v \in (I - P)(H)$, then $v = (I - P)p = p - Pp =$

$= p - q = w \in M^{\perp}$ and so we see that the converse inclusion also holds.

Ad (ii). Let $M \ni x = x_1 + x_2$, $x_1 \in M$, $x_2 \in M^{\perp}$. Then necessarily $x_2 = o$, $x_1 = x$ and, consequently, $Px = x_1 = x$. Suppose conversely that $x = Px$. Then $x = Px \in P(H) = M$. //

We are now in a position to give a characterization of the partial ordering on $\underline{P}$.

<u>Theorem 4.5</u>. Let P, Q be two projections on a Hilbert space H. Then the following conditions are equivalent:

(i) $P \leqslant Q$; (ii) $P = PQ$; (iii) $P = QP$;

(iv) $Q - P$ is a projection;

(v) $\langle (Q - P)x, x \rangle \geqslant 0$ for every $x \in H$;

(vi) $\|Qx\| \geqslant \|Px\|$ for every $x \in H$.

<u>Proof</u>. Ad (i) $\Rightarrow$ (ii). Let $M = P(H)$. Take $p \in H$. By Theorem 3.7, $p = q + w$ where $q \in M$ and $w \in M^{\perp}$. Therefore, since $M \subset Q(H)$, $Q(H)^{\perp} \subset M^{\perp}$ and so $Qp - q \in M^{\perp}$. By Lemma 4,

$$(PQ)p = P(Qp) = Pq = q = Pp.$$

Ad (ii) $\Rightarrow$ (iii). If $P = PQ$, then $P = P^{*} = (PQ)^{*} = Q^{*}P^{*} = QP$.

Ad (iii) $\Rightarrow$ (i). Here we have

$$P(H) = (QP)(H) = Q(M) \subset Q(H).$$

Ad (i) $\Rightarrow$ (iv). Clearly, $Q - P$ is continuous, linear and $(Q - P)^{*} = Q^{*} - P^{*} = Q - P$. Furthermore,

$$(Q - P)^2 = Q^2 - QP - PQ + P^2 = Q - P - P + P = Q - P,$$

by (ii) and (iii).

Ad (iv) $\Rightarrow$ (v). Evidently,

$$\langle (Q - P)x, x \rangle = \langle (Q - P)^2 x, x \rangle = \langle (Q - P)x, (Q - P)^{*}x \rangle = \\ = \langle (Q - P)x, (Q - P)x \rangle \geqslant 0.$$

Ad (v) $\Rightarrow$ (vi). Since $\langle (Q-P)x,x\rangle \geqslant 0$, $0 \leqslant \langle Qx - Px,x\rangle = \langle Qx,x\rangle - \langle Px,x\rangle$. Hence $\langle Qx,x\rangle \geqslant \langle Px,x\rangle$. Now

$$\langle Qx,Qx\rangle = \langle Q^2x,x\rangle \geqslant \langle P^2x,x\rangle = \langle Px,Px\rangle,$$

i.e., $\|Qx\| \geqslant \|Px\|$.

Ad (vi) $\Rightarrow$ (i). Let $N = Q(H)$. If $\|Qx\| \geqslant \|Px\|$, then $Qx = o$ implies $Px = o$. By Lemma 4 (iii), every element x of $N^{\perp}$ belongs to $M^{\perp}$. In other words, $N^{\perp} \subset M^{\perp}$ and, consequently, $M \subset N$ which shows that $P \leqslant Q$. //

The lattice $(\underline{P}, \leqslant)$ is called the <u>projection lattice</u>. It is a lattice with the least element O (the null operator) and with the greatest element I (the identity operator). These assertions are true since $H = I(H) \supset P(H) \supset O(H) = \{o\}$ for every projection P on H. The following theorem shows a characterization of the join and the meet in this lattice, provided the operators commute.

<u>Theorem 4.6</u>. Let P,Q be two projections of a Hilbert space. If $PQ = QP$, then $P \wedge Q = PQ$ and $P \vee Q = P + Q - PQ$.

<u>Proof</u>. 1. We know that $(PQ)^* = Q^*P^* = QP = PQ$ and that $(PQ)^2 = PQPQ = PPQQ = PQ$. Therefore, the mapping PQ is a projection. We shall show that it has the properties of the infimum of $\{P,Q\}$ in $(\underline{P}, \leqslant)$.

By Theorem 5 (iii), since $PQP = QPP = QP = PQ$, $PQ \leqslant P$. Similarly, $PQ \leqslant Q$. Given a projection R such that $R \leqslant P$ and $R \leqslant Q$, $R = RP$ and $R = RQ$. Therefore $RPQ = RQ = R$ which implies $R \leqslant PQ$.

2. It is clear that

$$(P + Q - PQ)^* = P^* + Q^* - (PQ)^* = P + Q - Q^*P^* = P + Q - QP = \\ = P + Q - PQ.$$

Further,

$(P + Q - PQ)(P + Q - PQ) = P^2 + QP - PQP + PQ + Q^2 - PQ^2 -$
$- P^2Q - QPQ + PQPQ = P + PQ - PPQ + PQ + Q - PQ - PQ - PQQ +$
$+ PPQQ = P + PQ - PQ + PQ + Q - PQ - PQ - PQ + PQ =$
$= P + Q - PQ.$

Hence $P + Q - PQ$ is a projection. To prove that it has the properties of the mentioned supremum, let us first verify that $P \leqslant P + Q - PQ$. But $P(P + Q - PQ) = P^2 + PQ - P^2Q = P + PQ - PQ = P$ which yields the desired statement. Finally, if $P \leqslant R$ and $Q \leqslant R$, we have to show that $P + Q - PQ \leqslant R$. Indeed,

$$(P + Q - PQ)R = PR + QR - PQR = P + Q - PQ. \quad //$$

<u>Remark 4.7</u>. It has been seen that $I - P \in \underline{P}$. Evidently

$$P(I - P) = P - P^2 = P - P = 0 = (I - P)P.$$

By Theorem 6, $P \wedge (I - P) = P(I - P) = 0$ and $P \vee (I - P) = P + I - P - P(I - P) = I$. If $P \leqslant Q$, then $(I - P)(I - Q) = I - P - Q + PQ = I - P - Q + P = I - Q$ and so $I - Q \leqslant I - P$. Moreover, $I - (I - P) = P$. Thus $(\underline{P}, \leqslant)$ is an ortholattice where $P^{\perp} = I - P$ for any $P \in \underline{P}$.

<u>Theorem 4.8</u>. The projection lattice $(\underline{P}, \leqslant)$ of all projections on a Hilbert space is an orthomodular lattice.

<u>Proof</u>. Suppose S,T are two projections such that $S \geqslant T^{\perp}$ and $S \wedge T = 0$. By Theorem 5, $T^{\perp} = T^{\perp}S = ST^{\perp}$ and so $S(I - T) = (I - T)S$. Thus $ST = TS$. Using Theorem 6, we get $ST^{\perp} = S(I - T) = S - ST = S$, i.e., $S \leqslant T^{\perp}$. In summary, we have $S = T^{\perp}$. //

EXERCISES

IN EXERCISES III;1 - III;9 LET $`L = (L, \vee, \wedge, ', 0, 1)$ BE AN ORTHOMODULAR LATTICE, $L^{\bullet} = (L, \dot{\vee}, \dot{\wedge}, ', 0, 1)$ THE BOOLEAN SKEW LATTICE ASSOCIATED WITH $`L$ AND a, b, c ELEMENTS OF L.

III;1. Show that the following conditions are equivalent:

(i) $a = a \dot{\wedge} b$;

(ii) $b = b \dot{\vee} a$.

III;2. Prove that for every a, b

(i) $(a \dot{\vee} b) \dot{\vee} b = a \dot{\vee} b$;

(ii) $(a \dot{\wedge} b) \dot{\wedge} b = a \dot{\wedge} b$.

III;3. Prove that for every a, b, $a \dot{\wedge} b \leqslant b$ and $b \leqslant a \dot{\vee} b$.

III;4. Show that in $L^{\bullet}$ holds the following implication: If $a \leqslant b$, then $a \dot{\vee} c \leqslant b \dot{\vee} c$ and $a \dot{\wedge} c \leqslant b \dot{\wedge} c$.

III;5. Show that the relations $c \leqslant a$ and $c \leqslant b$ imply $c \leqslant a \dot{\wedge} b$.

III;6. Prove that

$$a \wedge b = (a \dot{\vee} b') \dot{\wedge} b, \quad a \vee b = (a \dot{\wedge} b') \dot{\vee} b.$$

III;7. Characterize the following conditions on a, b:

(i) $a = (b \dot{\wedge} a) \dot{\vee} (b \dot{\wedge} a')$;

(ii) $a = (b \dot{\wedge} a') \dot{\vee} (b \dot{\wedge} a)$;

(iii) $a = (a \dot{\wedge} b') \dot{\vee} (a \dot{\wedge} b)$.

III;8. Prove that the following conditions on a, b are equivalent:

(i) aCb;

(ii) $a \dot{\wedge} (b \dot{\vee} a') = a \dot{\wedge} b$;

(iii) $(b \dot{\vee} a') \dot{\wedge} a = a \dot{\wedge} b$;

(iv) $(b \dot{\vee} a') \dot{\wedge} a = b \dot{\wedge} a$.

III;9. Characterize the following conditions on elements a, b:

(i) $a \dot{\wedge} (a' \dot{\vee} b) = a \dot{\wedge} b$;

(ii) $(a' \dot{\vee} b) \dot{\wedge} a = b \dot{\wedge} a$.

III;10. Let $`L$ be an ortholattice and let $L^{\bullet}$ be the skew lattice associated with $`L$. Prove that the following conditions are equivalent:

(i) $`L$ is orthomodular;

(ii) $(a \dot{\wedge} b') \dot{\vee} b = (b \dot{\wedge} a') \dot{\vee} a$ for any $a, b \in L$;

(iii) for all $a, b \in L$, $a \leqslant b \Rightarrow (b \dot{\wedge} a') \dot{\vee} a = b$.

III;11. Let a, b be elements of an orthomodular lattice $`L$. Prove that the interval algebras

$$`L[a \wedge (a' \vee b) \wedge (a' \vee b'), a \vee (a' \wedge b) \vee (a' \wedge b')],$$

$$`L[b \wedge (a \vee b') \wedge (a' \vee b'), b \vee (a \wedge b') \vee (a' \wedge b')]$$

are orthoisomorphic.

IN EXERCISES III;12 - III;18 LET A, B, C, D, P AND Q BE PROJECTIONS ON A HILBERT SPACE H.

III;12. Let $M = P(H)$, $N = Q(H)$. Prove that

$$M \subset N^{\perp} \Leftrightarrow PQ = 0.$$

III;13. Suppose that P, Q satisfy one of the conditions

(i) $P \leqslant Q$;

(ii) $P \leqslant Q^{\perp}$;

(iii) $P \geqslant Q$;

(iv) $P \geqslant Q^{\perp}$.

Prove that $PQ = QP$.

III;14. Show that

(i) $P \wedge Q \leqslant Q$ and $P \wedge Q^{\perp} \leqslant Q^{\perp}$;

(ii) $(P \wedge Q)(P \wedge Q^{\perp}) = (P \wedge Q^{\perp})(P \wedge Q)$;

(iii) $(P \wedge Q)Q = Q(P \wedge Q)$;

(iv) $(P \wedge Q^{\perp})Q = Q(P \wedge Q^{\perp})$.

III;15. For A,B,C such that $AB = BA$, $BC = CB$ and such that $A + C$ is a projection, show that $B(A + C)$ and $(A + C)B$ are the same projections.

III;16. Let A,B,C,D be such that $A \leqslant B$, $C \leqslant D$, $AC = CA$ and $BD = DB$. Prove that AC and BD are projections and $AC \leqslant BD$.

III;17. Show that

(i) $(P \wedge Q)(P \wedge Q^{\perp}) = 0$;

(ii) $(P \wedge Q) \vee (P \wedge Q^{\perp}) = (P \wedge Q) + (P \wedge Q^{\perp})$.

III;18. Prove that P,Q satisfy the relation $PQ = QP$ if and only if $P = (P \wedge Q) \vee (P \wedge Q^{\perp})$.

III;19. Show directly that the lattice $(\underline{F}, \vee, \wedge)$ of all closed subspaces of a Hilbert space (cf. Theorem 3.13) is an orthomodular lattice with respect to the orthocomplementation $^{\perp}: K \rightarrow K^{\perp}$.

Chapter

AMALGAMS IV

1. AMALGAMS OF POSETS

As a preparation for what follows, we introduce first concepts which will be prominent in our investigation.

Let $`P = (P, \leq)$ be a poset and let M, N be nonvoid subsets of P. The set consisting of all the intervals $[x,y] = \{z \in P;\ x \leq z \ \&\ z \leq y\}$ where $x \leq y$ and either $x \in M, y \in N$ or $x \in N, y \in M$ will be denoted by $[M,N]$.

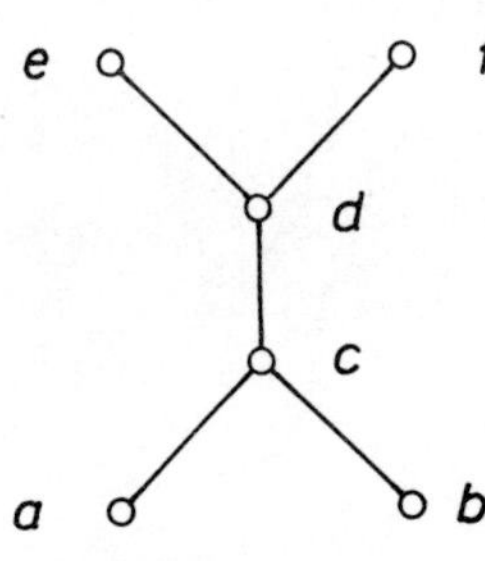

Fig. 24

Example 1.1. If $`P$ is the poset sketched in Figure 24 and if $M = \{b,e\}$, $N = \{a,d\}$, then $[M,N] = \{[a,e],[d,e],[b,d]\} = \{\{a,c,d,e\},\{d,e\},\{b,c,d\}\}$. In the case $M_1 = \{a\}$, $N_1 = \{b\}$ we have $[M_1,N_1] = \emptyset$.

Convention 1.2. Throughout the fourth chapter $(`S_i;\ i \in I)$ denotes a family of posets $`S_i = (S_i; \leq_i)$ where $I \neq \emptyset$. The subscripts distinguishing different partial orderings will be at times omitted. Suppose that a nonvoid subset S_i^o of the set S_i is determined for every $i \in I$. The associated subposet of the poset $`S_i$ will be denoted by $`S_i^o$. Next assume that an order-isomorphism $f_{i/j}$ from $`S_i^o$ onto the poset $`S_j^o$ is given for every $i,j \in I$.

We now adopt the following basic definition:

A poset $`S$ is called an <u>amalgam</u> (cf. [1], [79], [10], [12], [14]) of the posets $`S_i$ relative to the isomorphisms $f_{i/j}$ if and only if there exist order-monomorphisms $f_i : `S_i \to `S$, $i \in I$, such that

(A 1) the union $\bigcup(f_i(S_i);\ i \in I)$ of the sets $f_i(S_i) = \{s \in S; \exists x \in S_i :: s = f_i(x)\}$ is equal to S;

(A 2) the intersection $T \cap f_i(S_i^o)$ is nonvoid for any two $i \neq j$ of I and every nonempty interval $T \in [f_i(S_i), f_j(S_j)]$;

(A 3) the diagram

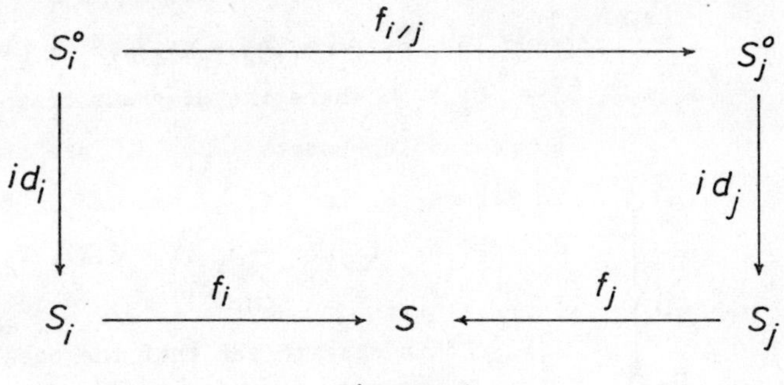

Diagram 1

is commutative for every $i,j \in I$, i.e., $f_{i/j} \circ id_j \circ f_j = id_i \circ f_i$. Here $id_i : S_i^o \to S_i$ and $id_j : S_j^o \to S_j$ are the identity mappings. When there is no danger of confusion we shall write merely id instead of id_j etc.

<u>Remark 1.3</u>. (A) To illustrate the definition in a suggestive way, consider Figure 25.

(B) The condition (A 2) has the following meaning: If $s \in S_i$ and $t \in S_j$, $i \neq j$, are such that $f_i(s) \leqslant f_j(t)$, then there exists $s_o \in S_i^o$ with $f_i(s) \leqslant f_i(s_o) \leqslant f_j(t)$.

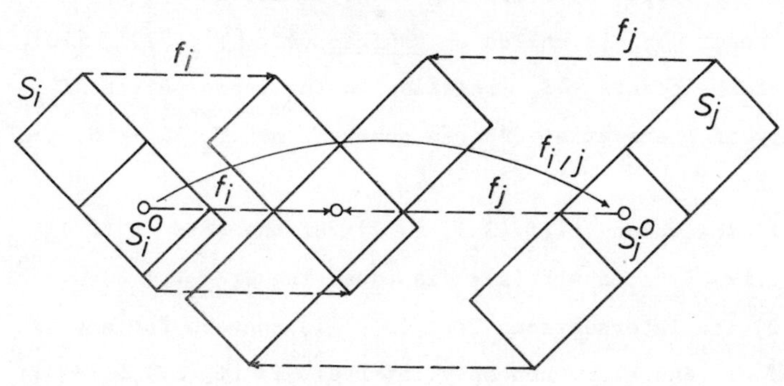

Fig. 25

(C) Let $I = \{1,2\}$ and let $S_1 = \{a_1, a_2, a_3, a_4\}$, $S_2 = \{b_1, b_2, b_3, b_4\}$, $S_1^o = \{a_2, a_3\}$, $S_2^o = \{b_2, b_3\}$ where the diagrams of the corresponding posets $`S_1$, $`S_2$ are shown in Figure 26. Let $f_{1/2}$ and $f_{2/1}$ be defined by $f_{1/2}: a_i \mapsto b_i$ $(i = 2,3)$, $f_{2/1} = f_{1/2}^{-1}$, $f_{1/1} = \mathrm{id}$ (on S_1^o), $f_{2/2} = \mathrm{id}$ (on S_2^o). It is easy to see that the poset $`P$ of Figure 24 is an amalgam of $`S_1$ and $`S_2$ relative to $f_{i/j}$. It is not difficult to choose the order-isomorphisms f_1, f_2 satisfying the conditions (A 1) - (A 3). To this end define the mappings f_1 and f_2 by

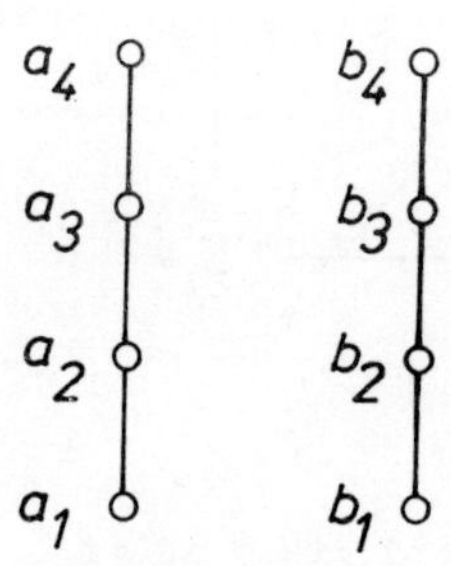

Fig. 26

$$f_1(a_1) = a,\ f_1(a_2) = c,\ f_1(a_3) = d,\ f_1(a_4) = e,$$
$$f_2(b_1) = b,\ f_2(b_2) = c,\ f_2(b_3) = d,\ f_2(b_4) = f.$$

The elements a_2,b_2 and a_3,b_3 are "pasted" by the isomorphism $f_{1/2}$. Similarly, b_i and a_i, $i = 2,3$, are "pasted" by the isomorphism $f_{2/1}$. The process visualized in this example justifies the following terminology:

The isomorphisms $f_{i/j}$ mentioned in the definition of an amalgam will be called the <u>pasting isomorphisms</u>. The set S_i^o and the poset $`S_i^o$ will be called a <u>pasted set</u> and a <u>pasted poset</u>, respectively. Similarly, an element $x \in S_i$ will be called a <u>pasted element</u> if and only if it belongs to S_i^o.

<u>Theorem 1.4</u>. (i) If $f_i(s) = f_j(t)$ and $i \neq j$, then s and t are pasted elements, i.e., $s \in S_i^o$, $t \in S_j^o$, and $t = f_{i/j}(s)$.

(ii) If $i \neq j$, then $f_i(S_i) \cap f_j(S_j) = f_i(S_i^o)$.

(iii) For every $i,j,k \in I$, $f_{i/k} = f_{j/k}f_{i/j}$.

(iv) The mapping $f_{i/i}$ is the identity mapping on S_i^o for every $i \in I$.

(v) For every $i,j \in I$, $f_{j/i} = f_{i/j}^{-1}$.

<u>Proof</u>. Ad (i). By Remark 3.B, $f_i(s) \leqslant f_i(s_o) \leqslant f_j(t)$ for an element $s_o \in S_i^o$. Since $f_i(s) = f_j(t)$, $f_i(s_o) = f_i(s)$ and so $s = s_o$. It follows from (A 3) that

$$f_j(f_{i/j}(s)) = f_i(s) = f_j(t).$$

But f_j is injective. Therefore $t = f_{i/j}(s) \in S_j^o$.

Ad (ii). If a belongs to the intersection $D = f_i(S_i) \cap \cap f_j(S_j)$, then $a = f_i(s) = f_j(t)$. By (i), a is of $f_i(S_i^o)$. Hence $D \subset f_i(S_i^o)$. Choose $b \in f_i(S_i^o)$. Then, a fortiori, $b \in \in f_i(S_i)$. Evidently, $b = f_i(s_o')$ where s_o' is a pasted element of S_i^o and so it is clear that $s_o' = f_{j/i}(t_o')$ with $t_o' \in S_j^o$. Consequently, $b = f_i(f_{j/i}(t_o')) = f_j(t_o') \in f_j(S_j)$, that is, $b \in D$. Thus $f_i(S_i^o) \subset D$.

Ad (iii). From the diagram

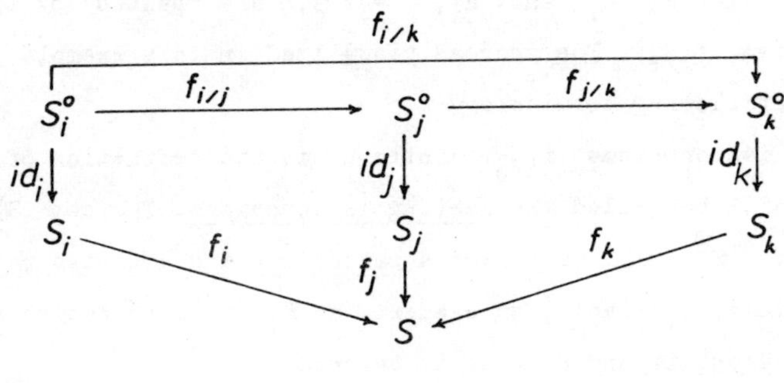

Diagram 2

we conclude that

$$f_k id f_{i/k} = f_i id = f_j id f_{i/j} = f_k id f_{j/k} f_{i/j}.$$

Since the mapping $f_k id$ is injective, we get $f_{i/k} = f_{j/k} f_{i/j}$.

Ad (iv). By (iii) we have $f_{i/i} id = f_{i/i} f_{i/i}$. The injectivity of $f_{i/i}$ implies $f_{i/i} = id$.

Ad (v). From (iii) and (iv) we infer that necessarily $id = f_{i/i} = f_{j/i} f_{i/j}$. //

<u>Theorem 1.5</u>. Suppose $({}^\backprime S_i;\ i \in I)$, ${}^\backprime S$, f_i and $f_{i/j}$ have the properties indicated in the definition of an amalgam. Let $f_i^*: {}^\backprime S_i \to {}^\backprime S^* = (S^*, \leq_*)$ be order-monomorphisms satisfying the three conditions of the definition of an amalgam. Then the poset ${}^\backprime S^*$ is isomorphic to the poset ${}^\backprime S$.

<u>Proof</u>. Let $s \in S$. Then, by (A 1), there exist $i \in I$, $t \in S_i$ such that $s = f_i(t)$. Define a mapping F from S into S^* (see Figure 27) by $F(s) = f_i^*(t)$. We shall prove that F is well defined. Indeed, if $s = f_j(u)$ where $j \neq i$ and $u \in S_j$, then $f_j(u) = f_i(t)$. By Theorem 4 (i) we see that $u = f_{i/j}(t)$. Thus

Thus $f_j^*(u) = f_j^*(f_{i/j}(t)) = f_i^*(t)$ in virtue of the condition (A 3) written for f_j^*.

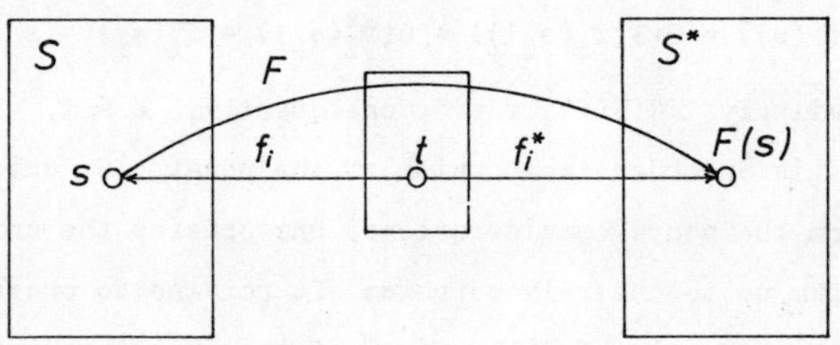

Fig. 27

Next we shall prove that F is a mapping of S onto S^*. Suppose $s^* \in S^*$. By (A 1), $s^* = f_i^*(v)$ where $i \in I$ and $v \in \in S_i$. Now it is immediate that $s^* = F(f_i(v))$.

We further wish to prove that the relation $s \leqslant t$ valid in $`S$ implies the relation $F(s) \leqslant_* F(t)$ valid in $`S^*$. In fact, if there exists $i \in I$ such that $s = f_i(s_i)$ and $t = f_i(t_i)$, then $s_i \leqslant t_i$ so that $F(s) = f_i^*(s_i) \leqslant_* f_i^*(t_i) = F(t)$. Suppose we have $s = f_i(s_i)$, $t = f_j(t_j)$ where $i \neq j$. By (A 2), there exists s_i^0 such that

$$s = f_i(s_i) \leqslant f_i(s_i^0) = f_j(s_j^0) \leqslant f_j(t_j)$$

where $s_j^0 = f_{i/j}(s_i^0)$. Thus

$$f_j^*(s_j^0) = f_j^*(f_{i/j}(s_i^0)) = f_i^*(s_i^0).$$

Since $s_i \leqslant s_i^0$, $s_j^0 \leqslant t_j$, we also have

$$F(s) = f_i^*(s_i) \leqslant_* f_i^*(s_i^0) = f_j^*(s_j^0) \leqslant_* f_j^*(t_j) = F(t).$$

Define next symmetrically a mapping $G: S^* \to S$ by $G(s^*) = = f_i(w)$ where $s^* = f_i^*(w)$, $w \in S_i$, $i \in I$. The assertion the proof of which we have just given shows that G is a mapping from $`S^*$

onto $`S$ such that the relation $s^* \leqslant_* t^*$ implies $G(s^*) \leqslant G(t^*)$.

Hence, if $F(s) \leqslant_* F(t)$, then, a fortiori, $G(F(s)) \leqslant G(F(t))$. Set $s = f_i(s_i)$. Here

$$G(F(s)) = G(F(f_i(s_i))) = G(f_i^*(s_i)) = f_i(s_i) = s$$

and, similarly, $G(F(t)) = t$. Consequently, $s \leqslant t$. This shows that F is an order-isomorphism of the poset $`S$ onto $`S^*$. //

From the above considerations, one obtains the uniqueness of an amalgam up to order-isomorphism. It remains to prove its existence. In preparation for the proof of the existence, we shall first introduce a new notation.

Suppose $i, j \in I$. Then the mapping $g_{i/j}$ is defined as follows:

$$g_{i/j} = \begin{cases} f_{i/j} & \text{in the case } i \neq j; \\ \text{identity mapping on } S_i & \text{in the case } i = j. \end{cases}$$

Given an element $s \in S_i$, we form a new symbol $s^{(i)}$. Let A denote the set of all the symbols $s^{(i)}$ which one obtains if i runs through I, i.e.,

$$A = \bigcup\{\bigcup\{s^{(i)};\ s \in S_i\};\ i \in I\}.$$

Remark 1.6. (A) It should be stressed that $g_{i/j}: S_i^o \to S_j^o$ whenever $i \neq j$ and $g_{i/i}: S_i \to S_i$.

(B) By the symbol $s^{(i)}$ we mean the ordered pair (s,i). In particular, $s^{(i)} = t^{(j)}$ if and only if $s = t$ and $i = j$.

Define a binary relation on A by

$$s^{(i)} R t^{(j)} \Leftrightarrow g_{i/j}(s) = t.$$

Theorem 1.7. (i) If $g_{i/j}(s) = t$ and $g_{j/k}(t) = u$, then $g_{i/k}(s) = u$.

(ii) The relation R is an equivalence relation on A.

Proof. Ad (i). We may suppose that $i \neq j$ and $j \neq k$. But then we can use Theorem 4 (iii).

Ad (ii). The transitivity of R follows by (i), the rest is obvious. //

The quotient set A/R will be denoted by $\underline{A}$. The equivalence class $[s]$ determined by $s \in S_i$ will also be denoted by $[s^{(i)}]$.

Remark 1.8. In what follows we assume that $(`S_i;\ i \in I)$ and $(f_{i/j};\ (i,j) \in I^2)$ are given and we do not use a more detailed notation for the set $\underline{A}$.

Example 1.9. The posets $`S_1, `S_2$ of Remark 3.C give

$$A = \left\{a_1^{(1)}, a_2^{(1)}, a_3^{(1)}, a_4^{(1)}, b_1^{(2)}, b_2^{(2)}, b_3^{(2)}, b_4^{(2)}\right\}.$$

The pasting isomorphisms $f_{i/j}$, $i,j \in \{1,2\}$ determine the following classes of the equivalence R:

$$[a_1^{(1)}] = \{a_1\},\ [a_2^{(1)}] = \{a_2, b_2\} = [b_2^{(2)}],\ [a_3^{(1)}] = \{a_3, b_3\} =$$
$$= [b_3^{(2)}],\ [a_4^{(1)}] = \{a_4\},\ [b_1^{(2)}] = \{b_1\},\ [b_4^{(2)}] = \{b_4\}.$$

This gives

$$\underline{A} = \left\{[a_1^{(1)}], [a_2^{(1)}], [a_3^{(1)}], [a_4^{(1)}], [b_1^{(2)}], [b_4^{(2)}]\right\}.$$

We shall now show that it is possible to define a natural partial ordering on $\underline{A}$.

Given $[m^{(i)}], [n^{(j)}] \in \underline{A}$, we define $[m^{(i)}] \leqslant [n^{(j)}]$ if and only if there exists $m_o \in S_i$ and $n_o \in S_j$ such that $m \leqslant_i m_o$, $n_o \leqslant_j n$ and $g_{i/j}(m_o) = n_o$.

This situation will often be described in a simple way as follows:

$$m \leqslant m_o \xrightarrow{g_{i/j}} n_o \leqslant n.$$

Sometimes it will be expressed diagrammatically in a more detailed form (see Diagram 3):

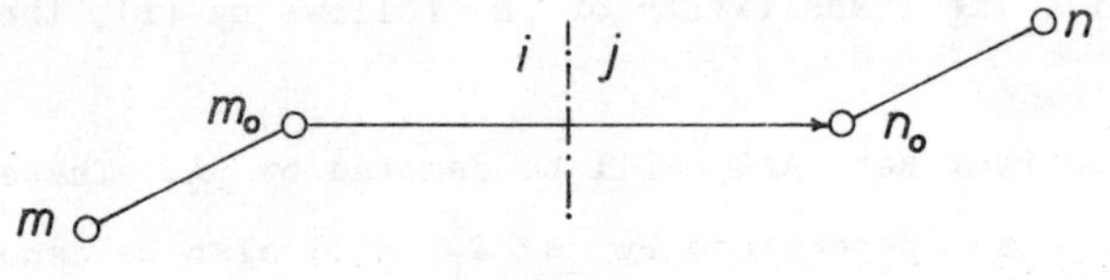

Diagram 3

If m_o, n_o belong to pasted sets, we shall denote them by shaded circlets in such a diagram.

Remark 1.10. Fix now our attention on the definition of the relation $\leq$. If $i = j$ and $g_{i/j}(m_o) = n_o$, then $m_o = n_o$. Hence, in this case $[m^{(i)}] \leq [n^{(j)}]$ is equivalent to $m \leq n$. If $i \neq j$, then $m_o \in S_i^o$, $n_o \in S_j^o$ and we have $m \leq m_o$, $f_{i/j}(m_o) = n_o$, $n_o \leq n$.

Lemma 1.11. Let $a \in S_i$, $b, c \in S_j$, $b \leq c$, $d \in S_k$ and let $g_{i/j}(a) = b$, $g_{j/k}(c) = d$. Then there exist $e \in S_i$, $f \in S_k$ such that $a \leq e$, $g_{i/k}(e) = f$ and $f \leq d$.

Proof. We shall distinguish the following typical situations (see Diagram 4):

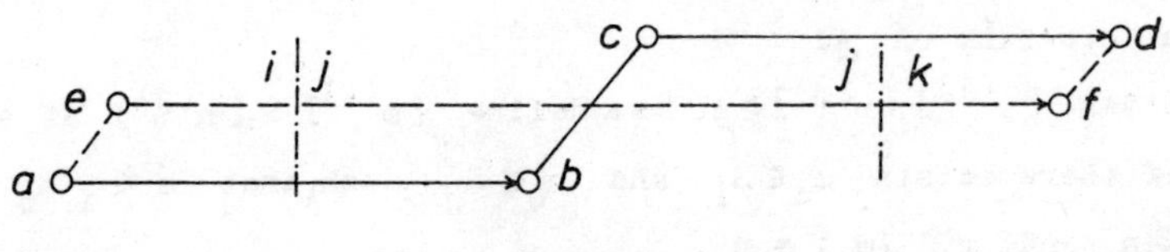

Diagram 4

Case I: $i = k$. Observe that $g_{i/j}(a) = b \leq c = g_{k/j}(d) = g_{i/j}(d)$. Thus $a \leq d$ and we can put $e = a = f$.

In the remaining cases we assume that $i \neq k$.

Case II: $j = k$. But then $c = d$ and we can set $e = a$, $f = b$.

Case III: $i = j$. Here we have $a = b$ and so we can write $e = c$, $f = d$.

Case IV: $i \neq j$ and $j \neq k$. Then $a \in S_i^o$, $f_{j/k}(b) \in S_k^o$. Put $e = a$, $f = f_{j/k}(b)$. Since $b \leqslant c$, we have $f = f_{j/k}(b) \leqslant f_{j/k}(c) = d$. Moreover, $b = f_{i/j}(a)$, $f = f_{j/k}(b)$ and, consequently, $g_{i/k}(e) = g_{i/k}(a) = f_{i/k}(a) = f_{j/k}(f_{i/j}(a)) = f_{j/k}(b) = f$, by Theorem 4 (iii). //

Lemma 1.12. The relation $\leqslant$ on $\underline{A}$ does not depend on the choice of the representatives $m^{(i)}$, $n^{(j)}$ in the classes $[m^{(i)}]$, $[n^{(j)}]$.

Proof. We wish to prove the following assertion: If

$$m \leqslant m_o \xrightarrow{g_{i/j}} n_o \leqslant n$$

and if $[m^{(i)}] = [m_1^{(i_1)}]$, $[n^{(j)}] = [n_1^{(j_1)}]$, then there exist p_o, q_o such that

$$m_1 \leqslant p_o \xrightarrow{g_{i_1/j_1}} q_o \leqslant n_1$$

(see Diagram 5).

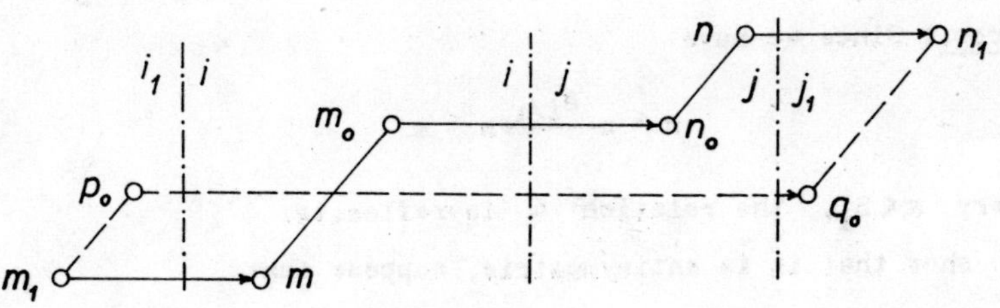

Diagram 5

We may first apply Lemma 11 to show that there exist E,F with the properties of Diagram 6.

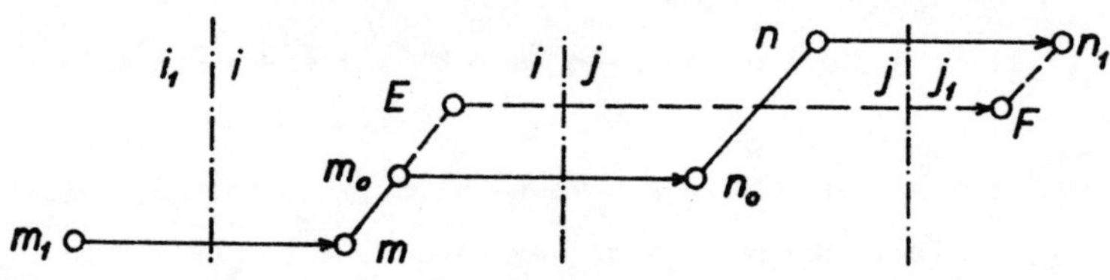

Diagram 6

Using again Lemma 11 with $a = m_1$, $b = m$, $c = E$, $d = F$, we conclude that there exist p_o,q_o having the properties of Diagram 7.

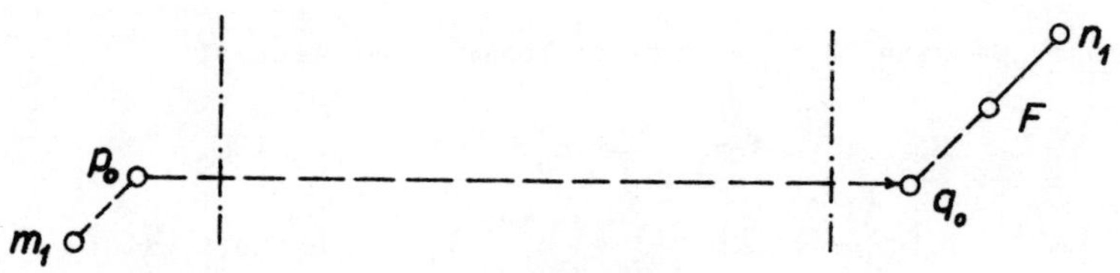

Diagram 7

The elements p_o,q_o satisfy the assertion stated above. //

<u>Lemma 1.13</u>. The relation $\leq$ is a partial ordering on $\underline{A}$.

<u>Proof</u>. Since we have

$$m \leq m \xrightarrow{g_{i/i}} m \leq m$$

for every $m \in S_i$, the relation $\leq$ is reflexive.

To show that it is antisymmetric, suppose that

$$[m^{(i)}] \leq [n^{(j)}] \quad [n^{(j)}] \leq [m^{(i)}].$$

If $i = j$, then this assumption is equivalent to $m \leq_i n$,

$n \leqslant_i m$, that is, $m = n$. Consider now the case $i \neq j$. Referring to Diagram 8, we see that $n_o \leqslant n_o'$ implies $m_o = f_{j/i}(n_o) \leqslant$

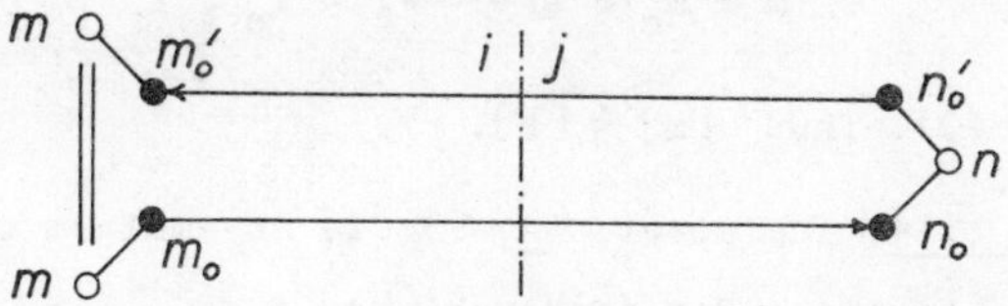

Diagram 8

$\leqslant f_{j/i}(n_o') = m_o'$. Thus $m \leqslant m_o \leqslant m_o' \leqslant m$ and, consequently, $m = m_o = m_o'$. From this it follows that $n_o' = n_o$. Since $n_o \leqslant n \leqslant n_o'$, we get $n_o = n = n_o'$. Hence $[n] = [n_o] = [m_o] = [m]$.

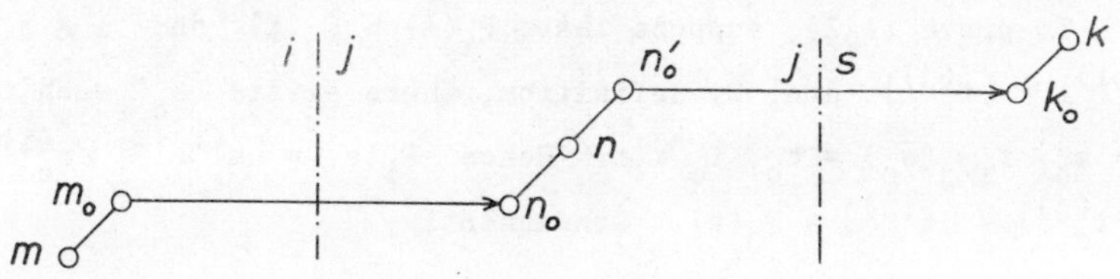

Diagram 9

We shall further confine our attention to the transitivity. If $[m^{(i)}] \leqslant [n^{(j)}]$ and $[n^{(j)}] \leqslant [k^{(s)}]$, consider Diagram 9.

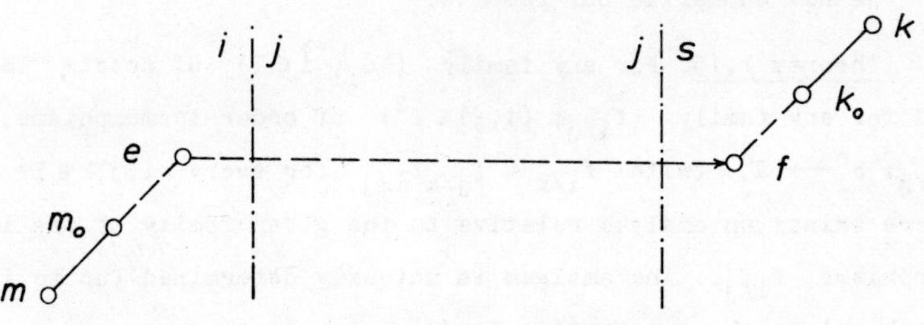

Diagram 10

Putting $a = m_o$, $b = n_o$, $c = n_o'$, $d = k_o$ in Lemma 11, we can conclude that there exist e,f with

$$m \leqslant m_o \leqslant e \xmapsto{g_{i/s}} f \leqslant k_o \leqslant k$$

(see Diagram 10). Thus $[m] \leqslant [k]$. //

<u>Theorem 1.14</u>. The poset $(\underline{A}, \leqslant)$ is an amalgam of the posets $`S_i$ relative to the order-isomorphisms $f_{i/j}$, $i,j \in I$.

<u>Proof</u>. Let $F_i : S_i \to \underline{A}$ be the mapping which assigns $[t^{(i)}]$ to $t \in S_i$. The condition (A 1) formulated for the mappings F_i, $i \in I$, is obviously valid.

Next consider the condition (A 3). For any $s \in S_i^o$ we have

$$F_j(f_{i/j}(s)) = [f_{i/j}(s)^{(j)}] = [s^{(i)}] = F_i(s).$$

To prove (A 2), suppose that $F_i(s) \leqslant F_j(t)$ and $i \neq j$. Then $[s^{(i)}] \leqslant [t^{(j)}]$ and, by definition, there exists s_o such that $s \leqslant s_o$, $f_{i/j}(s_o) = t_o$, $t_o \leqslant t$. Hence $F_i(s) = [s^{(i)}] \leqslant [s_o^{(i)}] = [t_o^{(j)}] \leqslant [t^{(j)}] = F_j(t)$. Consequently,

$$[s_o^{(i)}] = F_i(s_o) \in [F_i(s), F_j(t)] \cap F_i(S_i^o).$$

By Remark 10, the relation $s \leqslant_i t$ is equivalent to $F_i(s) = [s^{(i)}] \leqslant [t^{(i)}] = F_i(t)$ for every $s,t \in S_i$. Thus F_i is an order-monomorphism from $`S_i$ into $\underline{A}$. //

We now summarize our results:

<u>Theorem 1.15</u>. For any family $(`S_i;\ i \in I)$ of posets $`S_i$ and for any family $(f_{i/j};\ (i,j) \in I^2)$ of order-isomorphisms $f_{i/j} : `S_i^o \longrightarrow `S_j^o$ (with $f_{i/k} = f_{j/k} f_{i/j}$ for every $i,j,k \in I$) there exists an amalgam relative to the given family of the isomorphisms $f_{i/j}$. The amalgam is uniquely determined (up to isomorphism) by the two families. //

Remark 1.16. (A) An amalgam of two posets $`S_1, `S_2$ is determined by the pasting isomorphism $f_{1/2}$.

(B) In what follows, by an amalgam of posets $`S_i$, $i \in I$, relative to pasting isomorphisms $f_{i/j}$ we mean the poset $\underline{A}$ discussed above where we write f_i instead of F_i.

Example 1.17. The partial ordering on the set $\underline{A}$ of Example 9 is such that $[a_1^{(1)}] \leqslant [b_4^{(2)}]$. Note that this is valid because of

$$a_1 \leqslant a_2 \xmapsto{f_{1/2}} b_2 \leqslant b_4 .$$

The corresponding amalgam is shown in Figure 28.

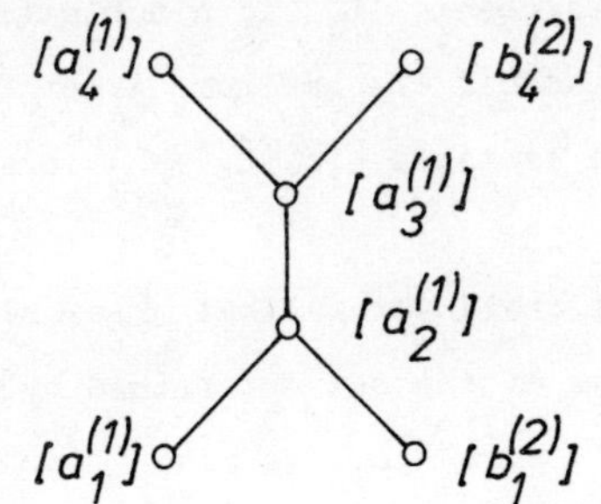

Fig. 28

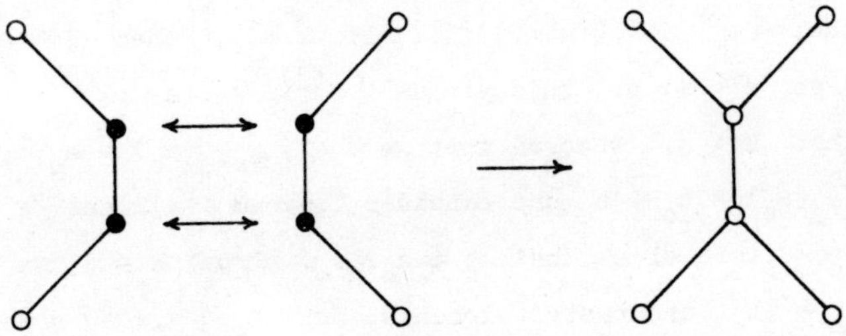

Fig. 29

The process of forming the amalgam is pictorially illustrated in Figure 29.

2. AMALGAMS OF LATTICES

Throughout this section we adopt the following convention: The letter `L having possibly subscripts will denote lattices.

We now turn our attention to the amalgams $\underline{A}$ in which the original family of posets $`S_i$, $i \in I$, is replaced by a family $(`L_i;\ i \in I)$ of lattices where I has more than one element.

Lemma 2.1. (i) If the pasted posets $`M_i^o$ are meet-subsemilattices of the lattices $`L_i$, then there exists the infimum $[a^{(j)}] \wedge [b^{(j)}]$ for every $[a^{(j)}], [b^{(j)}] \in \underline{A}$ and every $j \in I$ and it is equal to $[(a \wedge b)^{(j)}]$.

(ii) If the pasted poset $`L_i^o$ is a sublattice of the lattice $`L_i$ for every $i \in I$ and if the amalgam $\underline{A}$ of the lattices $`L_i$ is a lattice, then the lattices $(f_i(L_i), \leqslant)$ are sublattices of the amalgam $\underline{A}$.

Proof. Ad (i). We shall verify that $[(a \wedge b)^{(j)}]$ has the property of the infimum of the set determined by the elements $[a^{(j)}], [b^{(j)}]$. Since $a \wedge b \leqslant a$, we have $[(a \wedge b)^{(j)}] \leqslant [a^{(j)}]$, by Remark 1.10.

Suppose that $[e^{(k)}] \leqslant [a^{(j)}]$ and $[e^{(k)}] \leqslant [b^{(j)}]$. We wish to prove that $[e^{(k)}] \leqslant [(a \wedge b)^{(j)}]$. If $k = j$, then $e \leqslant a$ and $e \leqslant b$ and so $e \leqslant a \wedge b$. This yields $[e^{(j)}] \leqslant [(a \wedge b)^{(j)}]$. Suppose now that $k \neq j$. Observe that $e \leqslant e_o$, $g_{k/j}(e_o) = a_o \leqslant a$, $e \leqslant e_o'$, $g_{k/j}(e_o') = b_o \leqslant b$ and consider Diagram 11. Since $e \leqslant e_o$ and $e \leqslant e_o'$, it is clear that $e \leqslant e_o \wedge e_o'$. From $k \neq j$ we infer that e_o, a_o, e_o', b_o are pasted elements. But $f_{k/j}$ is an order-isomorphism of $`M_k^o$ onto $`M_j^o$. An argument analogous to that of the proof of Iséki's Theorem I.11 shows that $f_{k/j}(e_o \wedge e_o') = a_o \wedge b_o$. Thus $e \leqslant e_o \wedge e_o'$, $g_{k/j}(e_o \wedge e_o') \leqslant a \wedge b$, i.e., $[e] \leqslant [a \wedge b]$.

Ad (ii). The statement follows from (i) and its dual. //

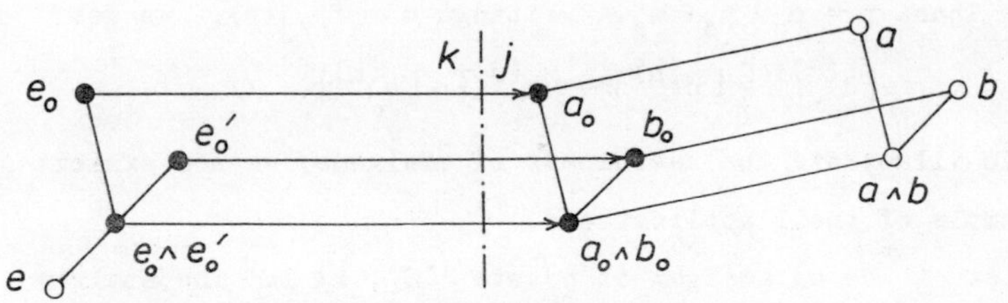

Diagram 11

Lemma 2.2. Let the pasted poset $`M_i^o$ be a meet-subsemilattice of the lattice $`L_i$ for every $i \in I$. Suppose the set $[a^{(j)}], [b^{(k)}]$ has the infimum $[a^{(j)}] \wedge [b^{(k)}]$ in the amalgam $\underline{A}$ of the lattices $`L_i$. Then there exists an element d belonging either to L_j or to L_k such that $[d] = [a^{(j)}] \wedge [b^{(k)}]$.

Proof. Let $[p^{(n)}] = [a^{(j)}] \wedge [b^{(k)}]$. If $n = j$ or $n = k$, the assertion is obvious. By Lemma 1, our statement holds also in the case $j = k$. Suppose now $j \neq k$, $n \neq j$ and $n \neq k$ (see Diagram 12).

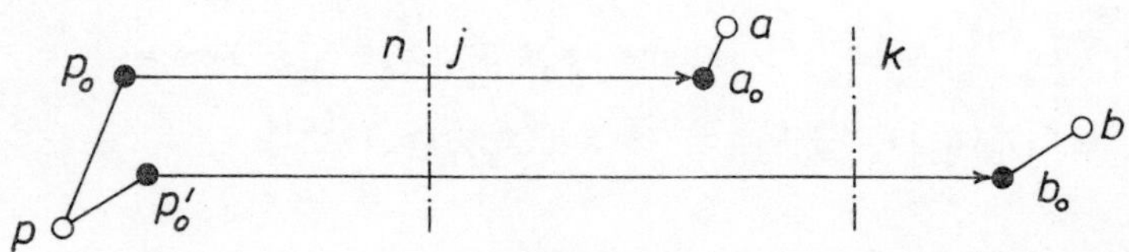

Diagram 12

The diagram suggests that p is a pasted element. We shall verify it by a simple calculation: First observe that $p \leqslant p_o$ and $p \leqslant p_o'$ so that $p \leqslant p_o \wedge p_o'$. Now, by Lemma 1 (i), we have

$$[(p_o \wedge p_o')^{(n)}] = [p_o^{(n)}] \wedge [p_o'^{(n)}] = [a_o^{(j)}] \wedge [b_o^{(k)}] \leqslant$$
$$\leqslant [a^{(j)}] \wedge [b^{(k)}] = [p^{(n)}].$$

However, the relation $[(p_o \wedge p_o')^{(n)}] \leq [p^{(n)}]$ implies $p_o \wedge p_o' \leq \leq p$. Thus $p = p_o \wedge p_o' \in M_n^o$. Setting $d = f_{n/j}(p)$, we get

$$[d^{(j)}] = [p^{(n)}] = [a^{(j)}] \wedge [b^{(k)}]. \quad //$$

To illustrate the usefulness of amalgams, we now exhibit an example of their applications.

Let $\underline{A}$ be an amalgam of posets $`S_i$, $i \in I$. The amalgam $\underline{A}$ is said to be convexly pasted if and only if it has the following property: If $[a_o^{(j)}] \leq [c^{(n)}] \leq [b_o^{(k)}]$ where a_o, b_o are pasted elements, then there exists a pasted element d_o such that $[c^{(n)}] = [d_o^{(m)}]$.

Lemma 2.3. The amalgam $\underline{A}$ of posets $`S_i$ is convexly pasted if and only if the pasted set $S_i^o$ is a convex subset of the poset $`S_i$ for every $i \in I$.

Proof. 1. Let $\underline{A}$ be a convexly pasted amalgam. Let $a_o \in S_k^o$, $b_o \in S_k^o$ and $a_o \leq c \leq b_o$. We shall show that $c \in S_k^o$. Clearly, we have $[a_o] \leq [c] \leq [b_o]$. Thus $[c^{(k)}] = [d_o^{(m)}] = [(f_{m/k}(d_o))^{(k)}]$, i.e., $c = f_{m/k}(d_o) \in S_k^o$.

2. Conversely, let S_i^o be convex subsets. Suppose that $[a_o^{(j)}] \leq [c^{(n)}] \leq [b_o^{(k)}]$ where $a_o \in S_j^o$, $b_o \in S_k^o$. Hence

$$[(f_{j/n}(a_o))^{(n)}] \leq [c^{(n)}] \leq [(f_{k/n}(b_o))^{(n)}]$$

and so

$$S_n^o \ni f_{j/k}(a_o) \leq c \leq f_{k/n}(b_o) \in S_n^o.$$

But S_n^o is convex in $`S_n$. Therefore, c is a pasted element. //

Before stating the next theorem we shall define a new notion. Given an element $[a] \in \underline{A}$, the set of all the indices j of I such that there exists an element b (depending on j) with $[a] = [b^{(j)}]$ will be denoted by $[a]^*$ and it will be called the set of upper indices of $[a]$. Every index of $[a]^*$ is said to be an <u>upper index</u> of $[a]$. Observe that if a is not a pasted element, then the set $[a]^*$ has only one element; if a is a pasted element, then $[a]^* = I$.

<u>Lemma 2.4</u>. Let $\underline{A}$ be a convexly pasted amalgam of posets $`S_i$. If $[u] \leq [v]$ and $[u]^* \cap [v]^* \neq \emptyset$, then $[u]^* \cap [w]^* \neq \emptyset$ for every w such that $[u] \leq [w] \leq [v]$.

<u>Proof</u>. Assume to the contrary that $[u]^* \cap [w]^* = \emptyset$. Then $[u]^* = [v]^* = \{i\} \neq I$ and so $[v]^* \cap [w]^* = \emptyset$. Since $[u] \leq [w]$ and $[w] \leq [v]$, there exist pasted elements u_o, v_o such that $[u_o] \leq [w] \leq [v_o]$. Since $\underline{A}$ is convexly pasted, $[w]^* = I$, a contradiction. //

Let $\underline{A}$ be an amalgam of lattices $`L_i$, $i \in I$. We shall say that $\underline{A}$ is a <u>convexly pasted lattice amalgam</u> if and only if it satisfies the following conditions:

(i) $\underline{A}$ is convexly pasted;

(ii) $\underline{A}$ is a lattice;

(iii) the pasted poset $`L_i^o$ is a sublattice of $`L_i$ for every $i \in I$.

<u>Lemma 2.5</u>. Let $\underline{A}$ be a convexly pasted lattice amalgam of modular lattices $`L_i$, $i \in I$. If $[x], [y]$ and $[z]$ are elements of $\underline{A}$ such that $[x] \leq [y]$, $[d] = [x] \wedge [z] = [y] \wedge [z]$, $[v] = [x] \vee [z] = [y] \vee [z]$ and $[d]^* \cap [v]^* \neq \emptyset$, then $[x] = [y]$.

<u>Proof</u>. By Lemma 4, there is an upper index j such that $[x] = [x^{(j)}]$, $[y] = [y^{(j)}]$ and $[z] = [z^{(j)}]$. Since the lattice $`L_j$ is modular, it is not hard to see that $[x^{(j)}] = [y^{(j)}]$. //

<u>Theorem 2.6</u>. Let $\underline{A}$ be a convexly pasted lattice amalgam of modular lattices $`L_s$, $s \in I$. Then $\underline{A}$ is a modular lattice.

<u>Proof</u>. Assume the $\underline{A}$ to be nonmodular. Then $\underline{A}$ contains as a sublattice a pentagon $[o^{(m)}] < [a^{(i)}] < [c^{(k)}] < [u^{(n)}], [b^{(j)}]$.

Because of Lemma 5, the elements o and u are not pasted. Hence $n \neq m$.

By Lemma 2 and by duality, it is possible to choose the indices m,n in such a way that $i = m \neq n = j$. Then, from $[c^{(k)}] \wedge [b^{(j)}] = [o^{(i)}]$ and Lemma 2 it follows that $i \in \{k,j\}$. Consequently, $k = i$.

Using $[c^{(i)}] \leqslant [u^{(j)}]$, we see that there exists a pasted element $d \in L_i^o$ such that $[c] \leqslant [d] \leqslant [u]$. Putting $[e^{(q)}] = [d] \wedge [b]$, $[h^{(r)}] = [a] \vee [e]$, $d_2 = f_{i/j}(d)$, we find that $[d_2] = [d]$. By Lemma 1, $[e^{(q)}] = [d_2^{(j)}] \wedge [b^{(j)}] = [(d_2 \wedge b)^{(j)}]$. Hence we can assume $q = j$. We shall show that e is a pasted element. Indeed, the relation $[o] \leqslant [e]$ implies that there exists a pasted element e_o such that $[o] \leqslant [e_o] \leqslant [e]$. Since $[e_o] \leqslant [e] \leqslant [d]$ where e_o and d are pasted elements and since $\underline{A}$ is convexly pasted, the element e is also pasted. Now

$$[u] = [b] \vee [a] \leqslant [b] \vee [h] \leqslant [b] \vee [d] \leqslant [u]$$

and $[e] \leqslant [h] \wedge [b] \leqslant [d] \wedge [b] = [e]$. This together with $[e]^* \cap [u]^* = I \cap \{j\} \neq \emptyset$ and Lemma 5 yields $[h] = [d]$.

Finally, we have $[d] = [h] = [a] \vee [e] \leqslant [c] \vee [e] \leqslant [d]$ and

$$[o] \leqslant [a] \wedge [e] \leqslant [c] \wedge [e] \leqslant [c] \wedge [b] = [o].$$

Moreover, $[o]^* \cap [d]^* = \{i\} \cap I \neq \emptyset$. Therefore, by Lemma 5, $[a] = [c]$, a contradiction. Thus $\underline{A}$ is modular. //

3. AMALGAMS OF ORTHOMODULAR LATTICES

We shall, in this section, develop the theory of amalgams in constructions of orthomodular lattices.

The basic idea can be briefly described by the following preliminary example: Consider two copies of the Boolean algebra $\underline{2}^3$ (see Figure 30). If we construct the amalgam in such a way

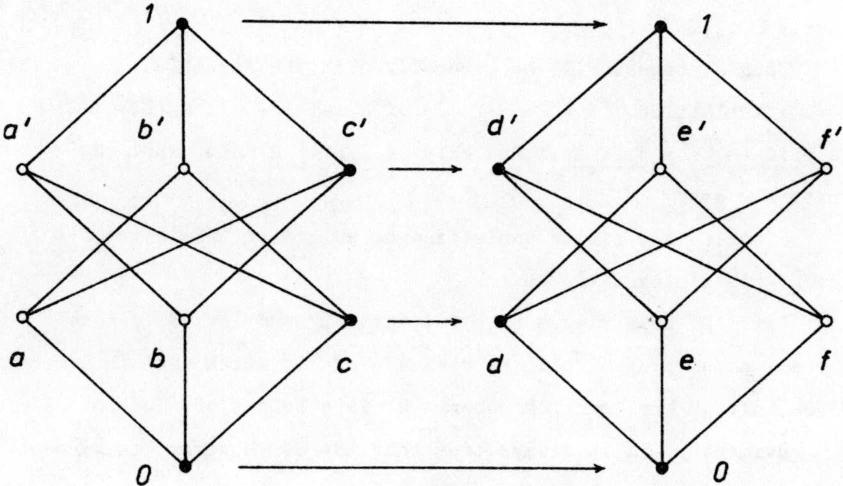

Fig. 30

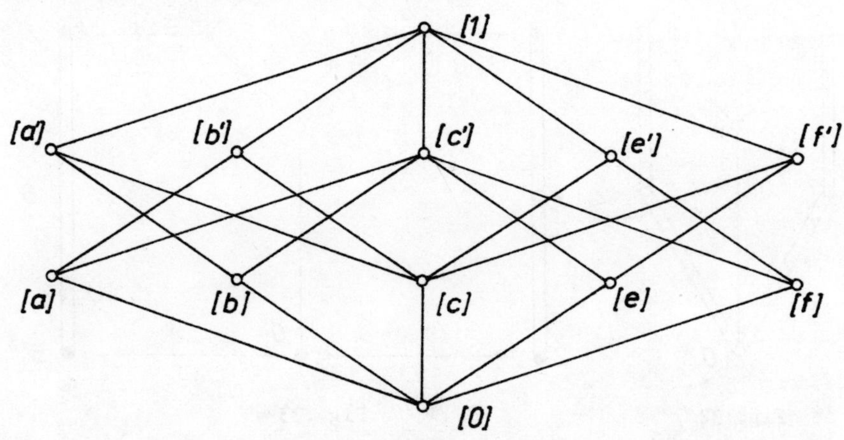

Fig. 31

that the elements 1 and 1, c' and d', c and d, 0 and 0 are pasted together, we obtain the lattice of Figure 31. It is easy to see that this lattice is orthomodular. Observe that we have pasted together the order-filters $[c'),[d')$ and the order-ideals $(c],(d]$.

Such a process will be generalized in the sequel.

A sublattice $`S = (S, \vee, \wedge)$ of a lattice $`L = (L, \vee, \wedge)$ is said to be a <u>plexus</u> if and only if $\inf_{`L} M$ and $\sup_{`L} M$ exist for every $M \subset S$.

Clearly, any finite sublattice of a lattice $`L$ with 0 and 1 is a plexus of $`L$.

Let $`S$ be a plexus of a lattice $`L$ and let $\overline{S}$ denote the set consisting of all the elements of L which are of the form $\inf_{`L} M$ or $\sup_{`L} M$ where M is a subset of S.

Evidently, it is always true that $S \subset \overline{S}$. However, in general

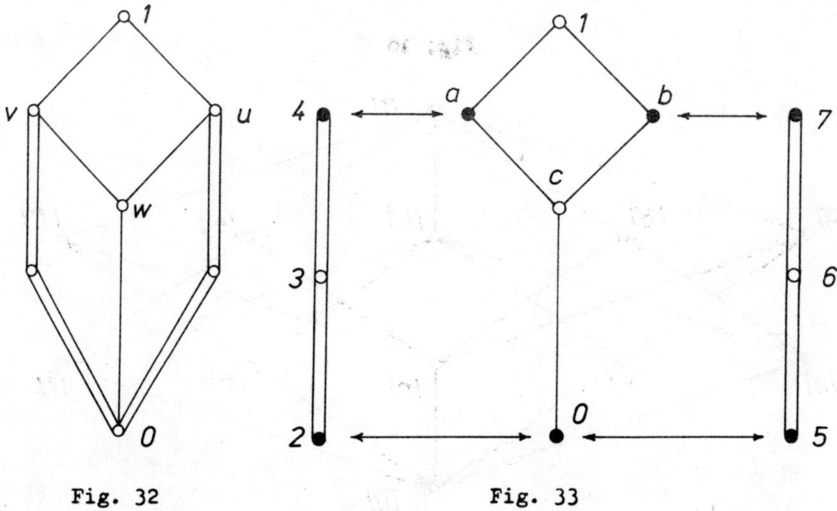

Fig. 32 Fig. 33

it is not possible to replace the inclusion indicated here by the equality sign. An example is furnished by the lattice of Figure 32. This lattice is obtained by pasting together the five-element lattice having the base set $\{0,a,b,c,1\}$ (see Figure 33) and two segments $[2,4]$, $[5,7]$ of real numbers. It is enough to consider the set S consisting of the open segments corresponding to $(2,4)$ and $(5,7)$, respectively, and of $0,1$. Notice that $\overline{S} = S \cup \{u,v\} \neq \overline{\overline{S}} = \overline{S} \cup \{w\}$.

A plexus $`S$ is said to be <u>closed</u> if and only if $S = \overline{S}$.

<u>Proposition 3.1</u>. A plexus $`S$ of a lattice $`L$ is closed if and only if the sublattice $`S$ is a complete lattice and $\sup_{`S} M = \sup_{`L} M$ as well as $\inf_{`S} M = \inf_{`L} M$ for every $M \subset S$.

<u>Proof</u>. 1. By definition, $\inf_{`S} M$ and $\sup_{`S} M$ belong to S for any $M \subset S$. Consequently, the given condition is sufficient for $`S$ to be closed.

2. If $`S$ is a closed plexus, then $\inf_{`L} M$ and $\sup_{`L} M$ are elements of $S$ whenever $M \subset S$. Thus $\inf_{`S} M$ exists and it is equal to $\inf_{`L} M$. An analogous assertion holds for the supremum $\sup_{`S} M$. Now, $1 = \inf_{`S} \emptyset \in S$. Because of I.13, $`S$ is a complete lattice. //

A sublattice $`T$ of a lattice $`L$ is called a <u>section</u> of $`L$ if and only if its base set satisfies the following conditions:

(i) $T = (s] \cup [t)$ for some $s,t \in L$;

(ii) $T \neq L$.

<u>Lemma 3.2</u>. Let $`L = (L, \vee, \wedge, ', 0, 1)$ be an ortholattice and let $`S$ be a subalgebra of $`L$. Then $(S, \vee, \wedge)$ is a section of $(L, \vee, \wedge)$ if and only if $S = (t'] \cup [t)$ for some $t \in L$.

<u>Proof</u>. 1. If S is a set of the given form, then it determines a section.

2. Now let $S = (s] \cup [t)$ be the base set of a subalgebra of $`L \neq `S$. Consequently, $t' \in S$. Suppose that $t' \in [t)$. Then $0 = t \wedge t' \in [t)$ and so $t = 0$. It follows that $S = L$, a contradiction. Thus $t' \in (s]$. Dually we get $s' \in [t)$. In other words, $t' \leqslant s$ and $s' \geqslant t$. Since $`L$ is an ortholattice, it follows that $s = t'$. //

An amalgam $\underline{A}$ of lattices $`L_i$, $i \in I \neq \emptyset$, is said to be <u>regularly pasted</u> if and only if

(i) the pasted sublattice $`L_i^o$ is a section in $`L_i$, i.e., $L_i^o = (s_i] \cup [t_i) \neq L_i$, for every $i \in I$;

(ii) the pasting isomorphisms $f_{i/j}$ are such that $f_{i/j}([t_i)) = [t_j)$ and $f_{i/j}((s_i]) = (s_j]$ for every $(i,j) \in I^2$.

Let $[a]$ and $[b]$ be two elements of a regularly pasted amalgam $\underline{A}$. Let $H_i(a,b) = \{h \in [t_i);\ [a] \leqslant [h]\ \&\ [b] \leqslant [h]\}$ and $D_i(a,b) = \{d \in (s_i];\ [d] \leqslant [a]\ \&\ [d] \leqslant [b]\}$.

<u>Lemma 3.3</u>. Let $\underline{A}$ be a regularly pasted amalgam of lattices $`L_i$, $i \in I$. Let the pasted sublattice $`L_i^o$ be a closed plexus of the lattice $`L_i$ for every $i \in I$. Suppose that $a \in L_j$, $b \in L_k$ $(j \neq k)$ are not pasted elements. Then

(i) $[a^{(j)}] \vee [b^{(k)}] = [u_i]$ where u_i is the infimum of the set $H_i(a,b)$ in $`L_i^o$;

(ii) $[a^{(j)}] \wedge [b^{(k)}] = [v_i]$ where v_i is the supremum of the set $D_i(a,b)$ in $`L_i^o$.

<u>Proof</u>. Let $[c^{(m)}]$ be such that $[c^{(m)}] \geqslant [a^{(j)}]$ and $[c^{(m)}] \geqslant [b^{(k)}]$. Without loss of generality we may assume that $m \neq j$. Hence there exist pasted elements a_o, c_o such that $a \leqslant a_o$, $f_{j/m}(a_o) = c_o$ and $c_o \leqslant c$. If $a_o \leqslant s_j$, then $a \leqslant s_j$ so that $a \in L_j^o$, a contradiction. Thus $a_o \geqslant t_j$ and, consequently, $c_o \geqslant t_m$. From $c \geqslant c_o \in [t_m)$ we see that c is

a pasted element. Therefore, $[c^{(m)}] = [c_1^{(i)}]$ where $c_1 = f_{m/i}(c) \in H_i(a,b)$. Since $`L_i^o$ is a closed plexus, the infimum u_i exists.

It remains to prove that $[a] \leqslant [u_i]$ and $[b] \leqslant [u_i]$. Clearly, $[a^{(j)}] \leqslant [u_j]$ and $[b^{(k)}] \leqslant [u_k]$. However, $f_{i/j}: :H_i(a,b) \mapsto H_j(a,b)$ and so, by I.7, $f_{i/j}(u_i) = u_j$, $f_{i/k}(u_i) = u_k$. Thus $[u_i] = [u_j] = [u_k]$.

The assertion (ii) follows by duality. //

<u>Theorem 3.4</u>. Let $\underline{A}$ be a regularly pasted amalgam of orthomodular lattices $`L_i$, $i \in I$. For any $i \in I$, let the pasted sublattice $`L_i^o$ determine a subalgebra $(L_i^o, \vee, \wedge, ', 0, 1)$ of the orthomodular lattice $`L_i = (L_i, \vee, \wedge, ', 0, 1)$ and let $(L_i^o, \vee, \wedge)$ be a closed plexus in the lattice $(L_i, \vee, \wedge)$. Suppose that the pasting isomorphisms $f_{i/j}: `L_i^o \twoheadrightarrow `L_j^o$ are orthoisomorphisms for every $(i,j) \in I^2$. Then $\underline{A}$ is an orthomodular lattice.

<u>Proof</u>. According to Lemma 2.1 and Lemma 3, $\underline{A}$ is a lattice which has $[0]$ and $[1]$ as its least and greatest element, respectively. Let $'$ be an operation on $\underline{A}$ defined by $[a^{(i)}]' = [a'^{(i)}]$. (Here $a'^{(i)}$ denotes the orthocomplement of $a \in L_i$ in $`L_i$.) Then it is an orthocomplementation on $\underline{A}$ which is well defined. In fact, if $[a^{(i)}] = [b^{(j)}]$ $(i \neq j)$, then the elements a, b are pasted and, furthermore, from the fact that $f_{i/j}$ is an orthoisomorphism, we conclude that $b' = f_{i/j}(a')$. Using Lemma 2.1 it is not difficult to see that $[x^{(i)}]'' = [x^{(i)}]$, $[x^{(i)}] \vee \vee [x^{(i)}]' = [1]$ and $[x^{(i)}] \wedge [x^{(i)}]' = [0]$ for any $[x]$ of $\underline{A}$. Assume $[a^{(i)}] \leqslant [b^{(j)}]$. If $i = j$, then $a \leqslant b$ and, consequently, $a' \geqslant b'$ which yields $[a^{(i)}]' \geqslant [b^{(j)}]'$. Now let $i \neq \neq j$. Then there exist $a_o \in L_i^o$, $b \in L_j^o$ such that $a \leqslant a_o$, $f_{i/j}(a_o) = b_o$ and $b_o \leqslant b$. Since $f_{i/j}$ is an orthoisomorphism, $f_{i/j}(a_o') =$

$= b_o'$. From $a' \geqslant a_o'$ and $b_o' \geqslant b$ we infer that $[a^{(i)}]' \geqslant$ $\geqslant [b^{(j)}]'$. In summary, we have shown that $\underline{A}$ is an orthocomplemented lattice.

There remains to show that $\underline{A}$ is orthomodular. Suppose the contrary. By Theorem II.5.4, $\underline{A}$ contains a subalgebra $`P$ isomorphic to the benzene ring, $P = \{[0],[x^{(i)}],[y^{(j)}],[x^{(i)}]',$ $[y^{(j)}]',[1]\}$. Evidently, $i = j$ leads to a contradiction of the fact that $`L_i$ is orthomodular. Assume $[x]^* \cap [y]^* = \emptyset$. Then $[x^{(i)}] < [y^{(j)}]$ implies $[x^{(i)}] < [d] < [y^{(j)}]$ for some pasted $d \in L_i$. Therefore, $[y^{(j)}]' > [d'] > [x^{(i)}]'$. Now $Q = \{[0],$ $[x^{(i)}],[d^{(i)}], [x^{(i)}]',[d^{(i)}]',[1]\}$ is the base set of a subalgebra isomorphic to the benzene ring, and the same contradiction as above ensues. //

4. ATOMIC AMALGAMS OF BOOLEAN ALGEBRAS

In order to compare and contrast our previous methods with

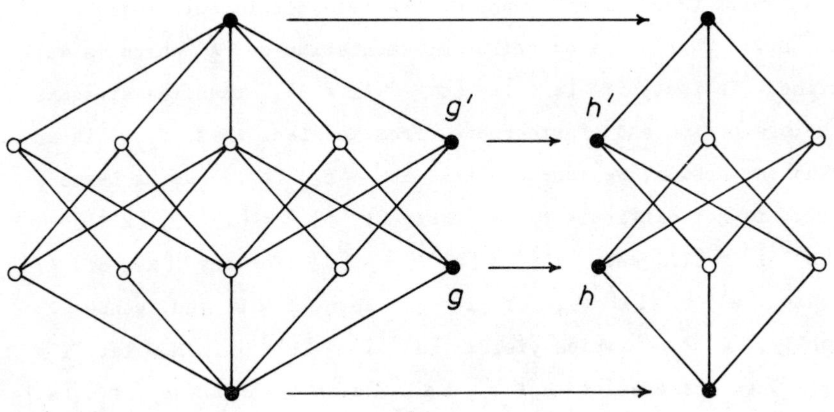

Fig. 34

the one given below, let us first consider the orthomodular lattice of Figure 31 where we for brevity write g for [f]. If we take a copy of the Boolean algebra $\underline{2}^3$ (see Figure 34), then we have another amalgam which we get by pasting together the both diagrams. The amalgam is usually called the <u>Dilworth's lattice</u> D_{16} and it is an orthomodular lattice, by Theorem 3.4. Its diagram is shown in Figure 35.

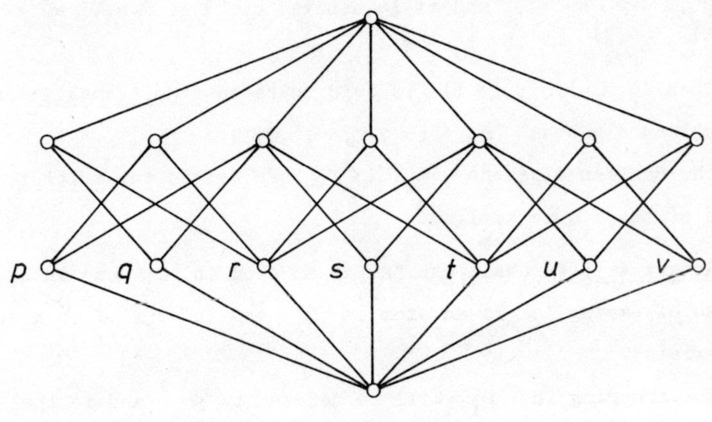

Fig. 35

The lattice D_{16} has been obtained by two amalgams using successively three copies of the Boolean algebra $\underline{2}^3$.

Although we can proceed similarly with many amalgams of this type, it is most convenient to seek a construction which uses a system of Boolean algebras and gives the resulting lattice instantaneously, i.e., without intermediate successively formed amalgams.

To this end we can use the following definition:

Let 0 and 1 be two fixed elements of a set B. For any i of a finite index set $I \neq \emptyset$ let B_i denote a finite Boolean algebra having the elements 0 and 1 fixed above as its least and greatest element, respectively. Suppose that the conditions (a 1) and (a 2) are satisfied:

(a 1) Any $`B_i$ has at least eight elements.

(a 2) For any $i \neq j$ of I, the intersection $B_i \cap B_j$ is either $\{0,1\}$ or $\{0,1,a,a'\}$ for some atom a of B_i and B_j. If $B_i \cap B_j = \{0,1,a,a'\}$, then it is supposed that the complement of a in $`B_i$, say $a^{\prime(i)}$, is equal to the complement of $a$ in $`B_j$, say $a^{\prime(j)}$, and it is denoted by a', i.e., $a' = a^{\prime(i)} = a^{\prime(j)}$.

Then $B = \bigcup(B_i;\ i \in I)$ is said to be an <u>atomic amalgam</u> of the Boolean algebras $`B_i = (B_i, \wedge, \vee, ', 0, 1)$.

The Boolean algebras $`B_i$, $i \in I$, are called the <u>initial blocks</u> of the atomic amalgam.

<u>Remark 4.1</u>. We shall say "a is an atom in B_i and in B_j" instead of saying "a is an atom in B_i and in B_j if such an atom exists".

The ordering in $`B_i$ will be denoted by $\leqslant_i$ and in the sequel it will be sometimes abbreviated to $\leqslant$.

Let B be an atomic amalgam. We shall now show that the relation $\leqslant$ and the operation $'$ defined on B by

$$a \leqslant b \text{ if and only if } a,b \in B_i \text{ for some } i \in I \text{ and } a \leqslant_i b \text{ in } `B_i,$$

and by

$$': a \to a^{\prime(i)} \quad \text{whenever } a \in B_i,$$

make B into a special poset. By an abuse of language, the structure $(B, \leqslant, ', 0, 1)$ will also be called, when no ambiguity arises, an <u>atomic amalgam</u> of the Boolean algebras $`B_i$.

A structure $(P, \leqslant, ', 0, 1)$ is called an <u>orthocomplemented</u>

<u>poset</u> if and only if $(P, \leqslant)$ is a poset and the unary operation $' : a \mapsto a'$ is such that

(i) $(a')' = a$ for every $a \in P$;

(ii) for every $a, b \in P$, the relation $a \leqslant b$ implies $a' \geqslant b'$;

(iii) $a \vee a' = 1$ for every $a \in P$;

(iv) $0 = 1'$.

<u>Theorem 4.2</u>. An atomic amalgam of Boolean algebras is an orthocomplemented poset.

<u>Proof</u>. 1. The relation $\leqslant$ introduced in the definition of an atomic amalgam $(B, \leqslant, ', 0, 1)$ is reflective. We shall show that it is also antisymmetric. However, if $x \leqslant y$ and $y \leqslant x$, then $x \leqslant_i y$ and $y \leqslant_j x$ for convenient indices $i, j \in I$. Hence, x and y belong to $B_i \cap B_j$. If $i = j$, then it is clear that $x = y$. Let $i \neq j$. By (a 2), $x, y \in \{0, 1, a, a'\}$ where a is an atom in B_i and in B_j. Suppose $x \neq y$. If $x = 0$, then $y \leqslant_j \leqslant_j x$ implies $y = 0$, a contradiction. If $x = 1$, then $x \leqslant_i y$ implies $y = 1$, a contradiction. Similarly we can prove that $y = 0, 1$ is not possible. Hence $x, y \in \{a, a'\}$. If $x = a$, then $x \leqslant_i y$ and $x \neq y \in \{a, a'\}$ imply $a = x \leqslant_i y = a'$. Thus $a = 0$ which contradicts the fact $a \succ 0$. Finally, if $x = a'$, then $x \leqslant_i y$ and $x \neq y \in \{a, a'\}$ imply $a' = x \leqslant y = a$. This yields $0 \prec a = 1$ which is impossible by (a 1).

Now we shall prove the transitivity of the relation $\leqslant$.

Suppose $x \leqslant_i y$ and $y \leqslant_j z$. If $i = j$, then obviously $x \leqslant_i z$ and we are done. If $i \neq j$, then, a priori, y is an element of $\{0, 1, a, a'\}$ where a is an atom in B_i and in B_j. We claim that in all cases $x \leqslant z$. This is clear if $x = y$ or $y = z$ or if $x, y \in \{0, 1\}$. Therefore, it remains to investigate only the case $y = a$ and the case $y = a'$. Let us start with

$y = a$. Then $x \leq_i y$ implies $x = a$ or $x = 0$ and we have the cases already discussed. Finally, suppose $y = a'$. Since $y \leq_j z$, $z = a'$ or $z = 1$. Hence we have again the cases discussed above.

Thus $(B, \leq)$ is a poset.

2. We next verify the properties (i)-(iv) indicated in the definition of an orthocomplemented poset. If $b \in B_i$, then $b' = b'^{(i)}$ by (a 2). Hence $(b')' = (b'^{(i)})'^{(i)} = b$. The validity of (ii) and (iv) is immediate. To prove (iii), suppose h is an upper bound of the set $\{x, x'\}$. Consequently, $x \leq_i h$ and $x' \leq_j h$ for some $i, j \in I$. If $i = j$, then we conclude that $h = 1$. Assume further that $i \neq j$. In this case the elements x, x', h belong to the set $\{0, 1, a, a'\}$ where a is an atom in B_i and in B_j. Since $\{0, 1, a, a'\} \subset B_i$, we have $x \leq_i h$ and $x' \leq_i h$. Thus $h = 1$. //

<u>Lemma 4.3</u>. Let $B = \bigcup(B_i;\ i \in I)$ be an atomic amalgam of Boolean algebras $`B_i$, $i \in I$. Let $\emptyset \neq J \subset I$. Then $K = \bigcup(B_i; i \in J)$ is also an atomic amalgam of Boolean algebras.

<u>Proof</u>. The proof of this statement is straightforward and left to the reader. //

<u>Theorem 4.4</u>. Let $B = \bigcup(B_i;\ i \in I)$ be an atomic amalgam of Boolean algebras.

(i) If $i \neq j$ and $i, j \in I$, then $B_i \cup B_j$ is the base set of an orthomodular lattice which is isomorphic to the amalgam pasting together the lattices $(B_i, \vee, \wedge)$ and $(B_j, \vee, \wedge)$ along the pasted sublattices $`B_i^o = `B_j^o = (B_i \cap B_j, \vee, \wedge)$.

(ii) Let $k \in I$. Then the poset $(B, \leq)$ is order-isomorphic to the amalgam pasting together the posets $`S_1 = (B_k, \leq)$, $`S_2 = (\bigcup(B_i;\ i \in I \setminus \{k\}), \leq)$ along the pasted subposets $`S_1^o = `S_2^o = (B_k \cap \bigcup(B_i;\ i \in I \setminus \{k\}), \leq)$.

Proof. Ad (i). Since $B_i \cap B_j$ is a finite lattice, the pasted sublattices $`B_i^o$, $`B_j^o$ are closed plexuses. Hence, by Theorem 3.4, the amalgam pasting together the two lattices along $`B_i^o$ and $`B_j^o$ is an orthomodular lattice. It is easily verified that this lattice is isomorphic to the lattice with the base set $B_i \cup B_j$.

Ad (ii). This can be obtained in a similar manner. //

From Theorem 4 we can derive the following conclusion which is intuitively clear: Any atomic amalgam can be constructed in a finite number of successive steps by pasting together smaller amalgams and/or some initial blocks; the result is not dependent on the order of steps.

As shown above, this remark can be applied to the construction of the Dilworth´s lattice D_{16}. Here we have started with three Boolean algebras. However, we can continue the process of

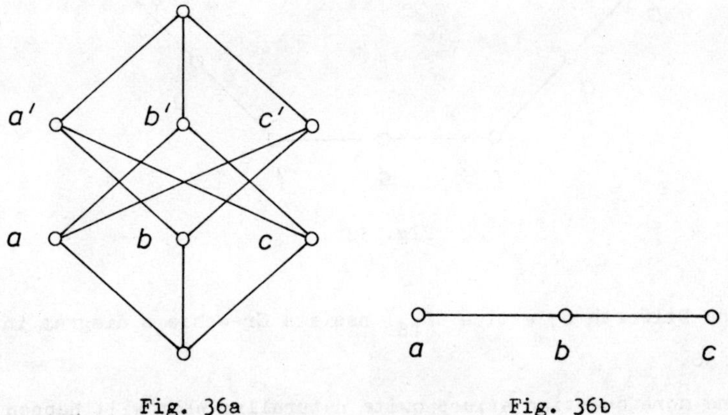

Fig. 36a Fig. 36b

successively enlarging the pasted bunch of Boolean algebras, but the resulting diagrams become extremely intricate, even if we paste a small number of Boolean algebras. Greechie has observed that the usually drawn diagrams may be simplified. The simplest example where we shall explain his idea is represented by the Boolean algebra $\underline{2}^3$. Now, instead of its usual diagram (see Figure 36a), we construct the new one (see Figure 36b) in such a way that we indicate only its atoms in a top view of the usual diagram. If we construct an amalgam, then we decline the segments in the pasted points. For instance, the orthomodular lattice of Figure 31 has its Greechie's diagram in Figure 37.

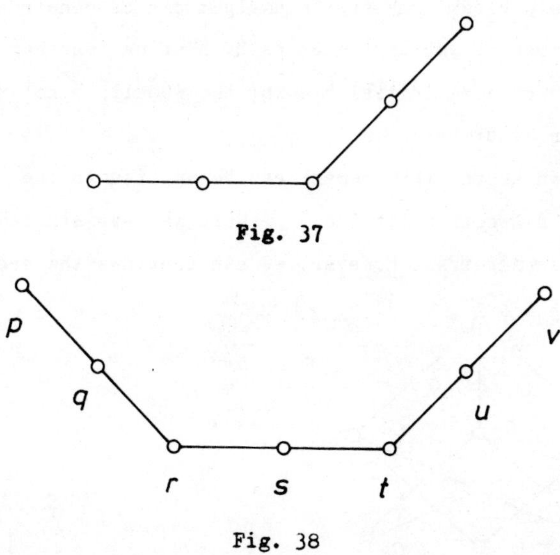

Fig. 37

Fig. 38

The Dilworth's lattice D_{16} has its Greechie's diagram in Figure 38.

One more question arises quite naturally: What will happen if the segments form a polygon?

Before proceeding to the more detailed study of these situations, we shall exhibit an application of such constructions known as the Janowitz's poset J_{18}. To get its diagram, we take two copies of the lattice sketched in Figure 31 and paste them together

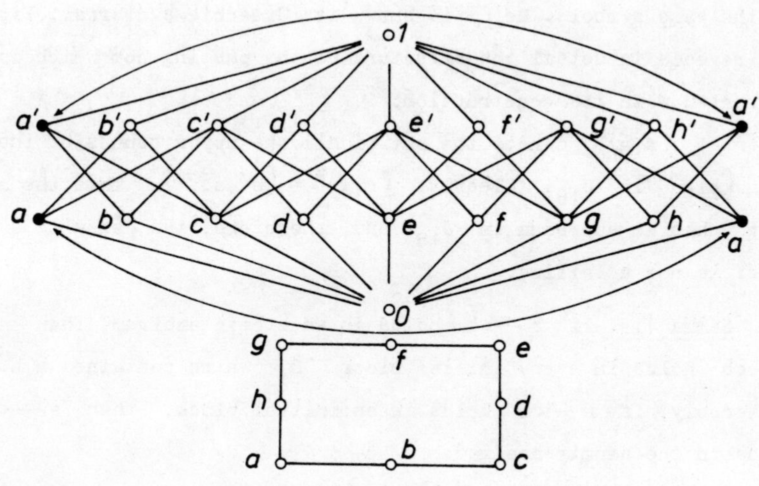

Fig. 39a

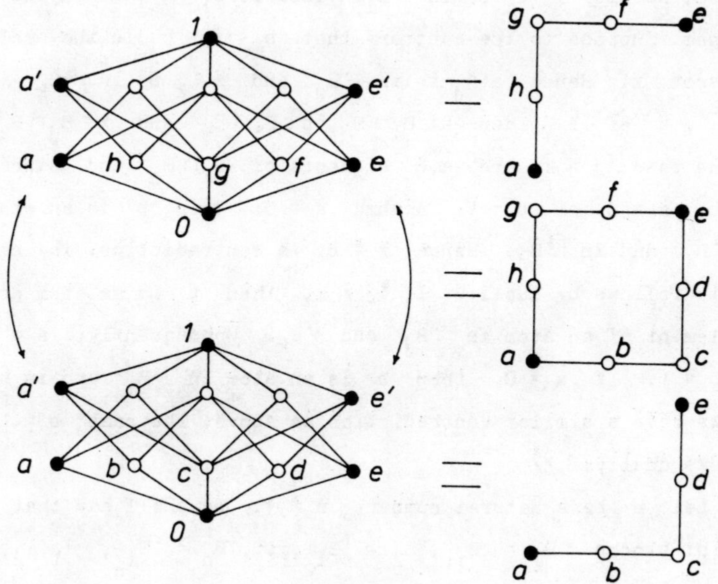

Fig. 39b

as shown in Figure 39a. The pasted elements are shaded and denoted by the same symbols. Below is shown its Greechie's diagram. Figure 39b reveals in detail the structure of the pasting job which is connected with its construction.

Let $\{c,g\}^*$ denote the set of all the upper bounds of the set $\{c,g\}$ in J_{18}. Clearly, $\{c,g\}^* = \{e',a',1\}$. Thus the set $\{c,g\}$ has no supremum in J_{18} and, therefore, the Janowitz's poset is not a lattice.

<u>Lemma 4.5</u>. If $a \prec b$ holds in an atomic amalgam, then $a \prec b$ holds in every initial block $`B_i$ which contains a,b. Conversely, if $a \prec b$ holds in an initial block, then $a \prec b$ holds in the atomic amalgam.

<u>Proof</u>. 1. Suppose $a \prec b$ in the amalgam. If $a <_i x <_i b$, then a fortiori, $a < x < b$, a contradiction.

2. Assume $a \prec_i b$ in $`B_i$. Therefore, $a < b$ in the amalgam. Suppose to the contrary that $a < x < b$ in the amalgam for some $x$. Hence $a <_j x$ in $`B_j$ and $x <_k b$ in $`B_k$ for some $j,k$ of $I$. Then $a \in B_i \cap B_j$, $b \in B_i \cap B_k$ and $x \in B_j \cap B_k$. In the case $j = k$ are $a,b$ elements of $B_i \cap B_j$ and, since $a < b$, $a = 0$ or $b = 1$. Assume $a = 0$. Then $b$ is an atom in $`B_i$ and in $`B_j$. Hence $x = b$, a contradiction. The case $b = 1$ follows by duality. If $j \neq k$, then $x$ is an atom or a complement of an atom in $`B_j$ and $`B_k$. Consequently, $a = 0$ or $b = 1$. If $a = 0$, then $b$ is an atom in $`B_i$ and in $`B_k$ and we have a similar contradiction as above. The case $b = 1$ follows dually. //

Let n be a natural number, $n \geqslant 3$. We shall say that the initial blocks $`B_1 = `B_{i_1}, `B_2 = `B_{i_2}, \dots, `B_n = `B_{i_n}$, $i_1, i_2, \dots$ $\dots, i_n \in I$, form an <u>atomic loop</u> of order n if and only if the following conditions are satisfied:

(L 1) for every $i = 1,2,\dots,n-1$, $B_i \cap B_{i+1} = \{0,1,a_i,a_i'\}$ where a_i is an atom in $`B_i$ and in $`B_{i+1}$; moreover, $B_n \cap B_1 = \{0,1,a_n,a_n'\}$ where a_n is an atom in $`B_n$ and in $`B_1$;

(L 2) $B_i \cap B_j = \{0,1\}$ for all indices $i \neq j$ not mentioned in (L 1);

(L 3) for any i,j,k such that $1 \leq i < j < k \leq n$, $B_i \cap B_j \cap B_k = \{0,1\}$.

<u>Remark 4.6</u>. (A) By (L 3),

$$\{0,1\} = B_i \cap B_{i+1} \cap B_{i+2} = (B_i \cap B_{i+1}) \cap (B_{i+1} \cap B_{i+2}) = \\ = \{0,1,a_i,a_i'\} \cap \{0,1,a_{i+1},a_{i+1}'\}.$$

Hence, a_i and a_{i+1} are distinct atoms. An entirely analogous argument leads to the conclusion that any two atoms of $a_1,\dots,a_n$ are distinct. This remark reveals the meaning of (L 3).

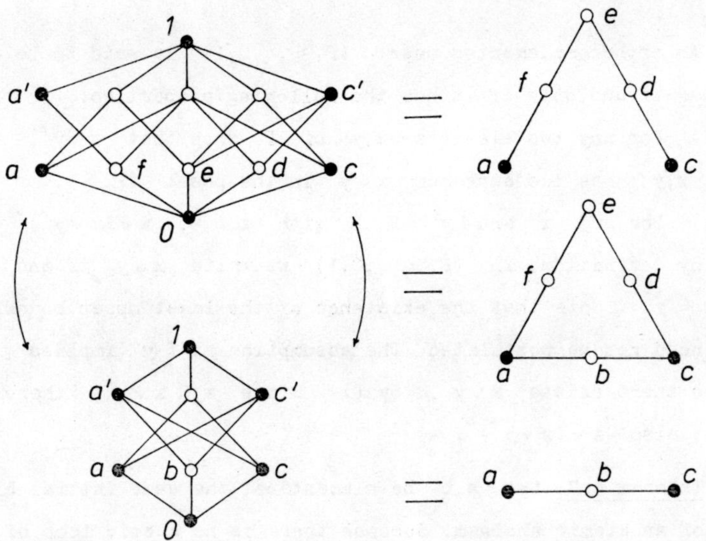

Fig. 40

(B) An atomic loop of order 3 constructed from three copies of the Boolean algebra $\underline{2}^3$ can be visualized using Figure 40.

(C) Consider the Janowitz's poset J_{18} defined above (see Figure 39a). Let $`B_1 = \langle a,b,c \rangle$ be the Boolean algebra generated by the atoms a,b,c so that it is an isomorphic copy of $\underline{2}^3$. Further, let $`B_2 = \langle c,d,e \rangle$, $`B_3 = \langle e,f,g \rangle$, $`B_4 = \langle g,h,a \rangle$. These Boolean algebras form an atomic loop of order 4 in which $a_1 = c$, $a_2 = e$, $a_3 = g$, $a_4 = a$.

(D) Strictly speaking, an atomic loop of order n is an ordered n-tuple $(`B_1, `B_2, \ldots, `B_n)$. Another n-tuple is considered to be equal to the first one if and only if it can be obtained by permuting cyclically the blocks $`B_1, `B_2, \ldots, `B_n$ in the original n-tuple.

In what follows, by an atomic loop $`B_i, `B_j, `B_k$, we mean an ordered triplet $(`A_1, `A_2, `A_3)$ where $`A_1 = `B_i$, $`A_2 = `B_j$, $`A_3 = `B_k$.

An orthocomplemented poset $(P, \leq, ', 0, 1)$ is said to be <u>orthomodular</u> if and only if it has the following properties:

(i) For any two elements x,y of P such that $x \leq y'$, the set $\{x,y\}$ has the supremum $x \vee y$ in the poset $(P, \leq)$.

(ii) For any x and y of P with $x \leq y$, $x \vee (x \vee y')' = y$.

By definition, in $(P, \leq, ', 0, 1)$ we write $x \perp y$ if and only if $x \leq y'$. Note that the existence of the least upper bounds in (ii) need not be postulated. The assumption $x \leq y$ implies $x \perp y'$ and so there exists $x \vee y'$, by (i). Since $x \leq x \vee y'$, there exists also $x \vee (x \vee y')'$.

<u>Theorem 4.7</u>. Let a,b be elements of the same initial block $`B_1$ of an atomic amalgam. Suppose there is no atomic loop of order 3 in this atomic amalgam. Then the set $\{a,b\}$ has the

supremum $a \vee b$ in the atomic amalgam and it is equal to the supremum of $\{a,b\}$ in $`B_1$.

Proof. Without loss of generality we may assume that a and b are incomparable in the amalgam.

Let c be an upper bound of the set $\{a,b\}$ in the atomic amalgam which does not belong to B_1. Since $a \leqslant c$ and $b \leqslant c$, there exist initial blocks $`B_2$, $`B_3$ such that $a \leqslant_2 c$ and $b \leqslant_3 c$. Then $a \in B_1 \cap B_2$, $b \in B_1 \cap B_3$ and $c \in B_2 \cap B_3$. If B_1, B_2 and B_3 are three different initial blocks, then they form an atomic loop of order 3. Consequently, $B_1 = B_2$ or $B_1 = B_3$ or $B_2 = B_3$. Hence $a \leqslant c$ and $b \leqslant c$ in $B_1 \cup B_2 \cup B_3$ which is an orthomodular lattice by Theorem 4 and, therefore, $a \vee_1 b \leqslant c$, by Lemma 2.1. //

Corollary 4.8. Let $b_1, b_2, \ldots, b_n$ be elements of the same initial block $`B_1$ of an atomic amalgam $B$. Suppose there is no atomic loop of order 3 in $B$. Then the supremum $\sup(b_i;\, i \in \in \{1,2,\ldots,n\})$ in the atomic amalgam exists and it is equal to the supremum $\sup(b_i;\, i \in \{1,2,\ldots,n\})$ in the initial block $`B_1$.

Proof. Use induction on n, Lemma I.2 and Theorem 7. //

Greechie's First Theorem 4.9 (cf. [82]). An atomic amalgam of Boolean algebras is an orthomodular poset if and only if it contains no atomic loop of order 3.

Proof. 1. Suppose the amalgam B contains initial blocks $`B_1, `B_2, `B_3$ forming an atomic loop of order 3, i.e., $B_1 \cap B_2 = \{0, a_1, a_1', 1\}$, $B_2 \cap B_3 = \{0, a_2, a_2', 1\}$, $B_3 \cap B_1 = \{0, a_3, a_3', 1\}$ where a_1, a_2, a_3 are distinct atoms in B (see Lemma 5). Clearly, $a_1 \leqslant a_3'$ in B. We shall show that the supremum $a_1 \vee a_3$ does not exist. Suppose that this is not the case and denote $s = a_1 \vee a_3$. Then the element $a_1 \vee_1 a_3$ (the supremum of $\{a_1, a_3\}$

in $`B_1$) and the element $a_2'$ are upper bounds of the set $\{a_1, a_3\}$ in the amalgam. But it follows that $s \leqslant a_2'$ and $s \leqslant a_1 \vee_1 \vee_1 a_3$. However, in $`B_1$ we have $a_3 \longrightarrow\!\!\!<_1 a_1 \vee_1 a_3$. By Lemma 5, we get $a_3 \longrightarrow\!\!\!< a_1 \vee_1 a_3$ in the amalgam. Similarly, $a_1 \longrightarrow\!\!\!< a_1 \vee_1 \vee_1 a_3$ and so $s = a_1 \vee_1 a_3 \leqslant a_2'$.

By the definition of the partial ordering in an atomic amalgam, there exists an initial block $`B_i$ with $a_1 \vee_1 a_3 \leqslant_i a_2'$. Let us consider the possible cases.

<u>Case I</u>: i = 1. But then $1 \succ\!\!\!\longrightarrow a_2'$ and a_2' belongs to $B_1 \cap \cap B_2 \cap B_3 = \{0,1\}$ which is absurd.

<u>Case II</u>: $i \neq 1$. Note that here $s \neq a_2'$, by (a 1). Hence, by Lemma 5, we have $0 \longrightarrow\!\!\!< a_1 \longrightarrow\!\!\!< a_1 \vee_1 a_3 = s < a_2' \longrightarrow\!\!\!< 1$ in B. By Lemma 5, this contradicts the fact that $s \in B_i \cap B_1$.

2. Suppose now that the atomic amalgam contains no atomic loop of order 3. By Theorem 2, it is an orthocomplemented poset. If $x \perp y$, then this means that $x \leqslant y'$ and so there exists an initial block $`B_1$ such that $x \leqslant_1 y'$. This latter fact also yields $x, y \in B_1$. By Theorem 7, there exists the supremum $x \vee y$ in the amalgam. To prove the second property formulated in the definition of an orthomodular poset, suppose $x \leqslant y$. Then again $x, y \in B_1$ for a convenient initial block $`B_1$. By Theorem 7, $x \vee y' = x \vee_1 y'$ and, similarly, $x \vee (x \vee y')' = x \vee_1 (x \vee_1 y')' = = y$ since the Boolean algebra $`B_1$ is evidently an orthomodular lattice. //

<u>Greechie's Second Theorem 4.10 (cf. [82])</u>. An atomic amalgam of Boolean algebras is an orthomodular lattice if and only if it contains no atomic loop of order 3 or 4.

<u>Proof</u>. 1. If the amalgam contains an atomic loop of order 3, then it is not a lattice by the first part of the proof of

Greechie's First Theorem.

2. Suppose that the considered amalgam B contains initial blocks $`B_1, `B_2, `B_3, `B_4$ which form ar atomic loop of order 4, i.e., $B_1 \cap B_2 = \{0, a_1, a_1', 1\}$, $B_2 \cap B_3 = \{0, a_2, a_2', 1\}$, $B_3 \cap B_4 = \{0, a_3, a_3', 1\}$, $B_4 \cap B_1 = \{0, a_4, a_4', 1\}$ where a_1, a_2, a_3 and a_4 are atoms. Suppose to the contrary that $s = a_1 \vee a_3$ exists in B. Note that $s \leq a_2'$ and $s \leq a_4'$. By Lemma 5 we therefore see that (i) $s \longrightarrow\!\!< 1$ is not possible in B; (ii) s is not an atom; (iii) $s \neq 0$ and (iv) $s \neq 1$. From (i)-(iv) it follows that s belongs to no two different initial blocks. However, $a_1 \leq_i s$ and $a_3 \leq_j \leq_j s$ for some $i, j \in I$. Hence $i = j$ and $a_1, a_3 \in B_i$. But then $`B_1, `B_i, `B_4$ form an atomic loop of order 3 and the conclusion follows from 1.

3. Suppose there is no atomic loop of order 3 or 4. By Greechie's First Theorem, the atomic amalgam is an orthomodular poset. It remains to prove that for any two elements a,b the set $\{a, b\}$ has the supremum in the amalgam. (The existence of the infimum then follows by De Morgan laws.)

Clearly, we may assume that the elements a,b are not comparable. In particular, we assume that $a, b \neq 0, 1$ and that $a \neq \neq b$. By Theorem 7, we can suppose that a,b do not belong to the same initial block.

Suppose $c_1 \neq 1$ and $c_2 \neq 1$ are two distinct upper bounds of the set $\{a, b\}$. Since $a \leq c_1$, $a \leq c_2$, $b \leq c_1$ and $b \leq c_2$, there exist initial blocks $`B_i, `B_j, `B_m, `B_n$ such that $\{a, c_1\} \subset \subset B_i$, $\{a, c_2\} \subset B_j$, $\{b, c_1\} \subset B_m$, $\{b, c_2\} \subset B_n$. Note that $c_1 \in B_i \cap B_m$, $b \in B_m \cap B_n$, $c_2 \in B_n \cap B_j$ and $a \in B_j \cap B_i$. Hence the initial blocks $`B_i, `B_m, `B_n, `B_j$ form an atomic loop of order 3 or 4, a contradiction. Consequently, the set $\{a, b\}$ has in B at most one upper bound which is not equal to 1. Thus $a \vee b$ evidently exists. //

The results on the Greechie's construction of atomic amalgams [82] will now be applied to an attractive example known as the <u>Greechie's lattice</u> G_{32}.

We start with its construction: The "bricks" of our "box of bricks" are represented by the following Boolean algebras all being copies of $\underline{2}^3$:

$`B_1 = \langle a,b,c\rangle$, $`B_2 = \langle c,d,e\rangle$, $`B_3 = \langle e,f,g\rangle$, $`B_4 = \langle g,h,i\rangle$, $`B_5 = \langle i,j,k\rangle$, $`B_6 = \langle k,l,a\rangle$, $`B_7 = \langle d,r,j\rangle$, $`B_8 = \langle b,s,h\rangle$, $`B_9 = \langle l,t,f\rangle$, $`B_{10} = \langle r,s,t\rangle$.

Here a,b,c; c,d,e; ... ; r,s,t are atoms in $`B_1$; $`B_2$; ... ; $`B_{10}$, respectively. We furthermore suppose that the greatest elements of $`B_1, `B_2, \ldots, `B_{10}$ are identified and denoted by 1 and that the same is true for their least elements 0. Moreover, we suppose that the complement of c in $`B_1$ is identified with its complement in $`B_2$. We adopt similar conventions for the other pasted elements (see the following figures).

Note that $`B_1, `B_2, \ldots, `B_6$ form an atomic amalgam (see Figure 41 which indicates how it could be constructed from two copies of the Dilworth' lattice D_{16}).

The Boolean algebras $`B_7, `B_8, `B_9, `B_{10}$ form also an atomic amalgam which is sketched in Figure 42a or, in a twisted form, in Figure 42b.

The Greechie's lattice G_{32} can now be defined as the atomic amalgam which we get in a diagrammatical form from the Boolean algebras $`B_1, `B_2, \ldots, `B_{10}$ by juxtaposing the diagrams of Figures 41 and 42b (see Figure 43).

It can be checked that there is no atomic loop of order 3 or 4 in G_{32}. Hence, by Greechie's Second Theorem, the lattice G_{32} shown in Figure 43 is orthomodular.

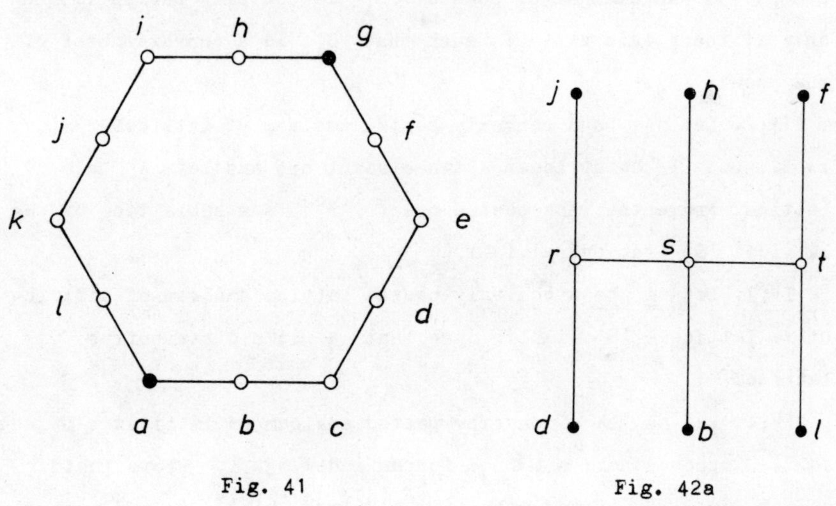

Fig. 41 Fig. 42a

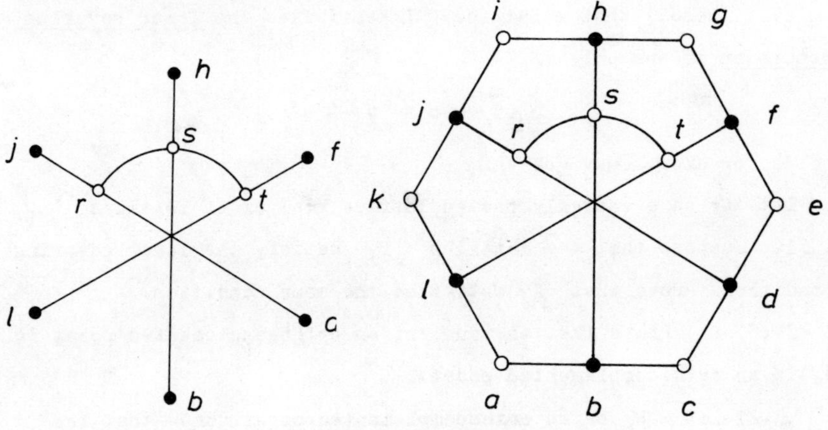

Fig. 42b Fig. 43

EXERCISES

IV;1. Discuss the validity of the following statement (compare Lemma 2.3): An amalgam of posets $`S_i$ is convexly pasted if and only if there exists $i \in I$ such that $S_i^o$ is a convex subset of the poset $`S_i$.

IV;2. Let $\underline{A}$ be a convexly pasted amalgam of lattices $`S_i$, $i \in I$. Let $I$ be at least a two-element set and let $\underline{A}$ be a lattice. Prove that the pasted poset $`S_i^o$ is a sublattice of the lattice $`S_i$ for every $i \in I$.

IV;3. Let $\underline{A}$ be a convexly pasted lattice amalgam of distributive lattices $`S_i$, $i \in I$. Show that $\underline{A}$ is a distributive lattice.

IV;4. Let $\underline{A}$ be a convexly pasted amalgam of lattices $`L_i$, $i \in I$. Suppose that $a, b \in L_j$ for an index $j \in I$. Prove that $a \prec b$ in $`L_j$ if and only if $[a^{(j)}] \prec [b^{(j)}]$ in the amalgam $\underline{A}$.

IV;5. Recall that a lattice $`L$ satisfies the <u>lower covering condition</u> if and only if

$$u \wedge v \prec u \Rightarrow v \prec u \vee v$$

holds for every u, v of $`L$.

Let $\underline{A}$ be a convexly pasted lattice amalgam of lattices $`L_i$, $i \in I$. Suppose that the lattices $`L_i$ satisfy the lower covering condition. Prove that $\underline{A}$ satisfies the same condition.

IV;6. (1) Prove that the dual of an orthocomplemented poset is again an orthocomplemented poset.

(2) Let $`P$ be an orthocomplemented poset. Show that the following conditions are satisfied for any $s, t \in P$:

(A) If there exists $s \vee t$, then there exists also $s' \wedge t'$

and $(s \vee t)' = s' \wedge t'$.

(B) If there exists $s \wedge t$, then there exists $s' \vee t'$ and $(s \wedge t)' = s' \vee t'$.

IV;7. Let `P be an orthocomplemented poset such that the following is true for every x, y of P:

(i) The set $\{x, y\}$ has the supremum $x \vee y$ whenever $x \leqslant y'$.

(a) Show that the condition (i′) below is true in `P:

(i′) The set $\{x, y\}$ has the infimum $x \wedge y$ whenever $x \geqslant y'$.

(b) Prove that if `P is an orthocomplemented poset satisfying the condition (i), then its dual is an orthocomplemented poset which satisfies the condition (i).

IV;8. Let $`P = (P, \leqslant, ', 0, 1)$ be an orthocomplemented poset which satisfies the condition (i) of Exercise IV;7 for every $x, y$ of $P$. Show that the following conditions on `P are equivalent:

(1) the poset `P is orthomodular;

(2) $(s \geqslant t' \ \& \ s \wedge t = 0) \Rightarrow s = t'$;

(3) $(s \leqslant t \leqslant u) \Rightarrow s \vee (t' \wedge u) = (s \vee t') \wedge u$;

(4) $s \leqslant t \Rightarrow s \vee (s' \wedge t) = t$;

(5) $s \geqslant t \Rightarrow s \wedge (s \wedge t')' = t$;

(6) $(s \perp t \ \& \ s \vee t = 1) \Rightarrow s = t'$;

(7) $s \geqslant t \Rightarrow s \wedge (s' \vee t) = t$.

IV;9. Let M be a finite set which has an even number of elements. Let $S(M)$ be the set of all the subsets of M which have an even number of elements. Prove that $`S(M) = (S(M), \subset, ', \emptyset, M)$ is an orthocomplemented poset. (Here $': H \mapsto H'$ where H' denotes the complement of the set $H \in S(M)$ in the set M, i.e., $H' = M \setminus H$.)

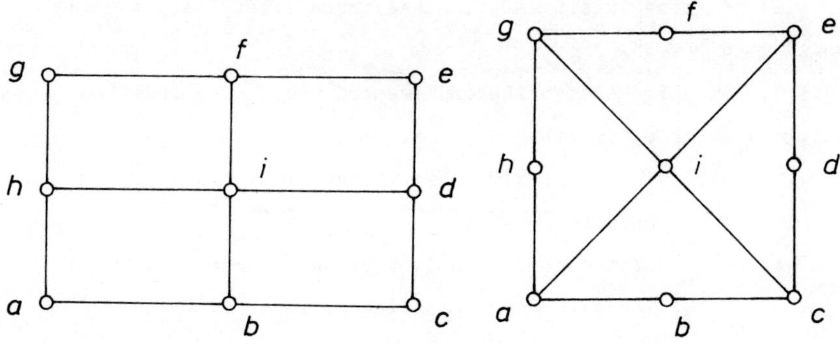

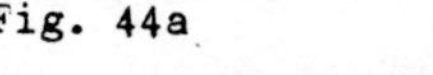
Fig. 44a

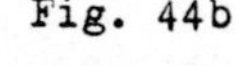
Fig. 44b

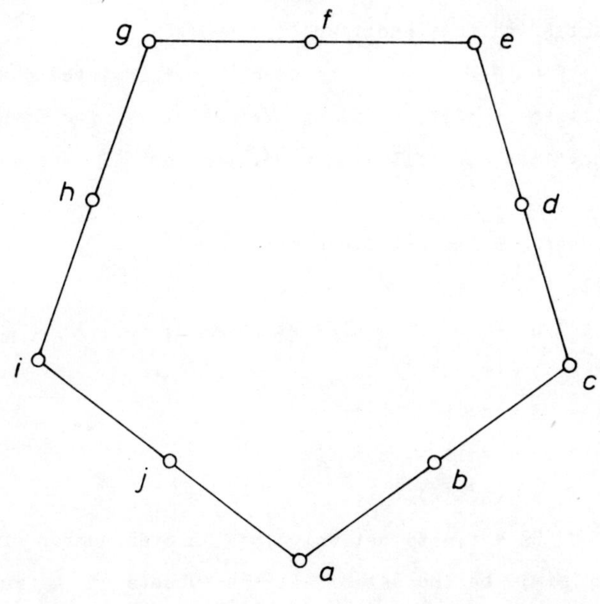

Fig. 44c

IV;10. Find the diagram of the poset $`S(M)$ defined in Exercise IV;9 in the case (a) $M = \{1,2\}$; (b) $M = \{1,2,3,4\}$; (c) $M =$

$= \{1,2,3,4,5,6\}$.

IV;11. Let M be a finite set having at least six elements. Prove that the poset `S(M) of Exercise IV;9 is not a lattice.

IV;12. Which of the Greechie's diagrams in Figures 44a-44c is an orthomodular poset (an orthomodular lattice)? Give the reasons for your answers.

GENERALIZED ORTHOMODULAR LATTICES

1. ORTHOGONALITY RELATION

Let `G = $(G, \vee, \wedge)$ be a lattice with the least element $0$. For any $a \in G$, define ${}^{P(a)}: [0,a] \to [0,a]$ to be a unary operation on $[0,a]$ such that ${}^{P(a)}: x \mapsto x^{P(a)}$. We shall say that `G is a generalized orthomodular lattice if and only if it satisfies the following conditions:

(G 1) The algebra $([0,a], \vee, \wedge, {}^{P(a)}, 0, a)$ is an orthomodular lattice for every $a \in G$.

(G 2) For any $x \leq a \leq b$ of G, $x^{P(a)} = x^{P(b)} \wedge a$.

The mapping P which assigns to each element $a \in G$ the unary operation ${}^{P(a)}: [0,a] \to [0,a]$ will be called the generalized orthogonality relation of `G.

Example 1.1. (A) Any orthomodular lattice is a generalized orthomodular lattice. By Corollary II.6.3, we namely see that $[0,a]$ is a base set of an orthomodular lattice, provided we define $x^{P(a)} = x' \wedge a$ for $x \in [0,a]$. The condition (G 2) is satisfied since $x^{P(b)} \wedge a = x' \wedge b \wedge a = x' \wedge a = x^{P(a)}$ for all $x \leq a \leq b$. From now on, therefore, we make the following convention: If $`G$ is an orthomodular lattice, then we always use $x^{P(a)}$ as an abbreviation for $x' \wedge a$.

(B) A generalized orthomodular lattice is an orthomodular lattice if and only if it has the greatest element.

(C) If $`G$ is a generalized orthomodular lattice and if I is an ideal of $(G, \vee, \wedge)$, then I is a generalized orthomodular lattice.

(D) Let $`L_i$ be an orthomodular lattice for every $i \in I \neq \emptyset$. Suppose that the set I is not finite and that no L_i has only one element. Let L consist of those sequences $\underline{a} = (a_i; i \in I)$ (with $a_i \in L_i$) having only a finite number of nonzero components $a_i \neq 0$. If $\underline{b} = (b_i; i \in I)$ is another sequence of L, we put $\underline{a} \vee \underline{b} = (a_i \vee b_i; i \in I)$, $\underline{a} \wedge \underline{b} = (a_i \wedge b_i; i \in I)$. The lattice $`L = (L, \vee, \wedge)$ is usually called the <u>discrete direct product</u> of the lattices $`L_i$. For later purposes note that if we modify this definition in accordance with I.16 by deleting the words "having only a finite number of nonzero components $a_i \neq 0$", we get the notion of the (<u>complete</u>, <u>unrestrected</u>) <u>direct product</u> $\prod(`L_i; i \in I)$ of these lattices. The least element of $`L$ is the null sequence $\underline{0} = (0; i \in I)$. Here $\underline{a} \leq \underline{b}$ if and only if $a_i \leq b_i$ for every $i \in I$. Using the notation of (A), we define $\underline{x}^{P(\underline{a})} = (x_i^{P(a_i)}; i \in I)$ where $\underline{x} \leq \underline{a}$ and $\underline{a} = (a_i; i \in I)$, $\underline{x} = (x_i; i \in I)$. Clearly, $\underline{x}^{P(\underline{a})} \in L$. Given $\underline{a} \in L$, the interval $[\underline{0}, \underline{a}]$ is a base set of an orthomodular lattice which we can describe as a direct product of a finite numeber of orthomodular lattices (cf. Remark II.3.3.E). Making use of (G 2), for $\underline{x} \leq \underline{a} \leq \underline{b}$, we find that

$$\underline{x}^{P(\underline{b})} \wedge \underline{a} = (x_i^{P(b_i)} \wedge a_i;\ i \in I) = (x_i^{P(a_i)};\ i \in I) = \underline{x}^{P(\underline{a})}.$$

Thus (G 2) holds also in $`L$. Evidently, no element of $`L$ plays the role of the greatest element in $`L$. In view of the earlier remarks it is now immediate that $`L$ is a generalized orthomodular lattice but it is not an orthomodular lattice.

Given a generalized orthomodular lattice $`G$ and a,b of G, we shall write $a \perp b$ if and only if $a \leqslant b^{P(a \vee b)}$.

Remark 1.2. If $`G$ is an orthomodular lattice, then $a \perp b$ if and only if $a \leqslant b'$, that is, if and only if a,b are orthogonal in the sense of the definition given in Section II.1. Indeed, the relation $a \perp b$ holds if and only if $a \leqslant b^{P(a \vee b)} = b' \wedge (a \vee b)$ which is equivalent to $a \leqslant b'$.

Lemma 1.3. Let $`G$ be a generalized orthomodular lattice. Then the following is true:

(i) $x \leqslant c \leqslant d \Rightarrow x^{P(c)} \leqslant x^{P(d)} \leqslant d$;

(ii) $x \leqslant y \leqslant c \Rightarrow y^{P(c)} \leqslant x^{P(c)}$;

(iii) $x \leqslant c \leqslant d \Rightarrow (x^{P(c)})^{P(d)} = x \vee c^{P(d)}$.

Proof. Ad (i). Here $x^{P(c)} = x^{P(d)} \wedge c \leqslant x^{P(d)}$.

Ad (ii). The assertion follows from (G 1).

Ad (iii). It is obvious that

$$(x^{P(c)})^{P(d)} = (x^{P(d)} \wedge c)^{P(d)} = (x^{P(d)})^{P(d)} \vee c^{P(d)} = x \vee c^{P(d)}. \ //$$

Lemma 1.4. For any generalized orthomodular lattice $`G$, the following conditions are equivalent:

(i) $a \perp b$;

(ii) $a \leqslant b^{P(c)}$ for any $c \in G$ such that $a \vee b \leqslant c$;

(iii) there exists $c \in G$ with $a \vee b \leqslant c$ and $a \leqslant b^{P(c)}$.

Proof. Ad (i) $\Rightarrow$ (ii). Let $a \perp b$ and let c be such that $a \vee b \leqslant c$. By Theorem 3 (i), $a \leqslant b^{P(a \vee b)} \leqslant b^{P(c)}$.

Ad (ii) $\Rightarrow$ (iii). Trivial.

Ad (iii) $\Rightarrow$ (i). If $a \leqslant b^{P(c)}$, then, a fortiori, $a \leqslant b^{P(c)} \wedge (a \vee b) = b^{P(a \vee b)}$ and so $a \perp b$. //

<u>Lemma 1.5</u>. Let a, b, c be elements of a generalized orthomodular lattice G. Then:

(i) $a \perp a \Leftrightarrow a = 0$;

(ii) $(\forall k \in G \;\; a \perp k) \Leftrightarrow a = 0$;

(iii) $a \perp b \Leftrightarrow b \perp a$;

(iv) $(a \perp b \;\&\; c \leqslant a) \Rightarrow c \perp b$;

(v) $(a \perp b \;\&\; a \perp c) \Leftrightarrow a \perp b \vee c$;

(vi) for every $a \leqslant b$ there exists c such that $b = a \vee c$ and $c \perp a$;

(vii) $a \leqslant b \Leftrightarrow [(\forall c \in G)\,(c \geqslant b \Rightarrow a \perp b^{P(c)})] \Leftrightarrow$
$\Leftrightarrow \{(\exists c \in G)\,[(c \geqslant a \vee b) \;\&\; (a \perp b^{P(c)})]\}$.

(viii) $(\forall c \in G)\,(c \geqslant a \Rightarrow a \perp a^{P(c)})$;

(ix) $x \leqslant y \Leftrightarrow (\forall z \in G)\,(z \perp y \Rightarrow z \perp x)$.

<u>Proof</u>. Ad (i). If $a \perp a$, then $a \leqslant a^{P(a)}$. Hence $a \leqslant a^{P(a)} \wedge a = 0$.

Ad (ii). The assertion follows from (i).

Ad (iii). Suppose that $a \leqslant b^{P(a \vee b)}$. By Lemma 3 (ii), $b = (b^{P(a \vee b)})^{P(a \vee b)} \leqslant a^{P(a \vee b)}$ so that $b \perp a$.

Ad (iv). Note that $c \leqslant a \leqslant b^{P(a \vee b)} = b^{P(a \vee b \vee c)}$. By Lemma 4 (iii) this implies that $c \perp b$.

Ad (v). Let $a \leqslant b^{P(a \vee b)}$ and $a \leqslant c^{P(a \vee c)}$. Then, by Lemma 3 (i), we have $a \leqslant b^{P(a \vee b \vee c)}$, $a \leqslant c^{P(a \vee b \vee c)}$ if and only if

$$a \leqslant b^{P(a \vee b \vee c)} \wedge c^{P(a \vee b \vee c)} = (b \vee c)^{P(a \vee b \vee c)}.$$

From Lemma 4 (iii) it is now clear that $a \perp b \vee c$ is equivalent to the assertion that $a \perp b$ and $a \perp c$.

Ad (vi). Put $c = a^{P(b)}$. By (G 1), $a \vee c = a \vee a^{P(b)} = b$. Now $c = a^{P(b)} = a^{P(c \vee a)}$. Thus $c \perp a$.

Ad (vii). If $a \leqslant b \leqslant c$, then $b^{P(c)} \leqslant a^{P(c)}$ and so $a \perp \perp b^{P(c)}$, by Lemma 4 and the symmetry of $\perp$.

If $c \geqslant a \vee b$ and $a \perp b^{P(c)}$, then $a \vee b^{P(c)} \leqslant a \vee c = c$ and this yields $a \leqslant (b^{P(c)})^{P(c)} = b$, by Lemma 4.

Ad (viii). Note that $a \leqslant a$ and use (vii).

Ad (ix). Suppose $x \leqslant y$ and $z \perp y$. By Lemma 4, $y \leqslant z^{P(c)}$ where $c \geqslant y$. From Lemma 4, (iii) and $x \leqslant z^{P(c)}$ it follows that $z \perp x$.

Conversely, assume the implication is true and put $z = = y^{P(c)}$ with $c \geqslant x \vee y$. By assumption, (viii) and (iii), $y^{P(c)} \perp \perp x$. It follows from (iii) and (vii) that $x \leqslant y$. //

Let M be a subset of the base set of a generalized orthomodular lattice `G. Define

$$M^{\perp} = \{x \in G;\ y \in M \Rightarrow x \perp y\}.$$

Remark 1.6. Let `G be an orthomodular lattice and let $x \in G$. Write as usual $(x] = \{y \in G;\ y \leqslant x\}$. Then $(x]^{\perp} = (x']$. In fact, $z \in (x]^{\perp}$ if and only if $z \perp y$ for every $y \in (x]$. This is equivalent to the assertion that $z \leqslant y'$ for every $y \in (x]$. Since $x' \leqslant y'$ for every $y \in (x]$, we see that the considered assertion is equivalent to $z \leqslant x'$. Hence $z \in (x]^{\perp}$ if and only if $z \leqslant x'$.

Lemma 1.7. Let `G be a generalized orthomodular lattice.

(A) The set $M^{\perp}$ is an ideal of the lattice $(G, \vee, \wedge)$.

(B) The mapping g of the set of all the subsets of G into the same set defined by $g: M \mapsto M^{\perp}$ determines a Galois connection on this set, that is,

(i) $M \subset M^{\perp\perp}$;

(ii) $M \subset N \Rightarrow N^{\perp} \subset M^{\perp}$.

Proof. Ad A. If $x, z \in M^{\perp}$, then $x \perp y$ and $z \perp y$ for every $y \in M$. By Lemma 5 (iii) and (v), we therefore have $x \vee z \perp y$ and so $x \vee z \in M^{\perp}$. If $u \leqslant x \in M^{\perp}$, then $x \perp y$ for every $y \in M$. From Lemma 5 (iv) it follows $u \perp y$ so that $u \in M^{\perp}$.

Ad B. (i) Let $y \in M$. By the definition of $M^{\perp}$, $y \perp x$ for every $x \in M^{\perp}$ and, therefore, $y \in M^{\perp\perp}$.

(ii) If $y \in M \subset N$ and $w \in N^{\perp}$, then $w \perp v$ for every $v \in N$. Hence $w \perp y$ and $w \in M^{\perp}$. Thus $N^{\perp} \subset M^{\perp}$. //

Lemma 1.8. Let `G be a generalized orthomodular lattice. Then:

(i) $t \perp x \Leftrightarrow t \in (x]^{\perp}$;

(ii) $t \in (x]^{\perp} \Leftrightarrow \exists X \in G :: x \leqslant X \ \& \ t \leqslant x^{P(X)}$;

(iii) $(x]^{\perp} = \bigcup \{(x^{P(X)}];\ X \in [x)\}$;

(iv) for any $X_0 \geqslant x$, $(x]^{\perp} = \bigcup \{(x^{P(X)}];\ X \geqslant X_0\}$.

Proof. Ad (i). This is straightforward.

Ad (ii). By (i), the relation $t \in (x]^{\perp}$ is equivalent to $t \perp x$. If there exists $X \in G$ with the mentioned property, then $t \leqslant x^{P(X)} \leqslant X$ so that $x \vee t \leqslant X$. By Lemma 4 (iii), we get $t \perp x$. Conversely, if $t \perp x$, then take $X = x \vee t$ and apply Lemma 4 (iii).

Ad (iii). Denote by L and R the left and the right side of (iii), respectively. By (ii), $L \subset R$. Let $z \in R$. Then $z \leqslant x^{P(Y)} \leqslant Y$ for an element Y such that $x \leqslant Y$. Since $z \vee x \leqslant Y$, $z \perp x$ by Lemma 4 (iii) and so $z \in (x]^{\perp}$. Thus $R \subset L$.

Ad (iv). Let $F(x_1) = \bigcup \{(x^{P(X)}];\ X \geqslant x_1\}$ whenever $x_1 \geqslant x$. By (iii), $(x]^{\perp} = F(x) \supset F(X_0)$. Let $t \perp x$ and $X_1 \geqslant X_0 \vee t$. Lemma 4 yields $t \leqslant x^{P(X_1)}$ and so $(x]^{\perp} \subset F(X_0)$. Thus $(x]^{\perp} = F(X_0)$. //

Lemma 1.9. In any generalized orthomodular lattice $`G$, $(x]^{\perp} = (y]$ if and only if there exists $A \in G$ such that $x \leqslant A$ and such that $x^{P(A)} = x^{P(B)}$ for every $B \geqslant A$. In this notation, $y = x^{P(A)}$.

Proof. 1. Let $(x]^{\perp} = (y]$ and let U denote the union of Lemma 8 (iii). Then $U = (y]$. However, this means that there exists A with $y \in (x^{P(A)}]$ and $x^{P(A)} \leqslant y$. Hence $y = x^{P(A)}$. If $B \geqslant A$, then $y = x^{P(A)} \leqslant x^{P(B)}$ by Lemma 3 (i). But $(y] = U \supset (x^{P(B)}]$. Thus $x^{P(B)} \leqslant y$ and $y = x^{P(A)} = x^{P(B)}$.

2. By Lemma 8 (i) and 4, $t \leqslant x^{P(A)}$ $(A \geqslant x)$ if and only if $t \in (x]^{\perp}$. Thus $(x]^{\perp} = (x^{P(A)}]$. //

Theorem 1.10. Let $`G$ be a generalized orthomodular lattice. Then the following conditions are equivalent:

(i) $`G$ is an orthomodular lattice;

(ii) there exist $x, y \in G$ such that $(x]^{\perp} = (y]$;

(iii) there exist elements $x, y \in G$ such that $(x]^{\perp} \subset (y]$.

If (ii) is valid, then y is the orthocomplement x' of the element x.

Proof. Ad (i) ⇒ (ii). By Remark 6, $(x]^{\perp} = (x']$ and so $y = x'$.

Ad (ii) ⇒ (iii). Trivial.

Ad (iii) ⇒ (i). Suppose $(x]^{\perp} \subset (y]$. By Lemma 8 (iii), $y \geqslant x^{P(X)}$ for every $X \geqslant x$. Hence $X = x \vee x^{P(X)} \leqslant x \vee y$. If t is an arbitrary element of G and if $X = x \vee t$, then $t \leqslant x \vee t \leqslant x \vee y$. Thus $x \vee y$ is the greatest element of the lattice $`G$. By Example 1.B, we therefore conclude that $`G$ is an orthomodular lattice. //

2. JANOWITZ'S EMBEDDING

Let $`G$ be a generalized orthomodular lattice. In what follows, we shall denote by D the set consisting of all the subsets I of G which are of the form $I = (x]$ or $I = (x]^{\perp}$ where x runs over G.

Theorem 2.1. For any $I \in D$, $I^{\perp\perp} = I$.

Proof. 1. Let $I = (x]$. By Lemma 1.7.B (i), we get $I \subset I^{\perp\perp}$. Clearly, $y \in (x]^{\perp\perp}$ if and only if $y \perp z$ for every $z \in (x]^{\perp}$. But $z \in (x]^{\perp}$ if and only if $z \perp x$, by Lemma 1.8 (i). Hence $z \perp x$ implies $z \perp y$. By Lemma 1.5 (ix), $y \leqslant x$, that is to say $y \in$ $\in (x]$. Thus $I^{\perp\perp} \subset I$ and, by what we have already proved above, $I = I^{\perp\perp}$.

2. Let $I = (x]^{\perp}$. Then $I^{\perp\perp} = ((x]^{\perp\perp})^{\perp}$ and, by 1, $(x]^{\perp} =$ $= (x]^{\perp\perp\perp}$ which yields $I^{\perp\perp} = I$. //

Recall (cf. [179]) that the ideals of a lattice $`L$ form a lattice $`I(`L) = (I(`L), \vee, \wedge)$ where for $I, J \in I(`L)$

$$I \vee J = \{x \in L; \exists i \in I \;\exists j \in J :: x \leqslant i \vee j\}, \qquad I \wedge J = I \cap J.$$

If $`L$ has the least element, then the lattice $`I(`L)$ is complete by I.13. The lattice $`I(`L)$ will be called the ideal lattice of $`L$.

Theorem 2.2. The set D is the base set of a sublattice $`D$ od the lattice $`I(`G)$ where $`G$ denotes a generalized orthomodular lattice. In addition, $I \wedge J = (I^{\perp} \vee J^{\perp})^{\perp} = I \cap J$ for $I, J \in D$.

Proof. 1. First we shall prove that D is closed under the operation $\vee$.

Case I: $I = (x]$ and $J = (y]$. Then $I \vee J = (x] \vee (y] =$ $= (x \vee y] \in D$.

Case II: $I = (x]^{\perp}$ and $J = (y]$. We shall show that $I \vee J = S$ where $S = (x \wedge y^{P(x \vee y)}]^{\perp}$. Since $(x \wedge y^{P(x \vee y)}] \subset (x]$, $(x]^{\perp} \subset S$ by Lemma 1.7. By Lemma 1.5 (viii), $y \perp y^{P(x \vee y)}$ and, by the same assertion, $(y] \subset (y^{P(x \vee y)}]^{\perp}$. However, $(x \wedge y^{P(x \vee y)}] \subset (y^{P(x \vee y)}]$. Consequently, $(y] \subset (y^{P(x \vee y)}]^{\perp} \subset S$. Thus S is an upper bound of $\{I, J\}$ in $`I(`G)$ and, therefore, $S \supset I \vee J$.

To prove the converse inclusion, choose $t \in S$. Then $t \perp x \wedge y^{P(x \vee y)}$. Setting $T = t \vee x \vee y$ and using Lemma 1.4 (ii) and Lemma 1.3 (iii), we obtain

$$t \leqslant (x \wedge y^{P(x \vee y)})^{P(T)} = x^{P(T)} \vee (y^{P(x \vee y)})^{P(T)} = x^{P(T)} \vee y \vee (x \vee y)^{P(T)}.$$

By Lemma 1.3 (ii), $(x \vee y)^{P(T)} \leqslant x^{P(T)}$. This yields $t \leqslant x^{P(T)} \vee y$. From Lemma 1.8 we see that $t \in I \vee J$.

Case III: $I = (x]^{\perp}$ and $J = (y]^{\perp}$. We shall prove that $I \vee J = (x \wedge y]^{\perp}$. By Lemma 1.7 we first note that $(x]^{\perp}, (y]^{\perp} \subset (x \wedge y]^{\perp}$. Hence $I \vee J \subset (x \wedge y]^{\perp}$. To prove the converse inclusion, consider $t \in (x \wedge y]^{\perp}$. By Lemma 1.4 (ii), $t \leqslant (x \wedge y)^{P(T)}$ where $t \vee x \vee y \leqslant T$. Therefore, $t \leqslant x^{P(T)} \vee y^{P(T)}$. By Lemma 1.8 (iii) and by the remark preceding Theorem 2, it follows that $t \in I \vee J$. Thus $I \vee J = (x \wedge y]^{\perp} \in D$.

2. In this step we shall verify that the set D is closed under $\wedge$.

Evidently, by Theorem 1 and by the first part of this proof, $K = (I^{\perp} \vee J^{\perp})^{\perp} \in D$ whenever $I, J \in D$.

Since $I^{\perp} \subset I^{\perp} \vee J^{\perp}$, we have $I = I^{\perp\perp} \supset K$ by Lemma 1.7.B (ii) and Theorem 1. Similarly, $J \supset K$. Thus K is a lower bound of the set $M = \{I, J\}$.

Next suppose E is an ideal of $I(`G)$ such that $I \supset E$ and $J \supset E$. By Lemma 1.7.B (ii), $I^{\perp} \subset E^{\perp}$, $J^{\perp} \subset E^{\perp}$. Hence $I^{\perp} \vee J^{\perp} \subset E^{\perp}$

and so $K = (I^{\perp} \vee J^{\perp})^{\perp} \supset E^{\perp\perp} \supset E$. Now it is immediate that $I \wedge J = (I^{\perp} \vee J^{\perp})^{\perp}$.

Since we know that the meet $I \wedge J$ is equal to $I \cap J$ in the lattice $`I(`G)$, the asserted identity of the theorem is clearly verified. //

The lattice $`D$ will be called the <u>Janowitz's hull</u> of the generalized orthomodular lattice $`G$ and at times it will be denoted more specifically by $`D(`G)$.

<u>Theorem 2.3 (cf. [102])</u>. The Janowitz's hull of a generalized orthomodular lattice is an orthomodular lattice.

<u>Proof</u>. 1. First observe that $(D, \vee, \wedge, ^{\perp}, (0], G)$ is an ortholattice. Indeed, by Theorem 1 and Lemma 1.7 we have $I^{\perp\perp} = I$ and $I^{\perp} \supset J^{\perp}$ whenever $I \subset J$ and $I, J \in D$. If $x \in I \wedge I^{\perp} = I \cap I^{\perp}$, then $x \perp x$ and so $x = 0$ by Lemma 1.5. Hence $I \wedge I^{\perp} = \{0\} = (0]$. By Theorems 2 and 1, $G^{\perp} = \{0\} = I \wedge I^{\perp} = (I^{\perp} \vee I^{\perp\perp})^{\perp} = (I^{\perp} \vee I)^{\perp}$ and, therefore, $G = G^{\perp\perp} = (I^{\perp} \vee I)^{\perp\perp} = I^{\perp} \vee I$.

2. Let $S, T \in D$ be such that $S \wedge T = (0]$ and $S \supset T^{\perp}$. We wish to prove that $S = T^{\perp}$.

<u>Case I</u>: $S = (x]$, $T = (y]$. Since $S = (x] \supset (y]^{\perp} = T^{\perp}$, $`G$ is an orthomodular lattice by Theorem 1.10.

<u>Case II</u>: $S = (x]^{\perp}$, $T = (y]^{\perp}$. By 1 and Theorem 1, $(0] = S \wedge T = (S^{\perp} \vee T^{\perp})^{\perp} = ((x] \vee (y])^{\perp} = (x \vee y]^{\perp}$. Again, $`G$ is an orthomodular lattice.

From Remark 1.6 we conclude that in the both cases, $S = (u]$ and $T = (v]$ for some $u, v \in G$. By assumption, $(0] = S \wedge T = (u \wedge v]$ and $S = (u] \supset T^{\perp} = (v']$. This yields $u \wedge v = 0$ and $u \geqslant v'$. Thus, by orthomodularity, $u = v'$ and so $S = (u] = T^{\perp} = (v']$.

<u>Case III</u>: $S = (x]$, $T = (y]^{\perp}$. In this case $(x] \wedge (y]^{\perp} = (0]$

and $(x]\supset(y]^{\perp\perp}=(y]$. Hence $y\leqslant x$ and, by Lemma 1.8,

$$(0]=(x]\wedge(y]^{\perp}=(x]\wedge\bigcup(y^{P(Y)}]\supset(x]\wedge(y^{P(x)}]=(x\wedge y^{P(x)}].$$

Since $y^{P(x)}\leqslant x$ by Lemma 1.3 (i), $0=x\wedge y^{P(x)}=y^{P(x)}$. From this we further find $x=y\vee y^{P(x)}=y\vee 0=y$. Thus $S=(x]=$ $=(y]=T^{\perp}$.

Case IV: $S=(x]^{\perp}$, $T=(y]$. This case is almost the same as Case III; we need only replace x by y and vice versa. //

Recall that $P\neq\emptyset$ is said to be a prime ideal of a lattice `L if and only if $P$ is an ideal of the lattice `L and the following implication holds for all elements a,b of L:

$$(a\wedge b\in P\ \&\ a\notin P)\Rightarrow b\in P.$$

Theorem 2.4. Let `G be a generalized orthomodular lattice and let $H=\{I\in D;\exists\, x\in G :: I=(x]\}$. Then $H$ is a prime ideal of the Janowitz's hull `D(`G).

Proof. 1. If $I,J\in H$, $I=(x]$, $J=(y]$, then $I\vee J=(x]\vee$ $\vee(y]=(x\vee y]\in H$. If $K\in D$ and $K\subset I\in H$, then either $K=(z]$ or $K=(u]^{\perp}$. In the former case, we evidently get $K\in H$; in the latter we find $(u]^{\perp}\subset I=(x]$. From Theorem 1.10 we infer that G is the base set of an orthomodular lattice. This yields $(u]^{\perp}=(u']\in H$. Thus H is an ideal in `D.

2. We shall prove that H is a prime ideal in `D. Given $I=(x]^{\perp}$, $J=(y]^{\perp}$, we have $I\wedge J=(x]^{\perp}\wedge(y]^{\perp}=((x]\vee(y])^{\perp}=$ $=(x\vee y]^{\perp}$, by Theorem 2. By Theorem 1.10, this means that $I\wedge J$ belongs to $H$ if and only if `G is an orthomodular lattice. However, then $I=(x']\in H$. Thus H is prime. //

Theorem 2.5. For any generalized orthomodular lattice `G there exists lattice-monomorphism $f$ of the lattice $(G,\vee,\wedge)$ into the orthomodular lattice $`D(`G)=(D(`G),\vee,\wedge)$ so that the image of G under f is a prime ideal of the lattice $`D(`G)$.

<u>Proof</u>. Let f be the mapping of G into $D(`G)$ which is defined by $f: x \mapsto (x]$. Then f is injective, $f(x \vee y) = f(x) \vee f(y)$ and $f(x \wedge y) = f(x) \wedge f(y)$ for any x,y of G. //

The mapping f of this proof is called the <u>Janowitz's embedding</u>.

<u>Remark 2.6</u>. (A) Let $`G$ be a generalized orthomodular lattice and let $f: `G \to `D(`G)$ be the corresponding Janowitz's embedding. In what follows, we identify x and $f(x) = (x]$ for every $x \in G$. In this way we can consider $`G$ to be a sublattice of $`D = `D(`G)$. Following the convention made above, we write $g = (g]$ for any $g \in G$. We know that for every $I \in D$ either $I = (x]$ or $I = (x]^{\perp}$ where x is an element of G. Hence, in particular, $0 = (0]$. Since the set G plays the role of the greatest element in $`D$, we shall denote it by 1. Note that the orthocomplement of $x = (x]$ in $`D$ is $(x]^{\perp}$. As usual, denote the orthocomplement of x by x'. Under these convention we have $x' = (x]^{\perp}$. Now it is clear that D consists of all the elements y which can be written either in the form $y = x$ or in the form $y = x'$ where $x \in G$. At the same time, $`D = (D, \vee, \wedge, ', 0, 1)$. In the following, we shall use all these conventions on the Janowitz's hull.

(B) Suppose $g, h \in G$, $g \leqslant h$. Let f be the Janowitz's embedding. Then $f(g^{P(h)}) = (g]^{\perp} \cap (h]$. Indeed, observe that $(g]^{\perp} = \{(g^{P(Q)}]; Q \geqslant h\}$ by Lemma 1.8 (iv). Hence

$$(g]^{\perp} \cap (h] = (\bigcup\{(g^{P(Q)}]; Q \geqslant h\}) \cap (h] = \bigcup\{(g^{P(Q)}] \cap (h]; Q \geqslant h\}.$$

However,

$$(g^{P(Q)}] \cap (h] = (g^{P(Q)} \wedge h] = (g^{P(h)}].$$

From this it follows immediately that $(g]^{\perp} \cap (h] = (g^{P(h)}]$. The above-mentioned convention gives $g^{P(h)} = (g]^{\perp} \cap (h] = (g]^{\perp} \wedge (h] = g' \wedge h$.

3. CONGRUENCE RELATIONS

Suppose `L is a given lattice. Recall that a binary relation T is called a congruence (or, more specifically, a lattice congruence relation) of `L if and only if T is a relation of equivalence on L such that the following is true for any $a, b, c \in L$: If $(a,b) \in T$ and $c \in L$, then also $(a \wedge c, b \wedge c)$ and $(a \vee c, b \vee c)$ belong to T. If T is a congruence relation and $(a,b) \in T$, $(c,d) \in T$, then $(a \wedge c, b \wedge c) \in T$ and $(b \wedge c, b \wedge d) \in T$. Hence $(a \wedge c, b \wedge d) \in T$ and, similarly, $(a \vee c, b \vee d) \in T$.

A <u>congruence relation of an ortholattice</u> $(L, \vee, \wedge, ', 0, 1)$ is any lattice congruence relation F of $(L, \vee, \wedge)$ which, moreover, satisfies the following implication for every a,b of L: If $(a,b) \in F$, then $(a', b') \in F$.

<u>Remark 3.1</u>. (A) By a congruence relation of an orthomodular lattice `L = $(L, \vee, \wedge, ', 0, 1)$ we of course mean any congruence relation of `L considered as an ortholattice.

(B) Note that there exist ortholattices with lattice congruence relations which are not congruence relations of the considered ortholattices. This is illustrated by the following example: In the lattice `L of Figure 9e, let $T = \{(a,a),(a,b),(b,a),(b,b),(a',a'),(b',b'),(0,0),(1,1)\}$. Then T is a lattice congruence relation of `L but it is not a congruence relation of the corresponding ortholattice.

(C) The set of all the congruence relations of a lattice, partially ordered under set inclusion, is a complete lattice. The same assertion holds for the set of all the congruence relations of an ortholattice.

Theorem 3.2. If $`L = (L, \vee, \wedge, ', 0, 1)$ is an orthomodular lattice, then any lattice congruence relation of $(L, \vee, \wedge)$ is also a congruence relation of the orthomodular lattice $`L$.

Proof. Let T be a lattice congruence relation. If $(a,b) \in T$ and $a \leqslant b$, then $(0, b \wedge a') = (a \wedge a', b \wedge a') \in T$. Hence $(b', b' \vee (b \wedge a')) \in T$. By orthomodularity and by the fact $b' \leqslant a'$, $a' = b' \vee (b \wedge a')$. Consequently, $(b', a') \in T$ and, since T is a symmetric relation, $(a', b') \in T$. Now choose $c, d \in L$ arbitrarily and suppose $(c,d) \in T$. It is well-known that for any lattice congruence T we have $(c,d) \in T$ if and only if $(c \wedge d, c \vee d) \in T$. Using the assertion we have already proved and putting $a = c \wedge d \leqslant c \vee d = b$, we obtain $((c \wedge d)', (c \vee d)') \in T$. Therefore, by De Morgan laws, $(c' \wedge d', c' \vee d') \in T$. By the remark above we conclude that $(c', d') \in T$. //

We note the following very useful theorem:

Theorem 3.3 (cf. G. Grätzer - E. T. Schmidt, [72]). Let T be a reflexive and symmetric relation on the base set L of a lattice $`L$. Then $T$ is a congruence relation of the lattice $`L$ if and only if it satisfies the following conditions:

(i) $(a,b) \in T \Leftrightarrow (a \wedge b, a \vee b) \in T$;

(ii) $[a \leqslant b \leqslant c \;\&\; (a,b) \in T \;\&\; (b,c) \in T] \Rightarrow (a,c) \in T$;

(iii) $[a \leqslant b \;\&\; (a,b) \in T] \Rightarrow [\forall c \in L \; (a \wedge c, b \wedge c) \in T \;\&\; (a \vee c, b \vee c) \in T]$.

For the proof see Exercise V;2. //

Recall the following conventions: Given two elements $a \leqslant b$

of a lattice `L, we write b/a instead of the ordered pair (a,b) and b/a will usually be called a quotient of `L. If, in particular, $a \prec b$, then b/a is called a prime quotient of `L. The set of all the prime quotients of `L will be denoted by P(`L). Two prime quotients b/a,d/c are said to be P-projective, and we write b/aPd/c, if and only if there exists $n \in \underline{N}$ and prime quotients b_i/a_i, $i = 1,2,\dots,n$, such that b/a = $= b_1/a_1$, $d/c = b_n/a_n$ and $b_i/a_i \sim b_{i+1}/a_{i+1}$ (see I.15) for every $i = 1,2,\dots,n-1$. The relation P we have just defined on the set P(`L) is obviously an equivalence relation on P(`L). The equivalence class of P determined by a prime quotient q/p will be denoted by [q/p].

We are now in a position to define the class of submodular lattices (cf. [180]).

A lattice `L is said to be submodular if and only if it satisfies the following dual conditions:

(S^*) If $u \wedge v \prec u$ and $v \leqslant p \prec q \leqslant u \vee v$, then $q/pPu/u \wedge v$.

(S^+) If $u \prec u \vee v$ and $u \wedge v \leqslant p \prec q \leqslant v$, then $q/pPu \vee v/u$.

Let $\tilde{c}: a = a_0 \prec a_1 \prec \dots \prec a_n = b$ be a maximal chain between a and b. Recall that n is called the length of this chain. We shall denote it by $l(\tilde{c})$, i.e., $n = l(\tilde{c})$.

We say that a lattice `L is an FC-lattice if and only if all the chains of the form $a = a_0 < a_1 < \dots$ and all the chains of the form $a = b_0 > b_1 > \dots$ are finite for every $a, a_i, b_j \in L$.

Let $\tilde{c}$ be a maximal chain, $\tilde{c}: a = a_0 \prec a_1 \prec \dots \prec a_n = b$. The symbol $P(\tilde{c})$ will denote the set of all the equivalence classes determined by the prime quotients of $\tilde{c}$, i.e.,

$$P(\tilde{c}) = \{[q/p];\ \exists i \quad 1 \leqslant i \leqslant n-1 :: q/pPa_{i+1}/a_i\}.$$

Example 3.4. In the lattice of Figure 45 we find

$$c/0 \nearrow a'/b \nearrow 1/c' \searrow d'/e, \quad b'/a \nearrow 1/c' \searrow e'/d.$$

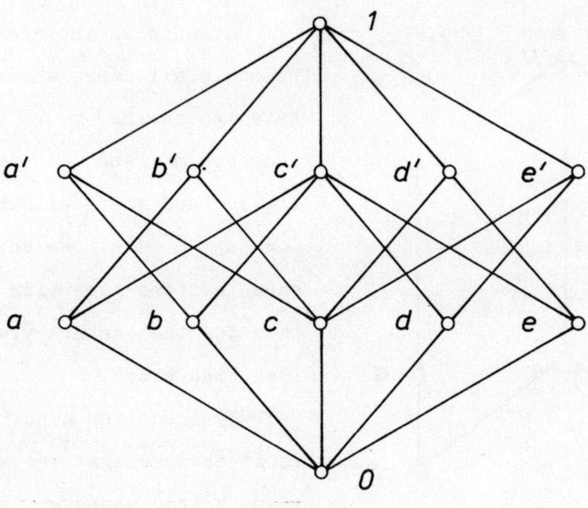

Fig. 45

Hence $[c/0] = \{c/0, b'/a, a'/b, d'/e, e'/d, 1/c'\}$. Since

$$c'/d \searrow a/0 \nearrow c'/e \searrow b/0 \nearrow c'/a \searrow e/0 \nearrow c'/b \searrow d/0 \nearrow e'/c \nearrow$$
$$\nearrow 1/a' \searrow d'/c \nearrow 1/b' \searrow a'/c \nearrow 1/e' \searrow b'/c \nearrow 1/d',$$

we also have

$$[a/0] = \{a/0, b/0, d/0, e/0, c'/a, c'/b, c'/d, c'/e, a'/c, b'/c, d'/c, e'/c, 1/a', 1/b', 1/d', 1/e'\}.$$

Let $\tilde{v}$ denote the chain $0 \prec b \prec c'$ of this lattice. Then $P(\tilde{v}) = \{[a/0]\}$. If $\tilde{w}$ denotes the chain $0 \prec a \prec b' \prec 1$, we find $P(\tilde{w}) = \{[a/0], [c/0]\}$.

Theorem 3.5. Let `L be a submodular FC-lattice and let

$$\tilde{v}: a \prec c \prec d \prec \dots \prec e \prec b, \quad \tilde{w}: a \prec p \prec q \prec \dots \prec r \prec b.$$

Then $P(\tilde{v}) = P(\tilde{w})$.

Proof. 1. The assertion clearly holds whenever the length m of $\tilde{v}$ is equal to 1 or the length n of $\tilde{w}$ is equal to 1.

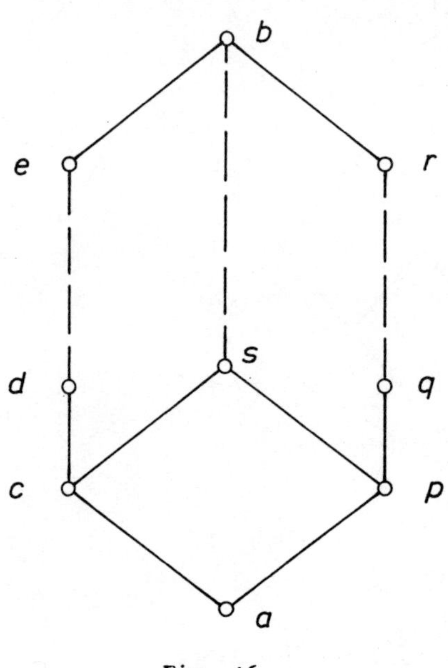

Fig. 46

2. Suppose the assertion holds in all cases where we have two chains $\tilde{v}', \tilde{w}'$ with lengths m' and n' respectively, and where either $m' < m$ or $n' < n$. We shall show that the assertion is true for the chains $\tilde{v}, \tilde{w}$ (see Figure 46).

By induction hypothesis it is obvious that the assertion is true whenever $p = c$. Next suppose that $p \neq c$. Consider the chains

$\tilde{w}_1': p \prec q \prec \ldots \prec r \prec b$,

$\tilde{w}_2': p \prec \ldots \prec s \prec \ldots \prec b$

where $s = c \vee p$. Here we can use the induction hypothesis, since $l(\tilde{w}_1') = l(\tilde{w}) - 1$. Hence

$$P(\tilde{w}_1') = P(\tilde{w}_2') = P(s \prec \ldots \prec b) \cup P(p \prec \ldots \prec s).$$

Let

$$\tilde{v}_1': c \prec d \prec \ldots \prec e \prec b, \quad \tilde{v}_2': c \prec s \prec \ldots \prec b.$$

Then $l(\tilde{v}_1') < l(\tilde{v})$ and we can apply the induction hypothesis to the chains $\tilde{v}_1', \tilde{v}_2'$. Therefore,

$$P(\tilde{v}_1') = P(\tilde{v}_2') = P(c \prec \ldots \prec s) \cup P(s \prec \ldots \prec b).$$

Moreover, by (S^*),

$P(p \prec \ldots \prec s) = P(a \prec c)$; $P(c \prec \ldots \prec s) = P(a \prec p)$.

Combining these facts, we obtain

$$P(\tilde{w}) = P(a \prec p) \cup P(\tilde{w}_1') =$$
$$= P(c \prec \ldots \prec s) \cup P(s \prec \ldots \prec b) \cup P(p \prec \ldots \prec s) =$$
$$= P(\tilde{v}_2') \cup P(p \prec \ldots \prec s) = P(\tilde{v}_1') \cup P(a \prec c) = P(\tilde{v}). \quad //$$

Theorem 5 justifies the following convention: Let $a < b$ be two elements of a submodular FC-lattice. We put $P(a,b) = P(\tilde{v})$ where $\tilde{v}$ is a maximal chain between a and b. If $a = b$, we set $P(a,b) = \emptyset$.

<u>Theorem 3.6</u>. Let $`L$ be a submodular FC-lattice. Suppose $y/x \nearrow h/g$ are two transposed quotients. Then $P(\tilde{v}) = P(\tilde{w})$ for every chain $\tilde{v}$ joining x to y and every chain $\tilde{w}$ joining g to h.

<u>Proof</u>. Let $b/a \nearrow d/c$ and let $a \prec e \leqslant c$. Observe that there exist $p \in [e,c]$ and $q \in [b,d]$ such that

(1) $b/a \nearrow q/e$ & $q/p \nearrow d/c$ & $c/p \nearrow d/q$ & $p/a \nearrow q/b$ & $P(e,q) \supset P(p,q)$.

Indeed, it suffices to put $q = e \vee b$, $p = q \wedge c$ and use Theorem 5.

We shall prove that

(2) $P(x,y) \supset P(g,h)$.

Clearly, by Theorem 5, we can suppose that y and g are incomparable. Now, $`L$ is an FC-lattice. Hence, by (1), we may assume that $x \prec g$.

Let $x \prec x_1 \prec \ldots \prec x_n = y$. Put $a = x$, $b = g$, $c = y$, $d = h$, $e = x_1$ and consider p,q of (1). Use induction on n. The case $n = 1$ is clear by (S^*).

<u>Case I</u>: $x_1 = p$. Then, by induction hypothesis, $P(x_1,y) \supset P(q,h)$ and from $x_1/x \nearrow q/g$ we see that $P(x,x_1) = P(g,q)$. Thus (2) holds.

Case II: $x_1 < p$. By (1), $g/x \nearrow q/x_1$ and $q/p \nearrow h/y$. Consequently, $P(x,y) \supset P(x_1,p) = P(y,h) = P(x,g)$. Hence, by Theorem 5, $P(q,h) \subset P(p,q) \cup P(q,h) = P(p,y) \cup P(y,h) \subset P(x,y)$ and, analogously as in Case I, (2) holds.

By duality, $P(x,y) \subset P(g,h)$ and this completes the proof. //

Theorem 3.7. Let `L be a submodular FC-lattice. Suppose B is a set consisting of some equivalence classes of P. Then there exists exactly one congruence relation T on `L such that the following is true for every $a,b \in L$: If $a \prec b$, then $(a,b) \in T$ if and only if $[b/a] \in B$.

Proof. 1. We shall prove that there exists at most one congruence relation T with the required property. Indeed, if such a congruence T exists, then, by Theorem 3, $(c,d) \in T$ holds if and only if $(c \wedge d, c \vee d) \in T$. Since `L is an FC-lattice, we have $c \wedge d = c_0 \prec c_1 \prec \ldots \prec c_{n-1} = c \vee d$ for some $n \in \underline{N}$. Therefore, $(c \wedge d, c \vee d) \in T$ if and only if $(c_i, c_{i+1}) \in T$ for every i. But it means that the property of T stated in the theorem completely determines whether (c,d) belongs to T or not.

2. Next we shall show the existence of such a congruence T by using Grätzer-Schmidt's Theorem 3. By the description of T we have just found in 1, T can be defined as follows:

$$(c,d) \in T \text{ if and only if } P(c \wedge d, c \vee d) \subset B.$$

The conditions (i) and (ii) of Theorem 3 are clearly satisfied.

Let $a \leqslant b$, $(a,b) \in T$ and $c \in L$. Then $c \wedge b/c \wedge a \nearrow [a \vee (c \wedge b)]/a$. Hence, by Theorem 6, $P(c \wedge a, c \wedge b) = P(a, a \vee (c \wedge b)) \subset P(a,b) \subset B$ and the condition (iii) also holds. //

Theorem 3.8. (i) Every relatively complemented FC-lattice is a submodular one.

(ii) Every orthomodular FC-lattice is submodular.

Proof. Ad (i). Let $a \prec b$, $c \leqslant y \prec d$ and let $b/a \nearrow d/c$. Then

(1) $b/a \nearrow d/y$.

Indeed, since $b < y$ does not hold, $a \leqslant b \wedge y < b$. Hence $b \wedge y = a$. At the same time, $d \geqslant b \vee y \geqslant b \vee c = d$.

Now suppose that $u \wedge v \prec u$ and $v \leqslant p \prec q \leqslant u \vee v$. Denote by r the relative complement of q in the interval $[p, u \vee v]$ and let s be such that $r \leqslant s \prec u \vee v$. By (1), $q/p \nearrow u \vee v/s$ and $u/u \wedge v \nearrow u \vee v/s$. Hence $q/pPu/u \wedge v$ and (S^{*}) holds. The condition (S^{+}) follows by duality.

Ad (ii). By Lemma II.6.2, every orthomodular lattice is relatively complemented. The remainder is a consequence of the statement (i). //

Remark 3.9. (A) By using Theorems 7 and 8 we can determine the congruence relations of orthomodular FC-lattices very easily. Such an approach is especially useful for the finite lattices. An illustrating example is furnished by the lattice of Figure 45 (see Example 4). Here we have four possibilities for the set B of Theorem 7: $B_1 = \emptyset$, $B_2 = \{[a/0]\}$, $B_3 = \{[c/0]\}$, $B_4 = \{[a/0], [c/0]\}$. It follows that this lattice has exactly four congruence relations. Let T_i denote the congruence relation which corresponds to B_i. Note that $T_i \subset T_j$ if and only if $B_i \subset B_j$. Therefore, the corresponding congruence lattice (that is to say the set of all congruence relations of the considered lattice partially ordered by the set inclusion) is isomorphic to the Boolean algebra of all the subsets of the set B_4. The same argument can be applied to any orthomodular FC-lattice. From this remark we can infer that the congruence lattice of any orthomodular FC-lattice `L is isomorphic to the Boolean algebra of all the subsets of the quotient set $P(`L)/P$ and, thus, it is a Boolean algebra.

(B) In general case we have the following situation: It is well-known [179, p. 182] that the congruence lattice of a lattice is distributive. Since the congruence lattice of an ortholattice $`L = (L, \vee, \wedge, ', 0, 1)$ is a sublattice in the lattice of congruence relations on $(L, \vee, \wedge)$, the congruence lattice of an ortholattice is distributive. A fortiori, the same is true for orthomodular lattices. However, the congruence lattice of an orthomodular lattice $`L$ need not be a Boolean lattice even if $`L$ is a Boolean algebra. Indeed, consider the direct product $`B = \prod(`L_i;\ i \in \underline{N})$ (see I.25) where $`L_i = \underline{2}$ for every $i \in \underline{N}$. Let I consist of all the sequences $\underline{a} = (a_i;\ i \in \underline{N})$ of B such that $a_i = 1$ only for a finite number of indices $i \in \underline{N}$. It can be readily shown that I is the kernel of a congruence relation S on B (see I.24 and Theorem 4.8). We shall prove that S has no complement in the congruence lattice. In fact, evidently, $S \neq B \times B$ and if $T \neq \Delta$ is a congruence relation on B, then there exists $\underline{r} \in B$ such that $\underline{r} \neq \underline{0}$ and $(\underline{0}, \underline{r}) \in T$. Consequently, there is a sequence $\underline{s} > \underline{0}$ with $\underline{r} \geqslant \underline{s} \in I$. Hence $(\underline{0}, \underline{s}) \in S \cap T$ and so $S \wedge T \neq \Delta$.

4. CONGRUENCE RELATIONS AND p-IDEALS

A subset $I \neq \emptyset$ of the base set L of an ortholattice $`L = (L, \vee, \wedge, ', 0, 1)$ is called a <u>p-ideal</u> if and only if the following conditions hold:

(i) I is an ideal of the lattice $(L, \vee, \wedge)$;

(ii) for any $i \in I$ and any $a \in L$, $(i \vee a') \wedge a \in I$.

Remark 4.1. In the theory of orthomodular lattices and Boolean skew lattices, a p-ideal is called also a right ideal (cf. [124]), since the condition (ii) may be rewritten in the form $i \dot{\wedge} a \in I$ (for details, see Exercises V;3 and V;4).

Theorem 4.2. A nonvoid subset $I \subset L$ is a p-ideal of an orthomodular lattice $`L$ if and only if I is an ideal and $(i \vee a) \wedge \wedge (i \vee a') \in I$ for every $i \in I$ and every $a \in L$.

Proof. 1. If I is a p-ideal, then $(i \vee a') \wedge a \in I$, by assumption. Since $i C i \vee a'$ and $a C i \vee a'$, we get $I \ni i \vee [(i \vee a') \wedge a] = = (i \vee a') \wedge (i \vee a)$, by Theorem II.3.10.

2. Suppose I has the mentioned properties. Then

$$(i \vee a') \wedge a \leq (i \vee a) \wedge (i \vee a') \in I$$

and so $(i \vee a') \wedge a \in I$. //

Corollary 4.3. An ideal I of an orthomodular lattice $`L$ is a p-ideal if and only if $\overline{\text{com}}\,(i,a) \in I$ for every $i \in I$ and every $a \in L$.

Proof. By the definition of $\overline{\text{com}}\,(i,a)$, $\overline{\text{com}}\,(i,a) \leq (i \vee a) \wedge \wedge (i \vee a')$. Now, if I is a p-ideal, then $(i \vee a) \wedge (i \vee a') \in I$ by Theorem 2. Hence, a fortiori, $\overline{\text{com}}\,(i,a) \in I$. Conversely, if $\overline{\text{com}}\,(i,a) \in I$, then $I \ni i \vee \overline{\text{com}}\,(i,a) = (i \vee a) \wedge (i \vee a')$. From this and from Theorem 2 we can deduce easily that I is a p-ideal. //

Example 4.4. Let F_2 be the free orthomodular lattice of Section III.2. We shall verify that the principal ideal $(\underline{\text{com}}\,(x,y)]$ is a p-ideal. Indeed, if $i \in (\underline{\text{com}}\,(x,y)]$ and $a \in F_2$, then $\overline{\text{com}}\,(i,a) = 0 \in (\underline{\text{com}}\,(x,y)]$ by Remark III.2.2 and Corollary III. 2.6. Hence, by Corollary 3, it is now obvious that $(\underline{\text{com}}\,(x,y)]$ is a p-ideal of F_2.

<u>Theorem 4.5</u>. Every prime ideal of an orthomodular lattice is its p-ideal.

<u>Proof</u>. Let $i \in I$ where I is a prime ideal of an orthomodular lattice $`L$. Since $a \wedge a' = 0$, either $a \in I$ or $a' \in I$. Hence $(i \vee a') \wedge a \in I$. //

<u>Remark 4.6</u>. (A) In the lattice of Figure 12, $I = \{0\}$ is a p-ideal and $d \wedge d' = 0$. However, neither d nor d' belongs to I. Hence, I is a p-ideal which is not a prime ideal.

(B) There exist ideals which do not contain only 0 and which have the same property as the ideal $\{0\}$ in the lattice of (A). Indeed, consider again the ideal $(\underline{\text{com}}\ (x,y)]$ of the lattice F_2 (see Example 4 and Figure 18). Then $0 = x \wedge x' \in (\underline{\text{com}}\ (x,y)]$. However, one can see at a glance that neither x nor x' belongs to $(\underline{\text{com}}\ (x,y)]$.

We shall say that an element a of a lattice with 0 is <u>perspective</u> to b via an element t if and only if

$$a \vee t = b \vee t \ \&\ a \wedge t = b \wedge t = 0.$$

In this case we write $a \underset{p}{\sim} b$ and say that a,b are <u>perspective</u>.

<u>Theorem 4.7</u>. Let $I \subset L$ be an ideal of an orthomodular lattice $`L$ such that the following implication is fullfilled for every i, j:

$$(i \in I \ \&\ i \underset{p}{\sim} j) \Rightarrow j \in I.$$

Then I is a p-ideal.

<u>Proof</u>. Choose $i \in I$ and $k \in L$. Write $a = i \wedge (i' \vee k)$, $b = k \wedge (k' \vee i)$, $s = (i \vee k) \wedge (i \vee k') \wedge (i' \vee k)$ and $t = i' \wedge s$. Then $a = i \wedge s$, $b = k \wedge s$. Clearly, $a \wedge t = 0$ and $b \wedge t \leqslant (i' \wedge k) \wedge (i \vee k') = 0$. By Remark III.2.2 and Theorem II.3.10, $a \vee t = (i \vee i') \wedge s = s$, $b \vee t = (k \vee i') \wedge s = s$. Now it is clear that

$a \in I$ and $a \underset{p}{\sim} b$. Hence the element $(i \vee k') \wedge k$ belongs to I. //

Recall that an ideal I is called the kernel of a lattice congruence relation T of a lattice `L with 0 if and only if the following is true for all i: An element i belongs to I if and only if $(i,0) \in T$.

<u>Theorem 4.8</u>. Let I be a p-ideal of an orthomodular lattice. Then I is the kernel of exactly one congruence relation T. If J is a p-ideal which is the kernel of a congruence relation F and if $I \subset J$, then $T \subset F$.

<u>Proof</u>. 1. Let `L be the mentioned orthomodular lattice. By Theorem II.6.2 we know that `L is a relatively complemented lattice. However, in any relatively complemented lattice with 0, an ideal is the kernel of at most one congruence relation T. In fact, by I.24, $(a,b) \in T$ if and only if $(0,r) \in T$ where r denotes a relative complement of $a \wedge b$ in the interval $[0, a \vee b]$. This means that $r \in I$ if and only if $(a,b) \in T$. Thus, if I is the kernel of T_1 and T_2, then $T_1 = T_2$. Clearly, the above argument shows that $T \subset F$.

2. We shall define a congruence relation T with required properties.

Let T be the relation defined on L by

$$(a,b) \in T \Leftrightarrow [a' \wedge (a \vee b) \in I \ \& \ b' \wedge (a \vee b) \in I].$$

It is easy to see that T is a reflexive and symmetric relation.

Next we note that the relations $a \leqslant b$ and $(a,b) \in T$ hold if and only if $a' \wedge b \in I$. Indeed, it is an easy consequence of $a' \wedge (a \vee b) = a' \wedge b$ and $b' \wedge (a \vee b) = b' \wedge b = 0$.

To prove that T is a congruence relation, we shall first verify the condition (i) of Grätzer-Schmidt's Theorem 3.3. Since $a \wedge b \leqslant a \vee b$, we get

$$(a \wedge b, a \vee b) \in T \iff I \ni (a \wedge b)' \wedge (a \vee b) = (a' \vee b') \wedge (a \vee b).$$

Now, suppose that $(a \wedge b, a \vee b) \in T$. Then $I \ni (a' \vee b') \wedge (a \vee b) \geqslant a' \wedge (a \vee b)$, i.e., $a' \wedge (a \vee b) \in I$ and, similarly, $b' \wedge (a \vee b) \in I$. From this we see that $(a,b) \in T$. Conversely, let $(a,b) \in T$. Then $a' \wedge (a \vee b) \in I$ and $b' \wedge (a \vee b) \in I$. Since I is an ideal and since $a' C a \vee b$, $b' C a \vee b$, we can conclude from Theorem II.3.10 that

$$I \ni [a' \wedge (a \vee b)] \vee [b' \wedge (a \vee b)] = (a' \vee b') \wedge (a \vee b)$$

and so $(a \wedge b, a \vee b) \in T$.

Further, let $a \leqslant b$, $b \leqslant c$ and let $(a,b) \in T$, $(b,c) \in T$. Then $a' \wedge b \in I$ and $b' \wedge c \in I$. Using Theorem II.3.10, orthomodularity, $b' \leqslant a'$ and $a' \wedge b \leqslant c$, we get

$$I \ni (b' \wedge c) \vee (a' \wedge b) = [b' \vee (a' \wedge b)] \wedge [c \vee (a' \wedge b)] = a' \wedge c.$$

Thus $(a,c) \in T$.

Finally, let $a \leqslant b$, $(a,b) \in T$ and $c \in L$. Then $a' \wedge b \in I$. Trivially, $a \wedge c \leqslant b \wedge c$. Hence, by our remark above, $(a \wedge c, b \wedge c) \in T$ is equivalent to $(a \wedge c)' \wedge (b \wedge c) \in I$. Since I is a p-ideal and $a' \wedge b \in I$, we obtain $I \ni [(a' \wedge b) \vee (b' \vee c')] \wedge (b \wedge c) = w$. But $(a' \wedge b) \vee b' = a'$, by orthomodularity. Therefore,

$$w = (a' \vee c') \wedge (b \wedge c) = (a \wedge c)' \wedge (b \wedge c) \in I.$$

Thus $(a \wedge c, b \wedge c) \in T$.

Furthermore, $a \vee c \leqslant b \vee c$. Hence $(a \vee c, b \vee c) \in T$ if and only if $I \ni (a \vee c)' \wedge (b \vee c) = (b \vee c) \wedge a' \wedge c'$. Since I is a p-ideal and $a' \wedge b \in I$, it is clear that the element $v = [(a' \wedge b) \vee (c' \wedge a')'] \wedge c' \wedge a'$ belongs to I. By orthomodularity, $(a' \wedge b) \vee a = b$. Consequently,

$$v = [(a' \wedge b) \vee c \vee a] \wedge c' \wedge a' = (b \vee c) \wedge c' \wedge a'$$

and it easily follows that $(a \vee c, b \vee c)$ belongs to T.

In view of $0 \leqslant a$, we conclude that

$$(0,a) \in T \Leftrightarrow 0' \wedge a \in I \Leftrightarrow a \in I,$$

i.e., I is the kernel of T. //

Theorem 4.9. Let $`L$ be a lattice with $0$. Suppose $I$ is an ideal of $`L$ which is the kernel of a congruence relation. Then I has the following property for every i,j:

$$(i \in I \ \& \ i \underset{p}{\sim} j) \Rightarrow j \in I.$$

Proof. Suppose that I is the kernel of a congruence T. Let $i \in I$ and let i be perspective to j via t. Then $T \ni (i,0)$. Furthermore, $T \ni (i \vee t, 0 \vee t) = (j \vee t, t)$. Hence $T \ni (j \wedge (j \vee t), j \wedge t) = (j, j \wedge t) = (j,0)$. Thus $j \in I$. //

Summarizing Theorems 7,8 and 9, we get the following theorem:

Theorem 4.10. In any orthomodular lattice, the following conditions on an ideal I are equivalent:

(i) I is a p-ideal;

(ii) I satisfies the implication

$$(i \in I \ \& \ i \underset{p}{\sim} j) \Rightarrow j \in I;$$

(iii) I is the kernel of a congruence relation. //

We shall now extend the notion of a p-ideal: A set $I \neq \emptyset$ is said to be a p-ideal of a generalized orthomodular lattice $`G$ if and only if it is an ideal of the lattice $(G, \vee, \wedge)$ which has the following property:

$$\forall i \in I \, \forall g \in G \quad (i \vee g^{P(i \vee g)}) \wedge g \in I.$$

Every p-ideal of a generalized orthomodular lattice is a generalized orthomodular lattice (cf. Example 1.1.C).

Observe that if $`G$ is an orthomodular lattice, then we have two definitions of a p-ideal which have the same meaning. Indeed, suppose I is an ideal such that $I \ni (i \vee g^{P(i \vee g)}) \wedge g = \{i \vee \vee [g' \wedge (i \vee g)]\} \wedge g = s$ for every $i \in I$ and $g \in G$. Since $i \vee g$ commutes with i and g', we have $s = (i \vee g') \wedge (i \vee g) \wedge g = = (i \vee g') \wedge g$, by Theorem II.3.10. Hence, we can conclude that $s \in I$ if and only if $I \ni (i \vee g') \wedge g$. This suffices to complete the proof of our remark.

Lemma 4.11. A nonvoid subset I of a generalized orthomodular lattice $`G$ is its p-ideal if and only if every set $I_h = I \cap (h]$, $h \in G$, is a p-ideal of the orthomodular lattice represented by the interval algebra $`G[0,h]$.

Proof. 1. Let I be a p-ideal in $`G$. Choose $i \in I_h$ and $k \in (h]$. We have to prove that $s = (i \vee k^{P(h)}) \wedge k \in I_h$. To establish this fact, it suffices to show that $s \in I$, since $s \leqslant \leqslant k \leqslant h$ is immediate. By assumption, $I \ni t = (i \vee k^{P(i \vee k)}) \wedge k$. But $i \leqslant h$, $k \leqslant h$ and so $i \vee k \leqslant h$. Therefore, $k^{P(i \vee k)} = = k^{P(h)} \wedge (i \vee k)$. The element $i \vee k$ commutes with $k^{P(h)}$ and i in the orthomodular lattice $`G[0,h]$. This yields

$$t = \{i \vee [k^{P(h)} \wedge (i \vee k)]\} \wedge k = (i \vee k^{P(h)}) \wedge (i \vee k) \wedge k = = (i \vee k^{P(h)}) \wedge k = s.$$

2. Clearly, $I = \bigcup \{I_h;\ h \in G\}$. It follows that I is an ideal of $`G$.

It remains to show that $(i \vee g^{P(i \vee g)}) \wedge g \in I$ for every $i \in I$ and $g \in G$. Set $h = i \vee g$. Since I_h is a p-ideal in $`G[0,h]$, it follows that $(i \vee g^{P(h)}) \wedge g \in I_h \subset I$. //

Remark 4.12. Let I be a p-ideal of a generalized orthomodular lattice $`G$ and let $i \in I$. If $h \geqslant g \in G$, then $I \ni$

$\ni (i \vee g^{P(h)}) \wedge g$. To prove this assertion, consider $H = i \vee g \vee h$. By Lemmas 1.3 (i), 11,

$$(i \vee g^{P(h)}) \wedge g \leqslant (i \vee g^{P(H)}) \wedge g \in I_H \subset I.$$

The transition from the results given in the previous theorems for orthomodular lattices to the corresponding results for generalized orthomodular lattices is now straightforward.

Theorem 4.13. An ideal I of a generalized orthomodular lattice $`G$ is its p-ideal if and only if $(i \vee k) \wedge (i \vee k^{P(h)}) \in I$ for every $h \in G$, $i \in I$, $i \leqslant h$ and every $k \in G$, $k \leqslant h$.

Proof. By Theorem 2, the mentioned condition is equivalent to the assertion that I_h is a p-ideal of $`G[0,h]$ for every $h \in G$. Hence, the result follows by Lemma 11. //

Theorem 4.14. If I is a p-ideal of a generalized orthomodular lattice $`G$ and if $J$ is a p-ideal of the generalized orthomodular lattice $I$, then $J$ is also a p-ideal of the generalized orthomodular lattice $`G$.

Proof. Choose $j \in J$ and $g \in G$. By assumption, $I \ni h = (j \vee g^{P(j \vee g)}) \wedge g \leqslant g$. From Lemma 1.3 (ii) it follows that $j \vee h^{P(j \vee g)} \geqslant j \vee g^{P(j \vee g)}$. In view of Remark 12 we have $J \ni (j \vee h^{P(j \vee g)}) \wedge h = h$. Thus J is a p-ideal of $`G$. //

Theorem 4.15. Every prime ideal of a generalized orthomodular lattice is its p-ideal.

Proof. Let $i \in I$ and $g \in G$. Then $g^{P(i \vee g)} \wedge g = 0$ implies $g^{P(i \vee g)} \in I$ or $g \in I$. Hence $(i \vee g^{P(i \vee g)}) \wedge g \in I$. //

Theorem 4.16. Let I be an ideal of a generalized orthomodular lattice $`G$. Suppose I has the following property for every i, j:

$$(i \in I \ \& \ i \underset{p}{\sim} j) \Rightarrow j \in I.$$

Then I is a p-ideal of the generalized orthomodular lattice `G.

Proof. Let i be an element of I which is perspective to j via t. Let $h = i \vee j \vee t$. Then $i \in I_h$. By the choice of h, we have $i \underset{p}{\sim} j$ in the orthomodular lattice `G[0,h]. Hence $j \in I_h$ and, by Theorem 7, $I_h$ is a p-ideal of `G[0,h]. By Lemma 11 this means that I is a p-ideal of `G. //

Theorem 4.17. Let I be a p-ideal of a generalized orthomodular lattice `G. Then I is the kernel of exactly one congruence relation of the lattice $(G, \vee, \wedge)$.

Proof. Choose $h \in G$. Denote by T_h the congruence relation of `G[0,h] having $I_h = (h] \cap I$ as its kernel. Clearly, if $g \leqslant h$, then $I_g \subset I_h$ and $T_g \subset T_h$. Recall [29; p. 186] that a directed set is a poset in which any two elements have an upper bound in the poset. Note that the congruence relations T_h where h runs over G form a directed set under $\subset$. Indeed, given $a, b \in G$, we have $T_a \subset T_{a \vee b}$ and $T_b \subset T_{a \vee b}$. Consequently, by a well-known argument, $T = \bigcup \{T_h;\ h \in G\}$ is a lattice congruence relation of $(G, \vee, \wedge)$.

Clearly, I is the kernel of T. From a remark made in the proof of Theorem 8 we can infer that there is no other congruence relation having the same property. //

Theorem 4.18. The following conditions are equivalent in any generalized orthomodular lattice:

(i) I is a p-ideal;

(ii) I satisfies the implication

$$(i \in I \ \&\ i \underset{p}{\sim} j) \Rightarrow j \in I;$$

(iii) I is the kernel of a congruence relation.

Proof. The theorem follows from Theorems 9,16 and 17. //

Theorem 4.19. Let $`D$ be the Janowitz's hull of a generalized orthomodular lattice $`G$. Then $I \subset G$ is a p-ideal of $`G$ if and only if it is a p-ideal of $`D$.

Proof. By definition, I is a p-ideal in $`G$ if and only if $I$ contains the element $u = (i \vee g^{P(i \vee g)}) \wedge g$ for every $i \in I$ and $g \in G$. From Remark 2.6.B we obtain $u = \{i \vee [g' \wedge (i \vee g)]\} \wedge g$ in $`D$. Since $i C i \vee g$ and $g' C i \vee g$, we have $u = (i \vee g') \wedge (i \vee \vee g) \wedge g = (i \vee g') \wedge g$, by Theorem II.3.10. Now, I is a p-ideal in $`D$ if and only if $u = (i \vee g') \wedge g \in I$. Hence, this is equivalent to the statement that $I$ is a p-ideal of $`G$. //

Theorem 4.20 (cf. [147]). Let $`D$ be the Janowitz's hull of a generalized orthomodular lattice $`G$. Suppose that $I \subset G$. Then the following conditions are equivalent:

(i) I is a p-ideal of $`G$.

(ii) I is a p-ideal of the orthomodular lattice $`D$.

(iii) I is the kernel of exactly one congruence relation of the lattice $(G, \vee, \wedge)$.

(iv) I is the kernel of exactly one congruence relation of the orthomodular lattice $`D$.

(v) For any i,j, the following implication is true:

$$(i \in I \ \& \ j \in G \ \& \ i \underset{p}{\sim} j) \Rightarrow j \in I.$$

(vi) For any i,j, the following implication holds:

$$(i \in I \ \& \ j \in D \ \& \ i \underset{p}{\sim} j) \Rightarrow j \in I.$$

Proof. To see this, it suffices to combine Theorems 18, 19 and 10. //

Remark 4.21. Let I be a p-ideal of a generalized orthomodular lattice $`G$. Then the quotient lattice $`G/I$ is a generalized orthomodular lattice.

Indeed, let $`D$ denote the Janowitz's hull of $`G$. From Theorem 2.5 and Theorem 15 we see that G is a p-ideal of $`D$. Hence, by Theorem 14, $I$ is a p-ideal of $`D$. Let $[x]$ denote an element of $`D/I$ which is an orthomodular lattice by Theorem 20 and Theorem 2.3. Take $\bar{G} = \{[x] \in D/I;\ x \in G\}$. Then $\bar{G}$ is an ideal of $`D/I$. In fact, it is easy to see that $[x] \vee [y] = [x \vee y] \in \bar{G}$ whenever $[x],[y] \in \bar{G}$. Moreover, if $[x] \leqslant [y]$ and $[y] \in \bar{G}$, then $[x] = [x \wedge y] \in \bar{G}$. The conclusion now follows from Example 1.1.C.

5. COMMUTATORS

Let us recall the definition of the upper commutator given in Section III.2 for two elements x,y. It is the element

$$\overline{\text{com}}\ (x,y) = (x \vee y) \wedge (x \vee y') \wedge (x' \vee y) \wedge (x' \vee y').$$

For convenience we sometimes write $\text{com}\ (x,y)$ for $\overline{\text{com}}\ (x,y)$.

Next we establish some properties of commutators.

Lemma 5.1. Let $`L$ be an orthomodular lattice and let $a \in L$, $k \in L$. Then

$$s_a(k) = (k \vee a') \wedge a \leqslant k \vee \text{com}\ (a,k).$$

Proof. Since k commutes with the elements $a \vee k$, $a \vee k'$, $a' \vee k$, $a' \vee k'$, we have

$$k \vee \text{com}\ (a,k) = k \vee [(a \vee k) \wedge (a \vee k') \wedge (a' \vee k) \wedge (a' \vee k')] =$$
$$= (a \vee k) \wedge (a' \vee k) \geqslant a \wedge (a' \vee k). \quad //$$

Lemma 5.2. Let $`G$ be a generalized orthomodular lattice and let $c,d \in [a,b]$, $b \leqslant B$. Then $\text{com}_{[a,b]}(c,d) = a \vee (\text{com}_{[0,B]}(c,d))$

where $\operatorname{com}_{[x,y]}(p,q)$ denotes the upper commutator of p,q in the orthomodular lattice $`G[x,y]$.

Proof. Denote by y^+ the orthocomplement of an element $y \in \in [a,b]$ in the interval $[a,b]$. Mark $x \in [0,B]$ with an asterisk to denote its orthocomplement in the interval $[0,B]$. By Corollary II.6.3 (ii), $y^+ = a \vee (y^* \wedge b) = (a \vee y^*) \wedge b$. From this we find

$$\begin{aligned}\bar{c} = \operatorname{com}_{[a,b]}(c,d) &= (c \vee d) \wedge (c \vee d^+) \wedge (c^+ \vee d) \wedge (c^+ \vee d^+) = \\ &= (c \vee d) \wedge \{c \vee [(a \vee d^*) \wedge b]\} \wedge \{[(a \vee c^*) \wedge b] \vee d\} \wedge \\ &\wedge \{[(a \vee c^*) \wedge b] \vee [(a \vee d^*) \wedge b]\}.\end{aligned}$$

Now, b commutes with a,c,d. Therefore by Theorem II.3.10

$$\begin{aligned} c \vee [(a \vee d^*) \wedge b] &= (c \vee a \vee d^*) \wedge b = (c \vee d^*) \wedge b, \\ [(a \vee c^*) \wedge b] \vee d &= (a \vee c^* \vee d) \wedge b = (c^* \vee d) \wedge b, \\ [(a \vee c^*) \wedge b] \vee [(a \vee d^*) \wedge b] &= (a \vee c^* \vee d^*) \wedge b. \end{aligned}$$

Application of $c \vee d \leqslant b$ gives

$$\bar{c} = (c \vee d) \wedge (c \vee d^*) \wedge (c^* \vee d) \wedge (a \vee c^* \vee d^*).$$

Therefore, since $a \leqslant c$ and $a \leqslant d$, we must have

$$\bar{c} = a \vee [(c \vee d) \wedge (c \vee d^*) \wedge (c^* \vee d) \wedge (c^* \vee d^*)],$$

by Theorem II.3.10. //

Remark 5.3. It follows from Lemma 2 that the smallest subset of a generalized orthomodular lattice $`G$ which contains all the commutators $\operatorname{com}_{[a,b]}(c,d)$ is the base set G. Indeed, $a = a \vee \vee \operatorname{com}_{[0,B]}(B,B)$ provided $B \geqslant a$. Hence the subset contains every $a \in G$.

Theorem 5.4. Let J be the ideal of an orthomodular lattice $`L$ which is generated by the set $K = \{m \in L; \exists\, x,y \in L :: m = = \operatorname{com}(x,y)\}$. Then

(i) The ideal J is a p-ideal of the orthomodular lattice $`L$.

(ii) The quotient lattice $`L/J$ is a Boolean algebra.

(iii) If I is a p-ideal such that $`L/I$ is a Boolean algebra, then $I \supset J$.

Proof. Ad (i). By definition, J is the smallest ideal which contains the set K. Hence $j \in J$ if and only if there exist $r \in \in \underline{N}$ and $k_1, k_2, \ldots, k_r \in K$ such that $j \leqslant k_1 \vee k_2 \vee \ldots \vee k_r$. This means that there are $x_i, y_i \in L$ with

$$j \leqslant \text{com}\,(x_1,y_1) \vee \text{com}\,(x_2,y_2) \vee \ldots \vee \text{com}\,(x_r,y_r).$$

By setting $k = \text{com}\,(x_1,y_1) \vee \ldots \vee \text{com}\,(x_r,y_r)$ in Lemma 1, we obtain

$$s_a(j) \leqslant s_a(k) \leqslant k \vee \text{com}\,(k,a) =$$
$$= \text{com}\,(x_1,y_1) \vee \ldots \vee \text{com}\,(x_r,y_r) \vee \text{com}\,(k,a).$$

Thus $(j \vee a') \wedge a \in J$ for every $j \in J$ and $a \in G$.

Ad (ii). The symbol $`L/J$ means the quotient lattice $`L/T$ where T is the uniquely determined congruence with the kernel J. Denote by $[a]$ the congruence class determined by $a \in L$. Then $[a] = [0]$ if and only if $a \in J$. Now, $`L/J$ is an orthomodular lattice by Exercise V;6. Hence by Theorem III.2.11 and Theorem II.4.5 it suffices to show that $\text{com}\,([x],[y]) = [0]$ for any elements $[x],[y]$ of the quotient lattice. But

$$\text{com}\,([x],[y]) = ([x] \vee [y]) \wedge \ldots \wedge ([x]' \vee [y]') =$$
$$= [x \vee y] \wedge \ldots \wedge [x' \vee y'] = [(x \vee y) \wedge \ldots$$
$$\ldots \wedge (x' \vee y')] = [\text{com}\,(x,y)] = [0]$$

because of $\text{com}\,(x,y) \in J$.

Ad (iii). Denote by $\bar{x}$ the congruence class of $`L/I$ determined by an element $x \in L$. Choose $j \in J$. By the same argument as before in the proof of (i) and (ii), $j \leqslant \text{com}\,(x_1,y_1) \vee \ldots$

and $\bar{j} \leqslant \mathrm{com}\,(\bar{x}_1,\bar{y}_1) \vee \ldots$. The distributivity of `L/I and Theorem III.2.11 now imply that every commutator on the right is equal to $\bar{0}$. Thus $\bar{j} = \bar{0}$ and so $j \in I$, i.e., $J \subset I$. //

Remark 5.5. Let I be a p-ideal of a generalized orthomodular lattice `G. According to Remark 4.21 we know that the quotient lattice `G/I is a generalized orthomodular lattice. Let $[x] \leqslant [a]$ be two elements of `G/I. It is useful to have a complete description of the generalized orthogonality relation $P([a]) : [x] \to [x]^{P([a])}$ on `G/I. From the considerations given in Remark 4.21 and from Example 1.1 it follows that $u = [x]^{P([a])} = [x]' \wedge [a]$ where $'$ denotes the orthogonality relation in the Janowitz's hull of `G. By Theorem 3.2, $u = [x'] \wedge [a]$. Using Remark 2.6 and the fact that $[a] = [a] \vee [x] = [a \vee x]$, we find $u = [x'] \wedge [a \vee x] = [x' \wedge (a \vee x)] = [(x]^{\perp} \wedge (a \vee x)]$. The proof of Theorem 2.3, Lemma 1.3 (i) and Remark 2.6 show that $(x]^{\perp} \wedge (a \vee x] = ((x]^{\perp\perp} \vee (a \vee x]^{\perp})^{\perp} = ((x] \vee (a \vee x]^{\perp})^{\perp} = ((a \vee x) \wedge x^{P(a \vee x)}]^{\perp\perp} = (x^{P(a \vee x)}] = x^{P(a \vee x)}$. Hence $u = [x^{P(a \vee x)}]$ and this yields the formula

$$(1) \quad [x]^{P([a])} = [x^{P(a \vee x)}]$$

valid for any $[x] \leqslant [a]$.

Now let $[b],[c],[d]$ be elements of `G/I such that $[c] \leqslant [b]$ and $[d] \leqslant [b]$. Since $[b] = [b] \vee [c] \vee [d] = [b \vee c \vee d]$, there is no loss in assuming that $c \leqslant b$ and $d \leqslant b$. To simplify notation, write $\bar{b},\bar{c},\bar{d},\bar{0}$ instead of $[b],[c],[d],[0]$, replace $[c]^{P([b])}, [d]^{P([b])}$ by $\bar{c}^{*}$ and $\bar{d}^{*}$, respectively, and, finally, put $c^{+} = c^{P(b)}$, $d^{+} = d^{P(d)}$. Observe that $\bar{c}^{*} = \overline{c^{+}}$, by (1). Hence,

$$\mathrm{com}_{[\bar{0},\bar{b}]}(\bar{c},\bar{d}) = (\bar{c} \vee \bar{d}) \wedge (\bar{c} \vee \bar{d}^{*}) \wedge (\bar{c}^{*} \vee \bar{d}) \wedge (\bar{c}^{*} \vee \bar{d}^{*}) =$$

$$= (\bar{c} \vee \bar{d}) \wedge (\bar{c} \vee \overline{d^+}) \wedge \ldots = \overline{(c \vee d)} \wedge \overline{(c \vee d^+)} \wedge \ldots =$$
$$= \overline{(c \vee d) \wedge (c \vee d^+) \wedge \ldots} = \overline{\mathrm{com}_{[0,b]}(c,d)},$$

i.e.,

$$(2) \quad \mathrm{com}_{[\bar{0},\bar{b}]}(\bar{c},\bar{d}) = \overline{\mathrm{com}_{[0,b]}(c,d)}$$

provided $c \leqslant b$ and $d \leqslant b$.

Theorem 5.6. Let $`G$ be a generalized orthomodular lattice and let G' denote the ideal generated by the set

$$M = \{g; \exists\, b,c,d \in G :: c,d \leqslant b \ \&\ g = \mathrm{com}_{[0,b]}(c,d)\}.$$

Then

(i) The ideal G' is a p-ideal.

(ii) The quotient lattice $`G/G'$ is distributive.

(iii) If I is a p-ideal of $`G$ such that $`G/I$ is distributive, then $I \supset G'$.

Proof. Ad (i). Choose $j \in G'$ and $g \in G$. Since $j \in G'$, there exist $b_1,c_1,d_1,\ldots,b_n,c_n,d_n$ such that

$$j \leqslant \mathrm{com}_{[0,b_1]}(c_1,d_1) \vee \ldots \vee \mathrm{com}_{[0,b_n]}(c_n,d_n) = q.$$

Let $`L = `G[0,q \vee g]$ and let J denote the p-ideal of $`L$ constructed in Theorem 4. By Lemma 1.3 (i),

$$(j \vee g^{P(j \vee g)}) \wedge g \leqslant (j \vee g^{P(q \vee g)}) \wedge g \in J \subset G'.$$

Ad (ii) and (iii). Clearly, the quotient lattice $`G/I$ is distributive if and only if every interval algebra of $`G/I$ is distributive. Hence, by Theorem III.2.11, $`G/I$ is distributive if and only if $\bar{0} = \mathrm{com}_{[\bar{0},\bar{b}]}(\bar{c},\bar{d})$ for every $b,c,d$ of $`G/I$ satisfying $\bar{c},\bar{d} \in [\bar{0},\bar{b}]$. From (2) we see that this is equivalent to $\mathrm{com}_{[0,b]}(c,d) \in I$ for every $b,c,d \in G$ such that $c,d \in [0,b]$. //

EXERCISES

V;1. Let u,v,x be elements of a generalized orthomodular lattice $`G$ such that $(u \vee v] = (x]^{\perp}$. Prove that there exist elements $s,t \in G$ satisfying $(u]^{\perp} = (s]$ and $(v]^{\perp} = (t]$.

V;2. Prove that the following conditions are equivalent for any relation T satisfying the hypotheses of Theorem 3.3:

(1) $(x,y) \in T$;

(2) $(u,v) \in T$ for every u,v such that $u,v \in [x \wedge y, x \vee y]$.

Using this result, prove Grätzer-Schmidt's Theorem 3.3.

V;3. Prove that a nonvoid subset I of L is a p-ideal of an orthomodular lattice $`L$ if and only if the following conditions hold:

(3) $i,j \in I \Rightarrow i \vee j \in I$;

(4) $(i \in I \ \& \ k \in L) \Rightarrow i \dot{\wedge} k \in I$.

V;4. A nonvoid subset J of L is called a <u>right ideal</u> of an orthomodular lattice $`L$ if and only if it satisfies the following conditions:

(5) $i,j \in J \Rightarrow i \dot{\vee} j \in J$;

(6) $(i \in J \ \& \ k \in L) \Rightarrow i \dot{\wedge} k \in J$.

Similarly, a nonvoid subset H L is said to be a <u>left ideal</u> of $`L$ if and only if it verifies the conditions (7),(8), below:

(7) $i,j \in H \Rightarrow i \dot{\vee} j \in H$;

(8) $(i \in H \ \& \ k \in L) \Rightarrow k \dot{\wedge} i \in H$.

Denote by $R(`L)$ the set of all right ideals, by $L(`L)$ the set of all left ideals and by $p(`L)$ the set of all p-ideals of $`L$. Prove that $p(`L) = R(`L) \subset I(`L) \subset L(`L)$ where $I(`L)$ denotes the set of all ideals of $`L$.

V;5. Find all p-ideals of the lattice $`L$ of Figure 45.

V;6. Let T denote a congruence relation of an orthomodular lattice `L. Show that the quotient lattice `L/T is an orthomodular lattice.

V;7. Prove that

$$\text{com}\,(a,b) = \text{com}\,(a',b) = \text{com}\,(a,b') = \text{com}\,(a',b')$$

in any ortholattice.

V;8. Find the set L' (see Theorem 5.6) in the case of the lattice of Figure 45.

V;9. Let `G be an orthomodular lattice. Prove that the ideal J of Theorem 5.4 is equal to the ideal G' of Theorem 5.6.

V;10. Show that the following conditions are equivalent in any orthomodular lattice:

(i) xCy;

(ii) $x \wedge \overline{\text{com}}\,(x,y) C y \wedge \overline{\text{com}}\,(x,y)$;

(iii) $x \vee \underline{\text{com}}\,(x,y) \leqslant \underline{\text{com}}\,(x,y)$.

SOLVABILITY OF GENERALIZED ORTHOMODULAR LATTICES

Chapter VI

1. *Reflective and coreflective congruences*
2. *Projective allelomorph*
3. *Commutator sublattices*
4. *Solvability in equational classes of lattices*

1. REFLECTIVE AND COREFLECTIVE CONGRUENCES

Suppose $\underline{K}$ is a given equational class of lattices (cf. I.26) and let $`L$ be a lattice. Denote by $E_{\underline{K}}(`L)$ the set formed by those congruences T on $`L$ for which $`L/T \in \underline{K}$. Since $`L/U \in \underline{K}$ for the universal congruence $U = L \times L$, $E_{\underline{K}}(`L) \neq \emptyset$. Define $C_{\underline{K}}(`L)$ to be the intersection $\bigcap\{T;\ \ T \in E_{\underline{K}}(`L)\}$. Then $C_{\underline{K}}(`L)$ is a congruence on $`L$. From I.25 we conclude that $`L/C_{\underline{K}}(`L)$ is isomorphic to a subdirect product of the lattices $`L/T$ where $T$ runs over $E_{\underline{K}}(`L)$. By I.26, $`L/C_{\underline{K}}(`L) \in \underline{K}$.

Hence, for any equational class $\underline{K}$ of lattices, there exists a congruence $C_{\underline{K}}(`L)$ on $`L$ which is the smallest one with regard to the property $`L/C_{\underline{K}}(`L) \in \underline{K}$.

The problem we wish to treat in this section is the following: given a lattice $`L$, determine the intrinsic structure of $C_{\underline{D}}(`L)$ where $\underline{D}$ denotes the equational class of distributive lattices. This can be done by means of "alles" defined and studied in [16].

Let b/a, d/c be two quotients of a lattice. Recall that

b/a is said to be weakly perspective down into d/c, written $b/a \searrow_w d/c$, if and only if $b = a \vee d$ and $a \wedge d \geq c$ (see Figure 47a). We say that b/a is weakly perspective up into d/c and write $b/a \nearrow^w d/c$ if and only if $a = b \wedge c$ and $b \vee c \leq d$ (see Figure 47b). In both cases we for convenience use the notation $b/a \sim_w d/c$ and say that b/a is weakly perspective into d/c.

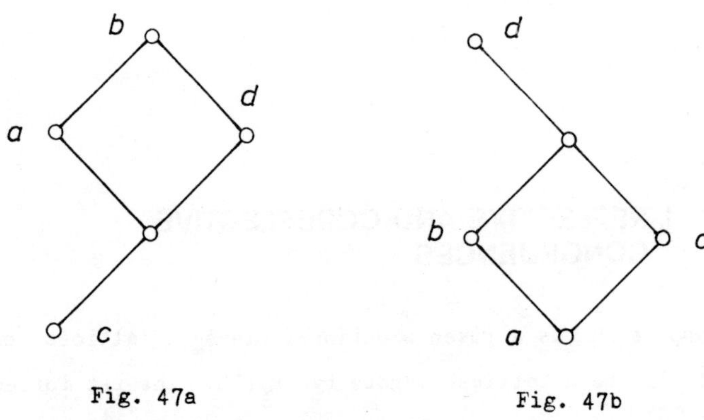

Fig. 47a Fig. 47b

Let q/p, s/r be two given quotients. If there exists a sequence

$$q_0/p_0 = q/p \sim_w q_1/p_1 \sim_w \cdots \sim_w q_n/p_n = s/r,$$

we say that q/p is weakly projective into s/r and we write $q/p \approx_w s/r$.

We are now ready to formulate the basic properties of these and similar notions introduced in Section I.15.

Lemma 1.1. In any lattice `L,

(i) $[(b/a \searrow_w d/c) \mathbin{\&} (d/c \searrow_w f/e)] \Rightarrow (b/a \searrow_w f/e)$;

(ii) $[(b/a \nearrow^w d/c) \mathbin{\&} (d/c \nearrow^w f/e)] \Rightarrow (b/a \nearrow^w f/e)$;

(iii) if $b/a \approx_w d/c$ and $(c,d) \in T$ where T is a congruence of `L, then also $(a,b) \in T$;

(iv) $b/a \searrow d/c \Rightarrow b/a \searrow_w d/c$; $b/a \sim d/c \Rightarrow b/a \sim_w d/c$;

$b/a \approx d/c \Rightarrow b/a \approx_w d/c$;

(v) $(b/a \sim d/c) \Leftrightarrow (b/a \sim_w d/c \;\&\; d/c \sim_w b/a)$;

(vi) $[p,q] \subset [r,s] \Rightarrow q/p \approx_w s/r$.

Proof. Ad (i). Note that $a \vee f \geqslant (a \wedge d) \vee f \geqslant c \vee f = d$, i.e., $a \vee f \geqslant a \vee d = b$. Hence $a \vee f = b$ and, furthermore, $a \wedge f \geqslant c \wedge \wedge f \geqslant e$.

Ad (ii). The assertion follows from (i) by duality.

Ad (iii). If $b/a \searrow_w d/c$ and $(c,d) \in T$, then $(a \wedge d, d) \in T$ by Exercise V;2. Since T is a congruence relation, $(a,b) = = (a \vee (a \wedge d), a \vee d) \in T$. Dually we get the validity of the implication

$$[(b/a \nearrow^w d/c) \;\&\; (c,d) \in T] \Rightarrow (a,b) \in T.$$

The statement (iii) now follows by induction.

Ad (iv). Trivial.

Ad (v). According to (iv) it suffices to prove that $b/a \sim_w d/c$ and $d/c \sim_w b/a$ imply $b/a \sim d/c$. However, if $b/a \searrow_w d/c$ and $d/c \searrow_w b/a$, then

$$b \geqslant d \;\&\; b \geqslant a \geqslant c \;\&\; d \geqslant b \;\&\; d \geqslant c \geqslant a$$

so that $b = d$ and $c = a$. Thus $b/a \sim d/c$ is in this case trivial. By symmetry and duality it remains to consider the case where $b/a \searrow_w d/c$ and $d/c \nearrow^w b/a$. But then from $d/c \nearrow^w b/a$ we conclude that $a \wedge d = c$ and so $b/a \searrow d/c$.

Ad (vi). Here we have $q/p \searrow_w q/r \nearrow^w s/r$. //

Lemma 1.2. The following two conditions are equivalent in any lattice:

(i) $q/p \approx_w s/r$;

(ii) there exist sequences $a_i \leqslant b_i$, $a_i' \leqslant b_i'$ such that

1. $b_i/a_i \sim b_i'/a_i'$ where $[a_i', b_i'] \subset [a_{i+1}, b_{i+1}]$ for every $i = 0, 1, \ldots, n-1$;

2. $b_0/a_0 = q/p$ & $b_n/a_n = s/r$.

Proof. If (i) is true, then also (ii) holds, by the definition of $\approx_w$. Conversely, if (ii) is valid, then the assertion follows from Lemma 1 (vi). //

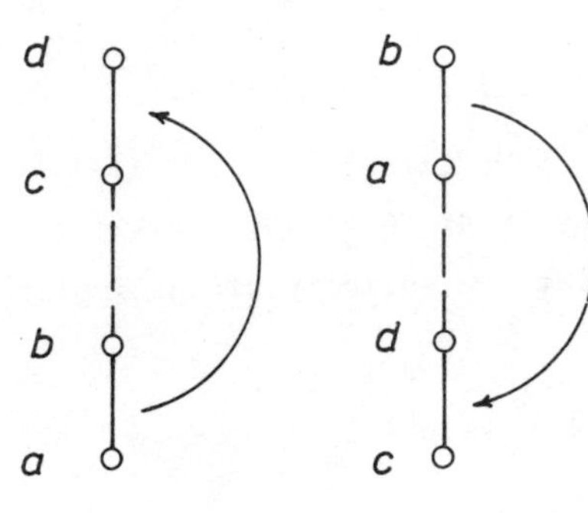

Fig. 48

A quotient b/a of a lattice `L is called an allele of `L if and only if there exists a quotient d/c such that $b/a \approx_w d/c$ and either $c \geqslant b$ or $a \geqslant d$ (see Figure 48). The set of all alleles of `L is denoted by A(`L) and it is called the allelomorph of the lattice `L. If $a \neq b$, then the allele b/a is said to be nontrivial.

Evidently, $a/a \in$ A(`L) for every $a \in L$.

Theorem 1.3. A lattice is distributive if and only if its allelomorph contains no nontrivial allele.

Proof. 1. Let `L be a lattice which contains as a sublattice the lattice `M_5 (see Figure 4). Then $a/o \nearrow i/c \searrow b/o \nearrow i/a$. Hence $a/o \in$ A(`L). Next suppose that `L contains `$N_5$ (see Figure 3) as a sublattice. Then $c/a \nearrow^{w} i/a \searrow b/o \nearrow i/c$ and so $c/a \in$ A(`L).

2. Let b/a be an allele of a distributive lattice `L. We shall prove that $a = b$. We may assume that $b/a \approx_w d/c$ and $a \leqslant b \leqslant c \leqslant d$ for the simple reason that in the case $c \leqslant d \leqslant a \leqslant b$ we can apply duality. By assumption, there exist a_i, b_i such that

$$b/a = b_0/a_0 \sim_w b_1/a_1 \sim_w b_2/a_2 \sim_w \ldots \sim_w b_n/a_n = d/c.$$

We shall show that $b_i/a_i \sim_w b_{i+1}/a_{i+1}$ implies $b_i \wedge b/a_i \wedge b \sim_w b_{i+1} \wedge$ $\wedge\, b/a_{i+1} \wedge b$.

First suppose that $b_i/a_i \nearrow^w b_{i+1}/a_{i+1}$. Then $b_i \wedge a_{i+1} = a_i$ and $b_i \vee a_{i+1} \leqslant b_{i+1}$. Since `L is distributive, we have

$$(b_i \wedge b) \vee (a_{i+1} \wedge b) = (b_i \vee a_{i+1}) \wedge b \leqslant b_{i+1} \wedge b.$$

In addition, $(b_i \wedge b) \wedge (a_{i+1} \wedge b) = a_i \wedge b$. Hence

$$b_i \wedge b/a_i \wedge b \nearrow^w b_{i+1} \wedge b/a_{i+1} \wedge b.$$

In the same way one can prove that

$$b_i/a_i \searrow_w b_{i+1}/a_{i+1}$$

implies

$$b_i \wedge b/a_i \wedge b \searrow_w b_{i+1} \wedge b/a_{i+1} \wedge b.$$

Therefore, by what we have just proved, $b/a = b \wedge b/a \wedge b \approx_w$ $\approx_w d \wedge b/c \wedge b = b/b$. By using Lemma 1 (iii) with T equal to the diagonal relation $\Delta = \{(a,a);\ a \in L\}$, we conclude that $a = b$. //

<u>Theorem 1.4</u>. Let $a < b$, $c < d$ be elements of a lattice. If $b/a \sim_w d/c$ and if $c = c_0 \leqslant c_1 \leqslant \ldots \leqslant c_n = d$, then there exist a_j such that $a = a_0 \leqslant a_1 \leqslant \ldots \leqslant a_n = b$ and such that $a_{j+1}/a_j \sim_w$ $\sim_w c_{j+1}/c_j$ for every $j = 0,1,\ldots,n-1$.

If $b/a \approx_w d/c$, then it is possible to choose the elements a_i so that $a_{j+1}/a_j \approx_w c_{j+1}/c_j$ for every j.

<u>Proof</u>. 1. Let $b/a \nearrow^w d/c$. Setting $a_j = b \wedge c_j$, we find that

$$a_{j+1} \vee c_j = (b \wedge c_{j+1}) \vee c_j \leqslant c_{j+1} \vee c_j = c_{j+1}.$$

Observe that

$$a_j \leqslant a_{j+1} \wedge c_j = (b \wedge c_{j+1}) \wedge c_j = b \wedge c_j = a_j,$$

i.e., $a_{j+1} \wedge c_j = a_j$. It follows that $a_{j+1}/a_j \nearrow^w c_{j+1}/c_j$. In the case $b/a \searrow_w d/c$ we can use a dual argument.

The remaining statement is a matter of an easy induction. //

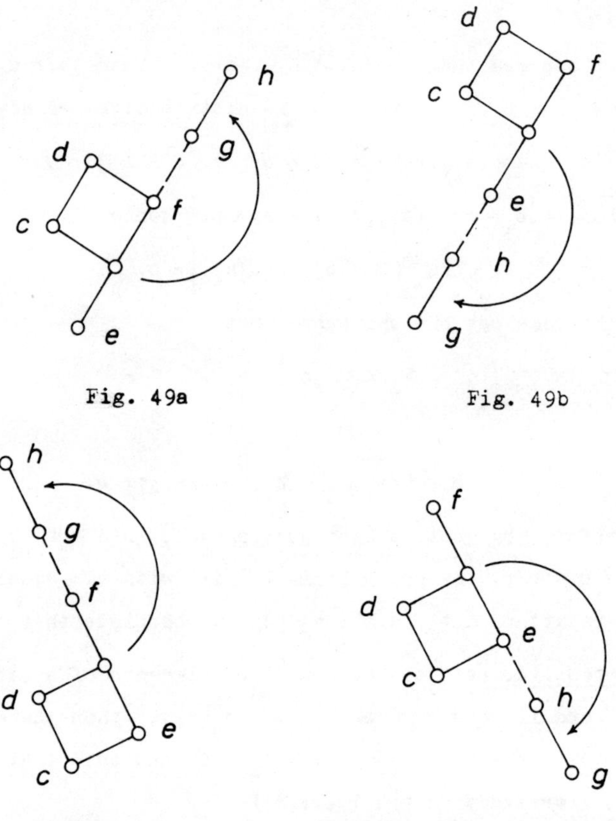

Fig. 49a

Fig. 49b

Fig. 49c

Fig. 49d

<u>Theorem 1.5</u>. Let $`L$ be a lattice and let $\hat{C}$ be the binary relation defined on L so that $(a,b) \in \hat{C}$ if and only if there exists $n \in \underline{N}$ and $a_1, a_2, \dots, a_n$ such that

$$a \wedge b = a_0 \leqslant a_1 \leqslant \dots \leqslant a_n = a \vee b$$

and such that $a_{j+1}/a_j \in A(`L)$ for every $j = 0,1,\dots,n-1$.

If $d/c \approx_w q/p$ and $(p,q) \in \hat{C}$, then $(c,d) \in \hat{C}$.

Proof. 1. Let $d/c \sim_w f/e$ where $f/e \in A(`L)$. We shall show that then necessarily $(c,d) \in \hat{C}$. Here we can consider four cases (see Figures 49a-d). The cases (a) and (d), and also (b) and (c) are dual. Hence it remains to investigate the cases (a) and (b).

If the case (b) occurs, then $d/c \searrow_w f/e \approx_w h/g$ and $h \leqslant e$. Consequently, $d/c \approx_w h/g$ and $c \geqslant c \wedge f \geqslant e \geqslant h$. It is now immediate that $d/c \in A(`L)$. Thus $(c,d) \in \hat{C}$.

Suppose the case (a) occurs. Here $d/c \searrow_w f/e \approx_w h/g$ and $f \leqslant g$. Let $r_0 = g$, $r_1 = g \vee (d \wedge h)$ and $r_2 = h$ (see Figure 50). Then

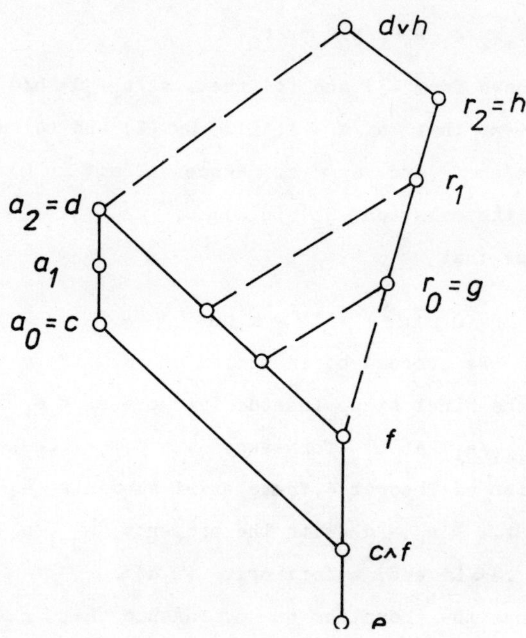

Fig. 50

(1) $r_1/r_0 \approx_w d/f$,

since $r_1/r_0 \searrow d \wedge h/d \wedge g$ and $d \wedge h/d \wedge g \approx_w d/f$. Using (1), together with $d/f \searrow c/c \wedge f$, we easily obtain the following formula

(2) $r_1/r_0 \approx_w c/c \wedge f$.

From $[r_1, r_2] \subset [d \wedge h, r_2]$ we infer that $r_2/r_1 \approx_w r_2/d \wedge h$. According to $r_2 \vee d = d \vee h$ and $r_2 \wedge d = d \wedge h$, it is evident that $r_2/d \wedge h \nearrow$ $\nearrow d \vee h/d$. Hence

(3) $r_2/r_1 \approx_w d \vee h/d$.

Next we see that $d/c \searrow_w f/e \approx_w h/g$ and so $d/c \approx_w h/g$. Clearly, $r_0 =$ $= g \leqslant r_1 \leqslant r_2 = h$. It follows from Theorem 4 that there are a_i such that $a_0 = c \leqslant a_1 \leqslant a_2 = d$ with

(4) $a_2/a_1 \approx_w r_2/r_1$ & $a_1/a_0 \approx_w r_1/r_0$.

We therefore have from (3) and (4) that $a_2/a_1 \approx_w d \vee h/d$ and $d \geqslant$ $\geqslant a_2$. This means that $a_2/a_1 \in A({}^{\backprime}L)$. By (2) and (4) we conclude that $a_1/a_0 \approx_w c/c \wedge f$ and $a_0 \geqslant c$. Hence $a_1/a_0 \in A({}^{\backprime}L)$. By the definition of the relation $\hat{C}$, $(c,d) \in \hat{C}$.

2. Suppose that

(5) $d/c = q_0/p_0 \sim_w q_1/p_1 \sim_w \ldots \sim_w q_m/p_m = q/p$

and $(p,q) \in \hat{C}$. We proceed by induction on m. If $m = 1$, then $(c,d) \in \hat{C}$ by the first step. Indeed, let $p = e_0 < e_1 < \ldots < e_k =$ $= q$ where $e_{i+1}/e_i \in A({}^{\backprime}L)$ for every $i = 0,1,\ldots,k-1$. By the second assertion of Theorem 4 there exist elements a_j such that $c = a_0 \leqslant a_1 \leqslant \ldots \leqslant a_k = d$ with the property $a_{j+1}/a_j \sim_w e_{j+1}/e_j$. By 1, $(a_j, a_{j+1}) \in \hat{C}$ and, a fortiori, $(c,d) \in \hat{C}$.

To continue the induction on m, assume that $m > 1$ and that the assertion of the theorem holds for any two quotients which are joined by a sequence of the form (5) having the length not greater than m-1. By the argument above, $(p_{m-1}, q_{m-1}) \in \hat{C}$. By the induction assumption, we then have $(c,d) \in \hat{C}$. //

Theorem 1.6. Let $`L$ be a lattice. The relation $\hat{C}$ defined in Theorem 5 is a congruence relation of the lattice $`L$.

Proof. We shall use Grätzer-Schmidt's Theorem V.3.3. The relation $\hat{C}$ is clearly reflexive and symmetric. Setting $d = a \wedge b$ and $s = a \vee b$, we get $d \wedge s = a \wedge b$ and $d \vee s = a \vee b$. It follows that $(a,b) \in \hat{C}$ if and only if $(a \wedge b, a \vee b) \in \hat{C}$. If $a \leqslant b \leqslant c$ and $(a,b) \in \hat{C}$, $(b,c) \in \hat{C}$, then $(a,c) \in \hat{C}$, by the definition of $\hat{C}$. Finally, suppose that $a \leqslant b$ and $(a,b) \in \hat{C}$. For any $c \in L$, $b \wedge c/a \wedge c \nearrow^{w} b/a$. Hence by Theorem 5, $(a \wedge c, b \wedge c) \in \hat{C}$. Dually we find $(a \vee c, b \vee c) \in \hat{C}$. //

The relation $\hat{C}$ will be called the coreflective congruence of the lattice $`L$. The quotient lattice $`L/\hat{C}$ is said to be the coreflection of the lattice $`L$ and it will be denoted by the abbreviation Coref $`L$.

Theorem 1.7. Let T be a congruence relation of a lattice $`L$. If $`L/T$ is a distributive lattice, then $T \supset \hat{C}$.

Proof. Let $(a,b) \in \hat{C}$. Then $a \wedge b = a_0 \leqslant a_1 \leqslant \ldots \leqslant a_n = a \vee b$ for suitable elements a_i such that $a_{i+1}/a_i \approx_w b_{i+1}/b_i$ where either $a_i \geqslant b_{i+1}$ or $b_i \geqslant a_{i+1}$. Denote by $[x]$ the congruence class of T determined by $x \in L$. Now

$$[a] \wedge [b] = [a_0] \leqslant [a_1] \leqslant \ldots \leqslant [a_n] = [a] \vee [b]$$

and $[a_{i+1}]/[a_i] \approx_w [b_{i+1}]/[b_i]$ in the lattice $`L/T$ where either $[a_i] \geqslant [b_{i+1}]$ or $[b_i] \geqslant [a_{i+1}]$. But it means that $[a_{i+1}]/[a_i]$ belongs to the allelomorph $A(`L/T)$ of the lattice $`L/T$. By Theorem 3, the distributivity of $`L/T$ implies that $[a_i] = [a_{i+1}]$ for every $i = 0,1,\ldots,n-1$. Hence $[a] \wedge [b] = [a] \vee [b]$ and this yields $[a] = [b]$, i.e., $(a,b) \in T$. Consequently, $\hat{C} \subset T$. //

See Exercise VI;9 for another proof.

Theorem 1.8 (cf. [188, Lemma 2, p. 95]). Let T be a congruence of a lattice $`L$. Suppose $[b]/[a] \approx_w [d]/[c]$ in the lattice $`L/T$. Then there are elements $`a, `b, `c, `d$ such that $`a \in [a]$, $`b \in [b]$, $`c \in [c]$, $`d \in [d]$ and such that $`b/`a \approx_w `d/`c$.

Proof. Since $[c] \leqslant [d]$, $[c \wedge d] = [c] \wedge [d] = [c]$. Hence we may assume without loss of generality that the element c is chosen in such a way that $c \leqslant d$. Let

$$[b]/[a] = [b_0]/[a_0] \sim_w [b_1]/[a_1] \sim_w \ldots \sim_w [b_n]/[a_n] = [d]/[c].$$

Define $`b_n = d = `d$, $`a_n = c = `c$. Suppose we have already constructed

$$`b_n, `b_{n-1}, \ldots, `b_{j+1}, `a_n, `a_{n-1}, \ldots, `a_{j+1}$$

such that $[`b_i] = [b_i]$ and $[`a_i] = [a_i]$ for every $i = n, n-1, \ldots$ $\ldots, j+1$, and such that the following implications hold for every $i = j+1, \ldots, n-1$:

$$([b_i]/[a_i] \nearrow^w [b_{i+1}]/[a_{i+1}]) \Rightarrow `b_i/`a_i \nearrow^w `b_{i+1}/`a_{i+1};$$

$$([b_i]/[a_i] \searrow_w [b_{i+1}]/[a_{i+1}]) \Rightarrow `b_i/`a_i \searrow_w `b_{i+1}/`a_{i+1}.$$

Our goal is to show that one can find elements $`b_j$ and $`a_j$ satisfying the just mentioned conditions with i equal to j. By duality, no generality is lost in assuming that $[b_j]/[a_j] \nearrow^w$ $\nearrow^w [b_{j+1}]/[a_{j+1}]$. Putting $`b_j = `b_{j+1} \wedge b_j$, $`a_j = `a_{j+1} \wedge b_j$, we get

$$`b_j \vee `a_{j+1} \leqslant (`b_{j+1} \wedge b_j) \vee `b_{j+1} = `b_{j+1},$$

$$`a_j \leqslant `b_j \wedge `a_{j+1} \leqslant b_j \wedge `a_{j+1} = `a_j.$$

This implies that $`b_j/`a_j \nearrow^w `b_{j+1}/`a_{j+1}$. Furthermore,

$$[`b_j] = [`b_{j+1} \wedge b_j] = [`b_{j+1}] \wedge [b_j] = [b_{j+1}] \wedge [b_j] = [b_j]$$

and, similarly,

$$[`a_j] = [`a_{j+1}] \wedge [b_j] = [a_{j+1}] \wedge [b_j] = [a_j].$$

To complete the proof, it suffices to put $`b_0 = `b$, $`a_0 = `a$. //

Theorem 1.9. Let T be a congruence relation of a lattice $`L$. Suppose $[b]/[a]$ is an allele of the lattice $`L/T$. Then there exist elements $`a, `b$ such that $`a \in [a]$, $`b \in [b]$ and $`b/`a \in A(`L)$.

Proof. By duality, we lose nothing in the way of generality by considering only the case where

$$(*)\ [b]/[a] = [b_0]/[a_0] \sim_w [b_1]/[a_1] \sim_w \dots \sim_w [b_n]/[a_n] = [d]/[c]$$

and where $[c] \geqslant [b]$. Arguing similarly as in the proof of Theorem 8, we see that we can assume that $a \leqslant b \leqslant c \leqslant d$ and that $b/a = b_0/a_0$. Next assume that the sequence $(*)$ is chosen in such a way that n is the smallest possible number. By Lemma 1 (i) and (ii), this assumption means that the symbols $\nearrow^w, \searrow_w$ alternate regularly. Let $\{`b_j/`a_j\}$ be the sequence constructed for the chain $(*)$ in the proof of Theorem 8. We shall distinguish between two cases:

Case I: $[b]/[a] \nearrow^w [b_1]/[a_1]$. By the construction of $`b_j$, $`a_j$, we have

$$`b_0 = `b_1 \wedge b_0 = `b_1 \wedge b \leqslant b, \quad `a_0 = `a_1 \wedge b_0 = `a_1 \wedge b \leqslant `b_1 \wedge b = `b_0.$$

Since $`b_0/`a_0 \approx_w `d/`c = `b_n/`a_n$ and since $`b_0 \leqslant b \leqslant c = `a_n$, it is now immediate that $`b_0/`a_0 \in A(`L)$.

Case II: $[b]/[a] \searrow_w [b_1]/[a_1]$ (see Figure 51). By the choice of the sequence above, $[b_1]/[a_1] \nearrow^w [b_2]/[a_2]$. Hence, the situation of the case I occurs for the quotients $[d]/[c]$, $[b_1]/[a_1]$. Therefore, $`b_1/`a_1 \approx_w `b_n/`a_n = d/c$ and $`b_1 \leqslant c$. By the construction of the elements $`a_0, `b_0$, we have $`a_0 = `a_1 \vee a$, $`b_0 = `b_1 \vee a$. Consequently,

$$`a_0 \wedge `b_1 = (`a_1 \vee a) \wedge `b_1 \geqslant `a_1 \wedge `b_1 = `a_1,$$

$$`a_0 \vee `b_1 = `a_1 \vee a \vee `b_1 = a \vee `b_1 = `b_0.$$

This implies that $`b_0/`a_0 \searrow_w `b_1/`a_1$. In summary, we find that $`b_0/`a_0 \searrow_w `b_1/`a_1 \approx_w d/c$, i.e., $`b_0/`a_0 \approx_w d/c$. Since $`b_1 \leqslant c$, $`b_0 = `b_1 \vee a \leqslant c \vee c = c$. Thus $`b_0/`a_0 \in A(`L)$. //

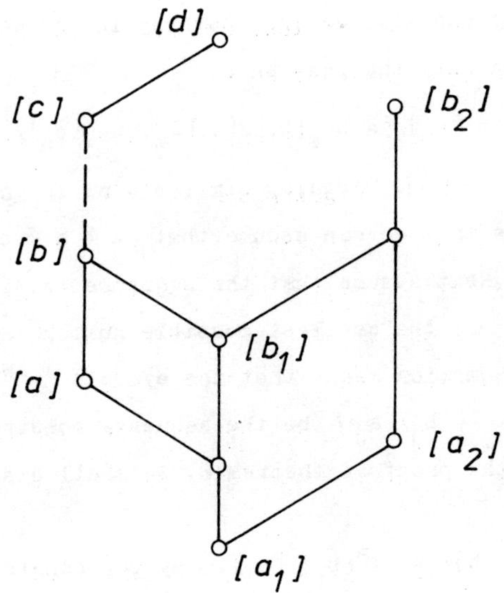

Fig. 51

Theorem 1.10 (cf. [16, Proposition 2.7, p. 334]). Let T be a congruence of a lattice $`L$. Then $`L/T$ is a distributive lattice if and only if $T \supset \hat{C}$.

Proof. By Theorem 7, it remains to show that $`L/T$ is distributive whenever $T \supset \hat{C}$. However, the lattice $`L/T$ is a homomorphic image of the lattice $`L/\hat{C}$. Hence, it suffices to prove that $`L/\hat{C}$ is a distributive lattice. To this end consider an allele $[b]/[a]\in A(`L/\hat{C})$. By Theorem 9, there exist elements $`a\in \in[a]$, $`b\in[b]$ such that $`b/`a$ belongs to the allelomorph $A(`L)$ of the lattice $`L$. It follows that $(`a,`b)\in\hat{C}$ so that $[a] = = [`a] = [`b] = [b]$. Thus, by Theorem 3, the lattice $`L/\hat{C}$ is necessarily distributive. //

Recall that an element a^* of a lattice $`L$ with 0 is called the pseudocomplement of $a\in L$ if and only if it has the following property for any $x\in L$:

$$a\wedge x = 0 \Leftrightarrow x \leqslant a^*.$$

We shall now construct (cf. [115; Lemma 14, p. 30], [116; 1;4, p. 51 and Theorem 3.5, p. 55], [16; Theorem 2.5, p. 333]) the pseudocomplement $\hat{C}^*$ of $\hat{C}$ in the lattice of all congruences on L. Let $\hat{B}$ be the binary relation defined on L by $(a,b)\in\hat{B}$ if and only if the following implication holds:

$$(d/c\approx_w a\vee b/a\wedge b \;\&\; d/c\in A(`L)) \Rightarrow d = c.$$

Theorem 1.11. In any lattice $`L$, the following is true:

(i) (cf. [115], [116], [16] loc. cit.) The relation $\hat{B}$ is a congruence relation of the lattice $`L$.

(ii) (Ibid.) Let D be a congruence relation of $`L$. Then $\hat{C}\cap D = \Delta$ if and only if $D\subset\hat{B}$. (Here $\Delta$ denotes the diagonal relation on $L$ and $\hat{C}$ is the coreflective congruence of $`L$.) In other words, $\hat{B} = \hat{C}^*$.

(iii) The lattice $`L$ is distributive if and only if $\hat{B} = L\times L$.

(iv) If $(a,b)\in\hat{B}$, then the interval $[a\wedge b, a\vee b]$ determines a distributive sublattice of $`L$.

Proof. Ad (i). We shall again use Grätzer-Schmidt's Theorem V.3.3. Clearly, $(a,b)\in\hat{B}$ if and only if $(a\wedge b, a\vee b)\in\hat{B}$. The relation $\hat{B}$ is reflexive and symmetric. Suppose that $a\leqslant b\leqslant c$,

$(a,b) \in \hat{B}$ and $(b,c) \in \hat{B}$. We shall prove that $(a,c) \in \hat{B}$ (see Figure 52). Let $f/e \in A(`L)$ be such that $f/e \approx_w c/a$. By the definition of an allele, there are $g \leqslant h$ such that $f/e \approx_w h/g$ and either $f \leqslant g$ (as it is shown in the figure) or $h \leqslant e$. Since

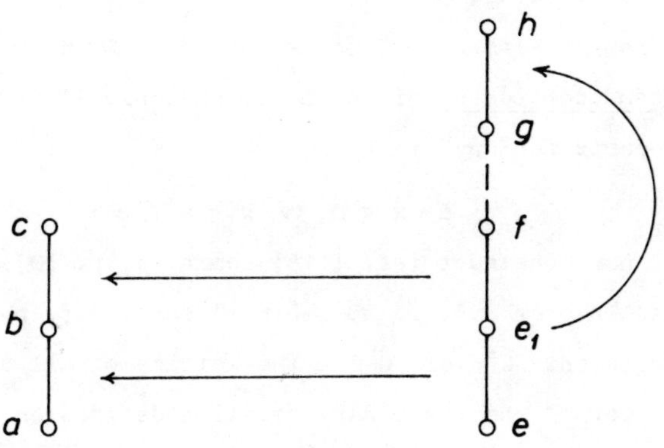

Fig. 52

$f/e \approx_w c/a$, we can conclude from Theorem 4 that there exist elements e_0, e_1, e_2 such that $e_0 = e \leqslant e_1 \leqslant e_2 = f$ and such that $e_1/e \approx_w$ $\approx_w b/a$, $f/e_1 \approx_w c/b$. Therefore, since trivially $e_1/e \approx_w f/e$ and $f/e \approx_w h/g$, we must have $e_1/e \in A(`L)$ and $e_1/e \approx_w b/a$. By the definition of $\hat{B}$. $e_1 = e$. Similarly we conclude that $f = e_1$. But this means that $e = f$ and so $(a,c) \in \hat{B}$.

Next suppose that $a \leqslant b$ and $(a,b) \in \hat{B}$. We shall prove that $(a \wedge c, b \wedge c) \in \hat{B}$ and $(a \vee c, b \vee c) \in \hat{B}$ for every $c \in L$. The first assertion can be proved in the following way: Let f/e be an allele such that $f/e \approx_w b \wedge c/a \wedge c$. Now $b \wedge c/a \wedge c \approx_w b/a$ and it follows that $f/e \approx_w b/a$. Since $(a,b) \in \hat{B}$, this yields $e = f$. The second statement follows by duality.

Ad (ii). Let D be such that $D \subset \hat{B}$. Then obviously $\hat{C} \cap D \subset$ $\subset \hat{C} \cap \hat{B}$. Choose $(a,b) \in \hat{C} \cap \hat{B}$, i.e., $(a,b) \in \hat{C}$ and $(a,b) \in \hat{B}$. By the definition of the coreflective congruence $\hat{C}$, there exist a_i such that

$$a \wedge b = a_0 \leq a_1 \leq a_2 \leq \dots \leq a_n = a \vee b$$

and such that $a_{i+1}/a_i \in A(`L)$ for every i. On the other hand, $(a,b) \in \hat{B}$ and $a_{i+1}/a_i \approx_w a \vee b/a \wedge b$, since $[a_i, a_{i+1}] \subset [a \wedge b, a \vee b]$. Hence, by the definition of the congruence $\hat{B}$ we have $a_i = a_{i+1}$ for every i. This shows that $a \wedge b = a \vee b$ and, a fortiori, $a = b$. Thus $\hat{C} \cap \hat{B} = \Delta$ and so $\hat{C} \cap D = \Delta$.

Conversely, suppose that $\hat{C} \cap D = \Delta$ and choose $(a,b) \in D$. Let $d/c \approx_w a \vee b/a \wedge b$ and $d/c \in A(`L)$. Since $D$ is a congruence and $(a,b) \in D$, $(a \vee b, a \wedge b) \in D$. By Lemma 1 (iii) this implies that $(c,d) \in D$. However, $d/c$ is an allele of the lattice `L. Hence $(c,d) \in \hat{C}$. By assumption, $\Delta = \hat{C} \cap D \ni (c,d)$ and so $c = d$. This proves that $(a,b) \in \hat{B}$, i.e., $D \subset \hat{B}$.

Ad (iii). The lattice `L is distributive if and only if the lattice $`L/\Delta \cong `L$ is distributive. Therefore, by Theorem 10, this is equivalent to $\Delta \supset \hat{C}$ which gives $\hat{C} = \Delta$. But $\hat{C} = \Delta$ if and only if $\hat{C} \cap (L \times L) = \Delta$. By (ii), this means that $\hat{B} = L \times L$.

Ad (iv). Suppose $(a,b) \in \hat{B}$ and let d/c be an allele of the sublattice determined by $[a \wedge b, a \vee b]$. Then $d/c \approx_w a \vee b/a \wedge b$ and so $c = d$. From Theorem 3 we see that the sublattice is in fact distributive. //

The congruence relation $\hat{B}$ is called the <u>reflective congruence</u> of the lattice `L. The quotient lattice $`L/\hat{B}$ is said to be the <u>reflection</u> of `L and it is denoted by Ref `L.

2. PROJECTIVE ALLELOMORPH

Let b/a be a quotient of a lattice $`L$. We shall write $b/a§d/c$ and say that $d/c$ is a close allele of $b/a$ if and only if $b/a \approx d/c$ and either $c \geqslant b$ or $a \geqslant d$. A quotient is said to be a projective allele of $`L$ if and only if it has a close allele.

The set $A(`L;\approx)$ of all the projective alleles of $`L$ will be called the projective allelomorph of $`L$.

Lemma 2.1. The following is true in any lattice $`L$:

(i) $[(b/a \searrow d/c) \ \& \ (d/c \searrow f/e)] \Rightarrow (b/a \searrow f/e)$;

(ii) $[(b/a \nearrow d/c) \ \& \ (d/c \nearrow f/e)] \Rightarrow (b/a \nearrow f/e)$;

(iii) the relation § is symmetric on the set of all quotients of $`L$;

(iv) for any congruence T of $`L$, if $b/a \approx d/c$, then $(c,d) \in T$ if and only if $(a,b) \in T$.

Proof. Ad (i). By Lemma 1.1 (i) and (iv), $b/a \searrow_w f/e$. In particular, we have $a \wedge f \geqslant e$. It remains to prove that $a \wedge f = e$. But $a \wedge f \leqslant a \wedge (c \vee f) = a \wedge d = c$ and so $a \wedge f \leqslant c \wedge f = e$.

Ad (ii). This statement follows by duality.

Ad (iii). This follows immediately from the definition of §.

Ad (iv). If $b/a \approx d/c$ and $(c,d) \in T$, then $(a,b) \in T$ by Lemma 1.1 (iii) and Lemma 1.1 (iv). Since $\approx$ is a symmetric relation, the converse implication is obvious. //

A lattice $`L$ is said to be weakly modular (cf. [49] and [72]) if and only if the following is true for every $m,n,p,q$ of $`L$: If n/m, q/p are quotients such that $m \neq n$ and such that n/m is weakly projective into q/p, then there exist elements p_1, q_1 satisfying $p \leqslant p_1 < q_1 \leqslant q$ and $q_1/p_1 \approx_w n/m$.

Theorem 2.2. (i) In any lattice $`L$, $A(`L;\approx) \subset A(`L)$.

(ii) Let `L be a lattice which is either modular or relatively complemented. Suppose that $a \leqslant {`a} \leqslant {`b} \leqslant b$, $c \leqslant d$ and $b/a \sim_w d/c$. Then there exist elements `c, `d such that $`b/`a \approx `d/`c$ and $c \leqslant$ $\leqslant `c \leqslant `d \leqslant d$.

(iii) Let

$$\bar{b}/\bar{a} = b_0/a_0 \sim_w b_1/a_1 \sim_w \ldots \sim_w b_n/a_n = \bar{d}/\bar{c}.$$

Under the hypotheses on `L of (ii), there exist elements $`a_i$, $`b_i$ $(i = 0,1,\ldots,n)$ such that $a_i \leqslant `a_i \leqslant `b_i \leqslant b_i$ and such that

$$\bar{b}/\bar{a} = `b_0/`a_0 \approx `b_1/`a_1 \approx \ldots \approx `b_n/`a_n.$$

(iv) The projective allelomorph is equal to the allelomorph whenever the lattice is modular or relatively complemented.

Proof. Ad (i). This follows from Lemma 1.1 (iv).

Ad (ii). Let $b/a \searrow_w d/c$.

1. Let `L be modular. Putting $`c = `a \wedge d$ and $`d = `b \wedge d$, it can be verified that $c \leqslant a \wedge d \leqslant `a \wedge d = `c \leqslant `d \leqslant d$. Further, $b = a \vee d \leqslant `a \vee d \leqslant b \vee d = b$ and so $b = `a \vee d$. By modularity, $`a \vee `d = `a \vee (`b \wedge d) = `b \wedge (`a \vee d) = `b \wedge b = `b$. At the same time, $`a \wedge `d = `a \wedge `b \wedge d = `a \wedge d = `c$. Hence $`b/`a \searrow `d/`c$ so that $`b/`a \approx `d/`c$.

2. Let `L be relatively complemented. Denote by $r$ the relative complement of `b in $[`a,b]$, i.e., $r \vee `b = b$ and $r \wedge `b = `a$. Setting $`c = r \wedge d$ and $`d = d$, we obtain $b \geqslant r \vee$ $\vee `d = r \vee d \geqslant `a \vee d \geqslant a \vee d = b$. Hence $b/r \searrow `d/`c$. Since $`b/`a \nearrow$ $\nearrow b/r$, we have $`b/`a \approx `d/`c$.

In the case where $b/a \nearrow^w d/c$ we can use a dual argument.

Ad (iii). The assertion follows from (ii) by induction.

Ad (iv). Suppose that $\bar{b}/\bar{a} \approx_w \bar{d}/\bar{c}$ where either $\bar{a} \geqslant \bar{d}$ or $\bar{b} \leqslant \bar{c}$. Then the chain constructed in (iii) verifies $\bar{c} \leqslant `a_n \leqslant$

$\leqslant {}^{`}b_n \leqslant \bar{d}$. Hence either $\bar{a} \geqslant {}^{`}b_n$ or $\bar{b} \leqslant {}^{`}a_n$. From this we conclude that $\bar{b}/\bar{a} \in A({}^{`}L;\approx)$ and so $A({}^{`}L) \subset A({}^{`}L;\approx)$. By (i) it follows that the projective allelomorph of `L is equal to the allelomorph of `L. //

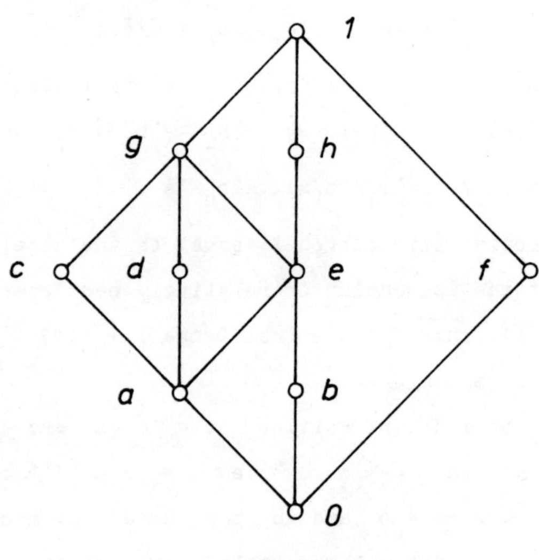

Fig. 53

Example 2.3. (A) From Theorem 2 (iii) we infer that every modular or relatively complemented lattice is weakly modular.

(B) The following relations hold in the lattice `L of Figure 53:

$$1/h \searrow f/0 \nearrow 1/g \searrow h/e; \quad 1/h \searrow g/e \searrow c/a \nearrow g/d \searrow e/a \nearrow g/c \searrow d/a;$$
$$g/c \searrow b/0 \nearrow 1/f \searrow a/0 \nearrow e/b.$$

If we have the situation as in the definition of a weakly modular lattice, then the assumption $m \neq n$ implies that in L there is a quotient n_1/m_1 such that $m \leqslant m_1 \prec n_1 \leqslant n$. By Lemma 1.1 (iii), $p \neq q$ and there exist p_1, q_1 satisfying $p \leqslant p_1 \prec q_1 \leqslant q$. By the argument above, $q_1/p_1 \approx n_1/m_1$. Since $n_1/m_1 \approx_w n/m$ by Lemma 1.1 (vi), $q_1/p_1 \approx_w n/m$. Thus the lattice of Figure 53 is weakly modular.

(C) By the part A, the class of weakly modular lattices contains the class of modular lattices and the class of relatively complemented lattices. In these two classes we have $A(`L;\approx) = A(`L)$. One might hope that this equality would remain true also for the class of weakly modular lattices. The lattice of Figure 53 eliminates that hope. Indeed, $g/a \nearrow x/y$ implies that $g \leqslant x$, $a \leqslant y$ and $g \wedge y = a$. A straightforward computation shows that $y = a$ and it follows that $x = g \vee y = g$. Since the lattice $`L$ is self-dual, we see that $g/a$ does not belong to $A(`L;\approx)$. On the other hand, $1/g$ is a close allele of g/a, since

$$g/a \nearrow^{w} 1/a \searrow f/0 \nearrow 1/g.$$

Thus $g/a \in A(`L)$. Note that $`L$ is neither modular nor relatively complemented.

Theorem 2.4 (cf. [116, Theorem 3.8, p. 56], [16, p. 333]). Let $\hat{B}$ be the reflective congruence of a weakly modular lattice. Then $(a,b) \in \hat{B}$ if and only if the following implication is true for every $m,n \in L$:

$$([m,n] \subset [a \wedge b, a \vee b] \ \& \ n/m \in A(`L)) \Rightarrow m = n.$$

Proof. 1. Suppose that $n/m \in A(`L)$, $(a,b) \in \hat{B}$ and $[m,n] \subset [a \wedge b, a \vee b]$. Lemma 1.1 (vi) gives $n/m \approx_w a \vee b/a \wedge b$. Hence, by the definition of $\hat{B}$, $n = m$.

2. Let $d/c \approx_w a \vee b/a \wedge b$ and $d/c \in A(`L)$. Let the condition mentioned in the theorem be valid. We claim that $c = d$. Suppose the contrary. Then by weak modularity of $`L$ we conclude that there exist elements c', d' such that $a \wedge b \leqslant c' < d' \leqslant a \vee b$ and

$d'/c' \approx_w d/c$. Therefore, by Theorem 1.5, $(c',d') \in \hat{C}$ and so

$$a \wedge b \leqslant c' = e_0 \leqslant e_1 \leqslant \ldots \leqslant e_t = d' \leqslant a \vee b$$

with $e_{i+1}/e_i \in A(`L)$ for $i = 0,1,\ldots,t-1$. By assumption, $c' = = e_0 = e_1 = \ldots = e_t = d'$, a contradiction. Thus $c = d$ and $(a,b) \in \hat{B}$. //

<u>Corollary 2.5</u>. Let `L be a modular or relatively complemented lattice. Then:

(i) The relation $(a,b) \in \hat{B}$ holds if and only if the following implication is valid for every m,n,a,b,p,q:

$$[([m,n] \subset [a \wedge b, a \vee b]) \ \& \ (n/m \S q/p)] \Rightarrow m = n.$$

In other words, $(a,b) \in \hat{B}$ if and only if the following implication is satisfied for every s,t,a,b:

$$[([s,t] \subset [a \wedge b, a \vee b]) \ \& \ (t/s \in A(`L))] \Rightarrow s = t.$$

(ii) The relation $(c,d) \in \hat{C}$ holds if and only if there exists k and elements $c_0, c_1, \ldots, c_k$ satisfying

$$c \wedge d = c_0 \leqslant c_1 \leqslant \ldots \leqslant c_k = c \vee d$$

and such that every c_{i+1}/c_i has a close allele.

<u>Proof</u>. The corollary follows from Theorems 4, 2 (iv) and Example 3 A. //

<u>Theorem 2.6</u>. A lattice `L belongs to a nontrivial equational class $\underline{K}$ of lattices if and only if its reflection belongs to $\underline{K}$.

<u>Proof</u>. 1. If `L belongs to $\underline{K}$, then also its quotient lattice $`L/\hat{B}$ = Ref `L belongs to $\underline{K}$.

2. Let Ref $`L \in \underline{K}$. By Theorem 1.11 (ii), $\hat{C} \cap \hat{B} = \Delta$. Hence, by I.25, `L is isomorphic to a subdirect product of the lattices $`L/\hat{C}$ = Coref `L and $`L/\hat{B}$ = Ref `L. By Theorem 1.10, Coref $`L \in \underline{D}$

where $\underline{D}$ denotes the class of distributive lattices. Now $\underline{D} \subset \underline{K}$, by I.27, and so $`L$ is isomorphic to a subdirect product of lattices belonging to $\underline{K}$. Thus $`L \subset \underline{K}$. //

3. COMMUTATOR SUBLATTICES

An element a of a lattice $`L$ having $0$ is said to be $\hat{C}$-accessible (or, more specifically, $\hat{C}(`L)$-accessible) if and only if $(a,0) \in \hat{C}$ where $\hat{C}$ is the coreflective congruence of $`L$. The set $\hat{C}(`L)$ of all $\hat{C}$-accessible elements of $`L$ is the kernel of the congruence $\hat{C}$.

An element b of such a lattice $`L$ is called fully $\hat{C}$-accessible (fully $\hat{C}(`L)$-accessible) if and only if $(b,0) \in \hat{C}([0,b])$ where $\hat{C}([0,b])$ is the coreflective congruence of the interval sublattice $`L[0,b]$ of the lattice $`L$. We shall denote by $F(`L)$ the set of all the fully $\hat{C}(`L)$-accessible elements of $`L$.

It is clear that $F(`L) \subset \hat{C}(`L)$.

Lemma 3.1. Let a,b be elements of an orthomodular lattice $`L$. Then $\mathrm{com}\,(a,b)$ is fully $\hat{C}$-accessible.

Proof. If $\mathrm{com}\,(a,b) = 0$, the assertion is trivial. Suppose $\mathrm{com}\,(a,b) \neq 0$. Then there exists an orthohomomorphism f of the free orthomodular lattice F_2 (see Section III.2) with two generators x,y such that $f(x) = a$ and $f(y) = b$. We have

$$f(\mathrm{com}\,(x,y)) = f((x \vee y) \wedge (x \vee y') \wedge (x' \vee y) \wedge (x' \vee y')) =$$
$$= (f(x) \vee f(y)) \wedge (f(x) \vee f(y')) \wedge \ldots = (f(x) \vee f(y)) \wedge (f(x) \vee$$
$$\vee [f(y)]') \wedge \ldots = (a \vee b) \wedge (a \vee b') \wedge \ldots = \mathrm{com}\,(a,b).$$

By Corollary III.2.4, (com (a,b)] contains a subalgebra, say $\{0,d,d^+,e,e^+,c = \text{com}(a,b)\}$, isomorphic to $`M_{02}$. Then

$$(*) \quad c/d \searrow e/0 \nearrow c/d^+ \searrow d/0.$$

Hence c/d and $d/0$ are alleles of the interval algebra $`L[0,\text{com}(a,b)]$. Thus $\text{com}(a,b) \in F(`L)$. //

Observe that the choice $d = a \wedge \text{com}(a,b)$, $d^+ = a' \wedge \text{com}(a,b)$, $e = b \wedge \text{com}(a,b)$, $e^+ = b' \wedge \text{com}(a,b)$ and a routine calculation using Theorem II.3.10 yield (*) and, therefore, give another more elementary proof of Lemma 1.

In analogy with group theory we now adopt the following definition: Let $`G$ be a generalized orthomodular lattice. Put $`G^{(0)} = `G$ and, with this start, suppose the generalized orthomodular lattices $`G^{(0)}, `G^{(1)}, \ldots, `G^{(n-1)}$ $(n \geq 1)$ have already been defined. We next define $`G^{(n)}$ by $`G^{(n)} = (`G^{(n-1)})'$. Then $`G^{(n)}$ is a generalized orthomodular lattice by Theorem V.5.6 (i) and Example V.1.1.C. It will be called the <u>n-th commutator sublattice</u> of $`G$. The first commutator sublattice $`G^{(1)} = `G'$ is called the <u>commutator sublattice</u> of $`G$. As customary, the second commutator sublattice $`G^{(2)}$ is also denoted by $`G''$ etc.

<u>Theorem 3.2</u>. Let $`G$ be a generalized orthomodular lattice. Then $G' = \hat{C}(`G)$. In other words, the base set of the commutator sublattice of a generalized orthomodular lattice is formed by the $\hat{C}$-accessible elements of that lattice.

<u>Proof</u>. By Theorem V.5.6 and V.4.20, G' is the kernel of exactly one congruence relation P of the lattice $(G, \vee, \wedge)$. Let I be the kernel of a congruence T. Suppose that $`G/T$ is distributive. By Theorem V.4.18, I is a p-ideal and, by Theorem V.5.6 (iii), $I \supset G'$. Hence, by Theorem V.4.8, $T \supset P$. Conversely,

suppose that $T \supset P$, i.e., $I \supset G'$. By Theorem V.5.6, $`G/G'$ and, a fortiori, $`G/I$ is distributive. In summary, $`G/T$ is distributive if and only if $T \supset P$. But in view of Theorem 1.10, $`G/T$ is distributive if and only if $T \supset \hat{C}$. Thus $\hat{C} = P$ and, consequently, these congruences have the same kernel, i.e., we have $\hat{C}(`G) = G'$. //

The following theorem gives a characterization of the coreflective congruence via relative complements.

Theorem 3.3. Let $`G$ be a generalized orthomodular lattice. Then $(a,b) \in \hat{C}$ if and only if a relative complement c of $a \wedge b$ in the interval $[0, a \vee b]$ belongs to G'.

Proof. Evidently, $a \vee b / a \wedge b \searrow c/0$. Thus $(a,b) \in \hat{C}$ if and only if $(0,c) \in \hat{C}$. Hence, by Theorem 2, $(a,b) \in \hat{C}$ if and only if c belongs to G'. //

Theorem 3.4. Let $`G$ be a generalized orthomodular lattice. Then the commutator sublattice $`G'$ is equal to the ideal of $`G$ generated by all the fully $\hat{C}(`G)$-accessible elements.

Proof. Denote by J the ideal of the theorem. Since $F(`G) \subset \subset \hat{C}(`G) = G'$, $J \subset G'$. By Lemma 1, $\mathrm{com}_{[0,b]}(c,d) \in F(`G[0,b]) \subset \subset F(`G)$ for any $b \in G$ and any $c,d \in [0,b]$. Therefore, by the definition of G', $G' \subset J$. Thus $G' = J$. //

Recall that a lattice $`L$ is said to be simple if and only if the diagonal relation Δ and the universal relation $L \times L$ are its only congruence relations.

Example 3.5. The lattice of Figure 53 is simple (cf. Example 2.3.B and Lemma 1.1 (iii)). Note that every element of this lattice is $\hat{C}$-accessible. This is a special case of the following more general statement: Every element of a simple nondistributive lattice $`L$ with $0$ is $\hat{C}(`L)$-accessible. Indeed, since $`L$ is nondistributive, $\hat{C} \neq \Delta$, by Theorem 1.3 and 1.6. From the simplicity of $`L$ it follows that $\hat{C} = L \times L$.

<u>Theorem 3.6</u>. Let $`G$ be a generalized orthomodular lattice. Suppose $(G, \vee, \wedge)$ is a simple nondistributive lattice. Then $`G = `G'$.

<u>Proof</u>. By Example 5, every element of the lattice $`G$ is $\hat{C}(`G)$-accessible. It follows from Theorem 2 that $`G = `G'$. //

The following result was obtained by E.L.Marsden (cf. [147, Thm 9]) in a different way.

<u>Theorem 3.7</u>. Let $`G$ be a generalized orthomodular lattice. Then $`G' = `G'' = \ldots$.

<u>Proof</u>. By Theorem 4, the ideal $(F(`G)]_{`G}$ generated in $`G$ by all fully $\hat{C}$-accessible elements is equal to $G'$. We therefore have $F(`G) \subset G'$ and it follows that $F(`G) \subset F(`G')$, since G' is an ideal. Consequently,

(1) $(F(`G)]_{`G} \subset (F(`G')]_{`G}$.

Now G' is an ideal in $`G$ and $F(`G') \subset G'$. Hence, by Theorem 4,

(2) $(F(`G')]_{`G} = (F(`G')]_{`G'} = G''$.

From (1) and (2) we infer that $G' = (F(`G)]_{`G} \subset G''$. However, it is always true that $G'' \subset G'$. Thus $`G' = `G''$. //

<u>Theorem 3.8</u>. Let $`G$ be a generalized orthomodular lattice under a generalized orthogonality relation $P$ and let $`H$ be a generalized orthomodular lattice under a generalized orthogonality relation Q. Let f be a homomorphism of the lattice $(G, \vee, \wedge)$ into the lattice $(H, \vee, \wedge)$. Let the commutator sublattices of $`G$ and $`H$ be denoted by $`G^P$ and $`H^Q$, respectively. Then $f(G^P) \subset H^Q$.

If, moreover, f is an isomorphism from $(G, \vee, \wedge)$ onto $(H, \vee, \wedge)$, then $f(G^P) = H^Q$.

Proof. Suppose f is a homomorphism. Let $g \in G$ be an element which is $\hat{C}((G, \vee, \wedge))$-accessible. Hence, by Theorem 2.2 (iv), there exist suitable elements g_i such that

$$0 = g_0 \leqslant g_1 \leqslant \ldots \leqslant g_n = g$$

and $g_{i+1}/g_i \S K_i/G_i$ for every $i = 0,1,\ldots,n-1$. Since f is a homomorphism, this implies that

$$0 = f(g_0) \leqslant f(g_1) \leqslant \ldots \leqslant f(g_n) = f(g)$$

and $f(g_{i+1})/f(g_i) \S f(K_i)/f(G_i)$. Now it is easily seen that $f(g)$ is an $\hat{C}((H, \vee, \wedge))$-accessible element. Thus the first assertion of the theorem follows from Theorem 2.

Next, suppose that f is an isomorphism. Here we can replace `G by `H and f by f^{-1}. By what we have already proved, $f^{-1}(H^Q) \subset G^P$, i.e., $H^Q \subset f(G^P)$ and so $f(G^P) = H^Q$. //

Theorem 3.9 (cf. [15, Thm 5]). Let `$G = (G, \vee, \wedge)$ be a lattice. Suppose there are two generalized orthogonality relations $P, Q$ defined on $G$ so that the corresponding generalized orthomodular lattices have `G^P and `$G^Q$ as commutator sublattices, respectively. Then `G^P = `G^Q.

Proof. The assertion of the theorem follows from Theorem 8 where we put $(G, \vee, \wedge) = (H, \vee, \wedge)$ and where f is equal to the identity mapping on G. //

Theorem 3.10. Let h be an endomorphism of a lattice $(G, \vee, \wedge)$. If `G is a generalized orthomodular lattice, then $h(G') \subset G'$. If, moreover, h is an automorphism of $(G, \vee, \wedge)$, then $h(G') = G'$.

Proof. Put $f = h$ and $H = G$ in Theorem 8. //

Theorem 3.11. Suppose a lattice $(G, \vee, \wedge)$ is the direct product of lattices $`H, `K$. Let $`G$ be a generalized orthomodular lattice with generalized orthogonality relation P. Let $g = (h,k) \in G$, $a = (b,c) \in G$ where $g \leqslant a$ and where $b, h \in H$, $c, k \in K$. Define $h^{Q(b)}$, $k^{R(c)}$ by $(h,k)^{P(a)} = (h^{Q(b)}, k^{R(c)})$. Then Q and R are generalized orthogonality relations on H and K, respectively.

For the proof see Exercise VI;10.

Theorem 3.12. Suppose that $`G$ is a generalized orthomodular lattice which is the direct product $`H \times `K$ of lattices $`H, `K$. Let $g = (g_1, g_2)$, $q = (q_1, q_2)$. Then

$$(\overline{\mathrm{com}}_{[0,q_1 \vee g_1]}(q_1, g_1), \overline{\mathrm{com}}_{[0,q_2 \vee g_2]}(q_2, g_2)) = \overline{\mathrm{com}}_{[0,q \vee g]}(q,g).$$

Proof. By Theorem 11,

$$\overline{\mathrm{com}}_{[0,q \vee g]}(q,g) = (q \vee g) \wedge (q \vee g^{P(q \vee g)}) \wedge (q^{P(q \vee g)} \vee g) \wedge (q^{P(q \vee g)} \vee$$
$$\vee g^{P(q \vee g)}) = (q_1 \vee g_1, \bullet) \wedge (q_1 \vee g_1^{Q(q_1 \vee g_1)}, \bullet) \wedge (q_1^{Q(q_1 \vee g_1)} \vee$$
$$\vee g_1, \bullet) \wedge (q_1^{Q(q_1 \vee g_1)} \vee g_1^{Q(q_1 \vee g_1)}, \bullet) =$$
$$= (\overline{\mathrm{com}}_{[0,q_1 \vee g_1]}(q_1, g_1), \bullet). \quad //$$

Theorem 3.13. Let $`G$ be a generalized orthomodular lattice which is isomorphic to the direct product $`H \times `K$ of lattices $`H$, $`K$. Then $`G' \cong `H' \times `K'$ and $(`H \times `K)' = `H' \times `K'$.

Proof. By Theorem 8, we need only show that $(`H \times `K)' = `H' \times `K'$. First observe that $`H' \times `K'$ is an ideal of the lattice $`H \times `K$. From Theorem 12 we conclude that this ideal contains every upper commutator $\overline{\mathrm{com}}_{[0,q \vee g]}(q,g)$ for any q, g of the direct product $`H \times `K$. Hence $(`H \times `K)' \subset `H' \times `K'$.

However, if t is an element of $`H' \times `K'$, then $t = (h', k')$ where

$$H' \ni h' \leq \bigvee_{i=1}^{m} \operatorname{com}_{[0,h_i \vee h_i^*]}(h_i, h_i^*),$$

$$K' \ni k' \leq \bigvee_{j=1}^{n} \operatorname{com}_{[0,k_j \vee k_j^*]}(k_j, k_j^*).$$

In these expressions we do not assume that the commutators on the right are distinct and so we can repeat them. Hence, we may suppose that $m = n$. By Theorem 12,

$$(h',k') \leq (\bigvee_{i=1}^{m} \overline{\operatorname{com}}_{\ldots}(h_i,h_i^*), \bigvee_{i=1}^{m} \overline{\operatorname{com}}_{\ldots}(k_i,k_i^*)) =$$

$$= \bigvee_{i=1}^{m} (\overline{\operatorname{com}}_{[0,h_i \vee h_i^*]}(h_i,h_i^*), \overline{\operatorname{com}}_{[0,k_i \vee k_i^*]}(k_i,k_i^*)) =$$

$$= \bigvee_{i=1}^{m} \overline{\operatorname{com}}_{[0,q_i \vee g_i]}(q_i,g_i) = r$$

where $q_i = (h_i,k_i)$, $g_i = (h_i^*,k_i^*)$ are elements of $`H \times `K$. At the same time, r is an element of the ideal $(`H \times `K)'$ and so, a fortiori, this ideal contains the element $t = (h',k')$. Now it is immediate that we have also proved the converse inclusion $`H' \times `K' \subset (`H \times `K)'$. //

An element a of a lattice $`L$ with $0$ is said to be <u>$\hat{B}$-accessible</u> if and only if $(a,0) \in \hat{B}$ where $\hat{B}$ is the reflective congruence of $`L$. The set of all $\hat{B}$-accessible elements of $`L$ will be denoted by $\hat{B}(`L)$.

Since $\hat{B}(`L)$ is the kernel of the congruence relation $\hat{B}$, it is obvious that this set is an ideal of the lattice $`L$.

<u>Theorem 3.14</u>. Let $`L$ be a relatively complemented lattice with $0$. Suppose that $`L$ satisfies the ascending chain condition in every interval $[0,a]$, $a \in L$. Then $`L$ is the direct sum (cf. I.17) of the sublattice $\hat{C}(`L)$ of all $\hat{C}$-accessible elements and the sublattice $\hat{B}(`L)$ of all $\hat{B}$-accessible elements.

<u>Proof</u>. Let $a \neq 0$. Since $[0,a]$ satisfies the ascending

chain condition, $(a]\cap\hat{C}(`L) = (c]$ with $c\in L$. Let $e\in[0,a]$, $(0,e)\in\hat{C}$ and let e^+ be a relative complement of e in $[0,a]$. Since $e/0\nearrow a/e^+$,

(1) $(e^+,a)\in\hat{C}$.

Let c^+ be a relative complement of c in $[0,a]$. If $0\leqslant u\leqslant v\leqslant c^+$ with $v/u\in A(`L)$ and if $r$ is a relative complement of $u$ in $[0,v]$, then $v/u\searrow r/0$ and so $(0,r)\in\hat{C}$. Hence $r\leqslant c$ and $r\leqslant c^+$, i.e., $r = 0$. Thus $u = v$ and we have $a = c\vee c^+$ with $c^+\in\hat{B}(`L)$.

If $a = c_1\vee d_1$ $(c_1\in\hat{C}(`L),\ d_1\in\hat{B}(`L))$, then $(0,c^+\vee d_1)\in\hat{B}$ and, by (1), $(c^+\wedge d_1,a)\in\hat{C}$. Hence $(c^+,d_1)\in\hat{B}\cap\hat{C} = \Delta$ and so $c^+ = d_1$. From $c_1\leqslant c$, $c^+/0\nearrow a/c_1$ it follows that $(c_1,c)\in\hat{C}\cap\hat{B} = \Delta$, i.e., $c_1 = d_1$. //

<u>Remark 3.15</u>. Note that, for any finitely generated orthomodular lattice, a result analogous to that of Theorem 14 is also true (see Theorem VII.2.6).

<u>Theorem 3.16</u>. Let $`L$ be a relatively complemented lattice with 0 which satisfies the ascending chain condition in every interval $[0,a]$, $a\in L$. Then

(i) $\hat{C}(`L)\cong `L/\hat{B} = \mathrm{Ref}\ `L$ & $\hat{B}(`L)\cong `L/\hat{C} = \mathrm{Coref}\ `L$;

(ii) $`L$ is isomorphic to the direct product of its reflection and coreflection.

<u>Proof</u>. By Theorem 14 and I.24,

$$`L = \hat{C}(`L)\times\hat{B}(`L)\cong `L/\hat{B}\times `L/\hat{C} = \mathrm{Ref}\ `L\times\mathrm{Coref}\ `L.\ //$$

<u>Theorem 3.17</u>. Let $`G$ be a generalized orthomodular lattice satisfying the ascending chain condition in every interval $[0,a]$, $a\in G$. Then

(i) the commutator sublattice G' is isomorphic to $\mathrm{Ref}\ `G$;

(ii) the lattice `G is the direct sum of `G′ and `H where `H is isomorphic to the coreflection Coref `G.

Proof. Use Theorem 16 and 2. //

In [49; Thm 4.4, p. 355], Dilworth showed that each relatively complemented lattice `L of finite length can be written as a direct sum of sublattices $`S_1, `S_2, \ldots, `S_k, `D_1, \ldots, `D_m$ where $`S_i$ and $`D_i$ are simple lattices ($`D_i$ distributive and $`S_i$ nondistributive), $k, m \in \underline{N}_0$. The reader is referred to the Szász′s monograph [179; Thm 102, p. 202] where he can find the corresponding proof. Using this result, we shall prove the following theorem:

Theorem 3.18. Let `L be an orthomodular lattice of finite length. Then its commutator sublattice `L′ is the direct sum of sublattices $`S_1, `S_2, \ldots, `S_k$ where $k \in \underline{N}_0$ and $`S_i$ are simple orthomodular lattices which are not distributive. In the case $k = 0$ we mean by the direct sum of the sublattices the one-element sublattice ⟨0⟩. Moreover, the lattice `L is the direct sum of `L′ and m two-element sublattices, $m \in \underline{N}_0$. In the case $m = 0$ this means that `L = `L′.

Proof. Let $`S_1, `S_2, \ldots, `S_k, `D_1, \ldots, `D_m$ be lattices mentioned above in the Dilworth′s theorem. By an argument analogous to the second part in the proof of Lemma I.17, the mapping f which assigns to the element $(a_1, \ldots, a_k, b_1, \ldots, b_m)$ of the direct product $`P = `S_1 \times `S_2 \times \ldots \times `S_k \times `D_1 \times \ldots \times `D_m$ the element $a_1 \vee \ldots \vee a_k \vee b_1 \vee \ldots \vee b_m$ of L is an isomorphism of the lattice `P onto the lattice `L. According to Example 5 and Theorem 4, $`S_i' = `S_i$. Since $`D_i$ are distributive lattices, we can conclude from Theorem 4 and 1.3 that $`D_i' = \langle 0 \rangle$. Hence, by Theorem 13, $`P' = `S_1 \times `S_2 \times \ldots \times \langle 0 \rangle$. Theorem 8 says that $f(P') = L'$. Under the isomorphism f, the direct product $`S_1 \times `S_2 \times \ldots \times \langle 0 \rangle$ is mapped onto a sublattice of the lattice L which is the direct sum of the sublattices $`S_1, `S_2, \ldots, `S_k$. Hence $`L'$ is the direct sum of the sublattices $`S_1, `S_2, \ldots, `S_k$. This implies that $`L$ is the direct sum of the sublattices $`L', `D_1, \ldots, `D_m$. By assumption, any $`D_i$ determines a finite Boolean algebra isomorphic to $\underline{2}^{n(i)}$. Since $`D_i$ is simple, $n(i) = 1$ for every $i$, by I.24. Consequently, $`D_i \cong \underline{2}$. //

4. SOLVABILITY IN EQUATIONAL CLASSES OF LATTICES

Recall (cf. [75, p. 138]) that an element d of a lattice $`L$ is said to be distributive if and only if $d \vee (x \wedge y) = (d \vee x) \wedge (d \vee y)$ for every $x, y \in L$. An element e is called dually distributive if and only if $e \wedge (x \vee y) = (e \wedge x) \vee (e \wedge y)$ for every x, y of L.

In accordance with this terminology, an ideal I of a lattice $`L$ is called dually distributive if and only if it is a dually distributive element in the lattice of all ideals of the lattice $`L$, i.e., in the ideal lattice $`I(`L)$.

Theorem 4.1. The commutator sublattice of a generalized orthomodular lattice is a dually distributive ideal.

Proof. We need only show that

$$G' \cap (I_1 \vee I_2) \subset (G' \cap I_1) \vee (G' \cap I_2)$$

for any two ideals I_1, I_2 of a generalized orthomodular lattice $`G$. Let $m \in G' \cap (I_1 \vee I_2)$, i.e., $m \in G'$ and $m \leqslant i_1 \vee i_2$ where $i_s \in I_s$ for $s = 1,2$. Denote by x' the orthocomplement of $x \in$

$\in [0, i_1 \vee i_2]$ in the orthomodular lattice $`G[0, i_1 \vee i_2]$. Let

$$k_1 = i_1 \wedge (i_1' \vee m \vee i_2) \wedge (i_1' \vee m \vee i_2'),$$

$$k_2 = i_2 \wedge (i_2' \vee m \vee i_1) \wedge (i_2' \vee m \vee i_1')$$

so that $k_s = i_s \wedge c \in I_s$ where

$$c = (i_1' \vee m \vee i_2) \wedge (i_1' \vee m \vee i_2') \wedge (i_2' \vee m \vee i_1).$$

Let $w = i_1 \wedge (i_1' \vee m)$ and $v = i_1 \wedge m$. Since $v \leqslant m \in G'$, $v \in G'$. Evidently,

$$w \wedge v' = i_1 \wedge (i_1' \vee m) \wedge (i_1' \vee m') = i_1 \wedge \text{com}\,(i_1, m) \in G'$$

and so $w = v \vee (w \wedge v') \in G'$.

Let $w^+ = w' \wedge k_1$. By Remark II.4.3, the element i_1 commutes with i_1' and $i_1 \wedge m'$. Hence $[i_1' \vee (i_1 \wedge m')] \wedge i_1 = i_1 \wedge m'$. Therefore,

$$\begin{aligned} w^+ &= [i_1' \vee (i_1 \wedge m')] \wedge i_1 \wedge (i_1' \vee m \vee i_2) \wedge (i_1' \vee m \vee i_2') = \\ &= i_1 \wedge m' \wedge (i_1' \vee m \vee i_2) \wedge (i_1' \vee m \vee i_2') \leqslant \\ &\leqslant (i_1' \vee m \vee i_2) \wedge (i_1' \vee m \vee i_2') \wedge [(i_1 \wedge m') \vee i_2] \wedge [(i_1 \wedge m') \vee i_2'] = \\ &= \text{com}\,(i_1' \vee m, i_2) \in G'. \end{aligned}$$

Consequently $w^+ \in G'$ and it follows that $k_1 = w \vee w^+ \in G' \cap I_1$. By symmetry, $k_2 \in G' \cap I_2$.

It is easily seen that $m \leqslant (i_1 \vee i_2) \wedge c$ and that every element i_s commutes with every element $(\ldots)$ of the definition of the element c. Hence, by Theorem II.4.2, i_1 and i_2 commute with c. From Theorem II.3.10 we conclude that

$$m \leqslant (i_1 \vee i_2) \wedge c = (i_1 \wedge c) \vee (i_2 \wedge c) = k_1 \vee k_2.$$

This shows that $m \in (G' \cap I_1) \vee (G' \cap I_2)$ which completes the proof of the theorem. //

<u>Lemma 4.2</u>. Let $`L$ be a relatively complemented lattice with

0 and let f be a lattice-homomorphism of $`L$ into a lattice $`T$ with 0. Suppose that $f(0) = 0$. Then f is injective if and only if the set

$$\mathrm{Ker}\, f = \{k \in L;\ f(k) = 0\}$$

is equal to $\{0\} \subset L$.

<u>Proof</u>. Evidently, we need only show the sufficiency of the condition. However, if $f(a) = f(b)$, then $f(a) = f(a) \wedge f(b) = f(a \wedge b)$ and, similarly, $f(a) = f(a \vee b)$. Hence $f(a \wedge b) = f(a \vee b)$. Let r be a relative complement of $a \wedge b$ in $[0, a \vee b]$. Then $a \vee b = (a \wedge b) \vee r$ and we have $f(a \wedge b) = f(a \vee b) = f(a \wedge b) \vee f(r) \geqslant f(r)$. Now $0 = f(0) = f((a \wedge b) \wedge r) = f(a \wedge b) \wedge f(r) = f(r)$. Consequently, $r \in \mathrm{Ker}\, f = \{0\}$ and this implies $a \wedge b = a \vee b$. Thus $a = b$. //

Given an element g of a generalized orthomodular lattice $`G$, we denote by $[g]$ the corresponding element of the quotient lattice $`G/\hat{B}$ where $\hat{B}$ denotes the reflective congruence relation of $`G$. Let $`I(`G')$ denote the ideal lattice of the lattice $`G$ (cf. V.2) and let f be a mapping of $`G/\hat{B}$ into $`I(`G')$ defined by $f: [g] \mapsto (g] \cap G'$. Note that f is well defined. In fact, suppose $[g] = [g_1]$ and choose $k \in (g] \cap G'$. Then $[g] = [g] \wedge [g_1] = [g \wedge g_1]$ and, therefore, $(g_1 \wedge g, g) \in \hat{B}$. Now $k \leqslant g$ and this yields $k/k \wedge g_1 \wedge g \nearrow^{w} g/g_1 \wedge g$. Hence, by Lemma 1.1 (iii), $(k \wedge g_1 \wedge g, k) \in \hat{B}$. At the same time, $k \in G'$. Consequently, we get $(k \wedge g_1 \wedge g, k) \in \hat{C}$. From Theorem 1.11 (ii) we can see that $k = k \wedge g_1 \wedge g \leqslant g_1$. Thus $(g] \cap G' \subset (g_1] \cap G'$ and, by symmetry, $(g] \cap G' = (g_1] \cap G'$.

Recall (cf. [74, Lemma 8, p. 34]) that for a lattice $`L$ belonging to an equational class <u>K</u> of lattices, the ideal lattice $`I(`L)$ belongs also to <u>K</u>.

Theorem 4.3 (cf. [16, Thm 3.5, p. 336]). Let $\underline{K}$ be an equational class of lattices which contains a lattice with more than one element. Then a generalized orthomodular lattice is a lattice of $\underline{K}$ if and only if its commutator sublattice belongs to $\underline{K}$.

Proof. 1. If $`G$ belongs to $\underline{K}$, then its sublattice $`G'$ belongs also to $\underline{K}$.

2. Conversely, assume that $`G' \in \underline{K}$ and consider the mapping $f: `G/\hat{B} \to `I(`G')$ defined above. By Theorem 1 it is readily shown that f is a lattice-homomorphism of $`G/\hat{B}$ into $`I(`G')$. Indeed, for any $[g],[h]$ of $`G/\hat{B}$,

$$f([g \vee h]) = (g \vee h] \cap G' = ((g] \vee (h]) \cap G' = $$
$$= ((g] \cap G') \vee ((h] \cap G') = f([g]) \vee f([h])$$

and

$$f([g \wedge h]) = (g \wedge h] \cap G' = (g] \cap (h] \cap G' = $$
$$= (g] \cap G' \cap (h] \cap G' = f([g]) \wedge f([h]).$$

Assume $f([x]) = (0]$ for some $x \in G'$. Then $(x] \cap G' = (0]$. We claim that $(x,0) \in \hat{B}$. To see it, observe first that from $0 \leqslant u \leqslant v \leqslant x$, $(u,v) \in \hat{C}$ and from the fact that for any relative complement r of u in $[0,v]$ we have $v/u \searrow r/0$ it follows that $(0,r) \in \hat{C}$. Consequently, $r \in G'$, by Theorem 3.2. Hence $r \in (x] \cap G'$ which yields $r = 0$ and $u = v$. Thus $(x,0) \in \hat{B}$ and $[x] = [0]$. From Lemma 2 we infer that $`G/\hat{B}$ is isomorphic to a sublattice of $`I(`G')$.

Now, by the remark above, $`I(`G') \in \underline{K}$. Since $\underline{K}$ is an equational class of lattices, it contains any isomorphic image of a sublattice of the lattice $`I(`G')$. Hence $`G/\hat{B} = \text{Ref } `G$ belongs to $\underline{K}$. It follows from Theorem 2.6 that $`G \in \underline{K}$. //

A generalized orthomodular lattice $`G$ is said to be

solvable in a class $\underline{K}$ of lattices if and only if there exists $n \in \underline{N}_0$ such that the n-th commutator sublattice $`G^{(n)}$ belongs to $\underline{K}$.

Theorem 4.4. A generalized orthomodular lattice is solvable in an equational class $\underline{K}$ of lattices which contains a lattice with more than one element if and only if it belongs to the class $\underline{K}$.

Proof. The theorem follows from Theorems 3 and 3.7. //

EXERCISES

VI;1. Find the commutator sublattice

(a) of the lattice in Figure 45;

(b) of the free orthomodular lattice F_2 (see Figure 18).

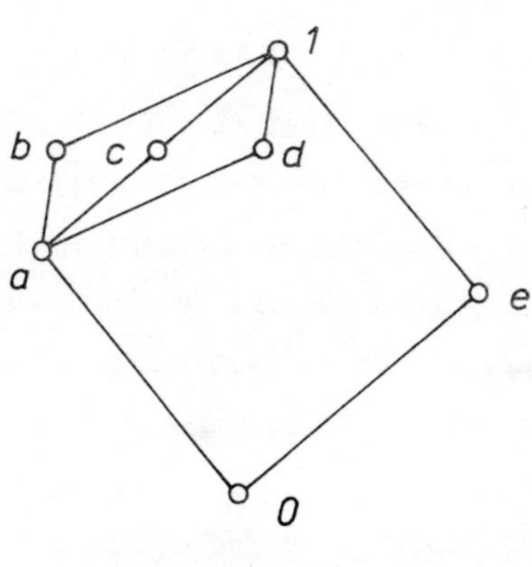

Fig. 54

VI;2. Find the projective allelomorph of the lattice shown in Figure 54.

VI;3. Find the allelomorph of the lattice in Figure 54.

VI;4. Prove that the following implication is true in any lattice $`L$ for every p,q,r:

$$(q/p \approx_w q/r \;\&\; p < r \leqslant q) \Rightarrow (r/p \in A(`L).$$

VI;5. If $`L$ is a modular lattice, then the following implication holds for every $p,q,r \in L$:

$$(q/p \approx_w q/r \;\&\; p < r \leqslant q) \Rightarrow (r/p \in A(`L;\approx)).$$

VI;6. Show that there exists a nonmodular lattice which satisfies the implication of Exercise VI;5. Find a nonmodular lattice which does not satisfy this implication.

VI;7. Find all the $\hat{C}$-accessible elements of the orthomodular lattice sketched in Figure 12.

VI;8. Find all the $\hat{B}$-accessible elements of the lattice in Figure 9g.

VI;9 (T. Katriňák). Let $\hat{C}$ denote the coreflective congruence of a lattice $`L$. Use the following sequence of statements to prove Theorem VI.1.7:

(i) Let P be a congruence of $`L$ such that $`L/P$ is distributive and let $Q = \hat{C} \cap P$. Then $`L/Q$ is isomorphic to a subdirect product of $`L/\hat{C}$ and $`L/P$.

(ii) The quotient lattice $`L/Q$ is distributive.

(iii) If $b/a \in A(`L)$, then $(a,b) \in Q$.

(iv) $\hat{C} \subset Q$.

VI;10. Prove that under the assumptions of Theorem 3.11 the following statements are true:

(i) If $(h,k)^{P(B,c)} = (x,y)$ and $(h,k)^{P(b,c)} = (u,v)$ where $(h,k) \leqslant (B,c)$ and $(h,k) \leqslant (b,c)$, then $y = v$.

(ii) If $(h_1,k)^{P(b,c)} = (s,t)$ and $(h,k)^{P(b,c)} = (u,v)$ where $(h_1,k) \leqslant (b,c)$ and $(h,k) \leqslant (b,c)$, then $t = v$.

(iii) The elements $k^{R(c)}$ and $h^{Q(b)}$ are well defined.

(iv) H is the base set of a generalized orthomodular lattice under the generalized orthogonality relation Q.

IN EXERCISES VI;11 - VI;14 $`G$ DENOTES A GENERALIZED ORTHOMODULAR LATTICE AND $G^{(n)}$ THE BASE SET OF THE n-TH COMMUTATOR SUBLATTICE $`G^{(n)}$.

VI;11. Prove that $(G, \vee, \wedge)$ is a distributive lattice if and only if $G' = \{0\}$.

VI;12. Define `G to be <u>solvable</u> (in the sense of Marsden (cf. [147])) if and only if there exists $n \in \underline{N}$ such that $G^{(n)} = \{0\}$.

Show that `G is solvable if and only if it is solvable in the class $\underline{D}$ of distributive lattices.

VI;13. Without assuming Theorem 4.4, prove directly that `G is solvable in the class of distributive lattices if and only if it is a distributive lattice.

VI;14. Without assuming Theorem 4.4, prove directly that `G is solvable in the class of modular lattices if and only if it is a modular lattice.

SPECIAL PROPERTIES OF ORTHOMODULARITY

Chapter **VII**

1. COMMUTATORS OF n ELEMENTS

The upper and lower commutator of two elements x,y have already been mentioned in Section III.2. We shall now generalize (cf. [20]) the original definition of the commutator $\mathrm{com}\,(x,y)$ given by Marsden (cf. [147]) for n elements.

Let $a_1,a_2,\dots,a_n \in [p,q]$ be elements of an ortholattice. We write $a_i^1 = a_i$, $a_i^{-1} = p \vee (a_i' \wedge q)$ and define the upper commutator of the elements $a_1,a_2,\dots,a_n$ in the interval $[p,q]$ as the element

$$\overline{\mathrm{com}}_{[p,q]}(a_1,a_2,\dots,a_n) = \bigwedge (a_1^{i_1} \vee a_2^{i_2} \vee \dots \vee a_n^{i_n})$$

where the superscripts $i_1,i_2,\dots,i_n$ run over the elements of $\{-1,1\}$. Dually we define the lower commutator as the element

$$\underline{\mathrm{com}}_{[p,q]}(a_1,a_2,\dots,a_n) = \bigvee (a_1^{i_1} \wedge a_2^{i_2} \wedge \dots \wedge a_n^{i_n}).$$

If $p = 0$ and $q = 1$, we omit the subscript $[0,1]$ and delete the words "in the interval $[0,1]$".

Remark 1.1. In special cases $n = 1$ and $n = 2$ we have

$$\overline{\text{com}}\,(a_1) = a_1 \wedge a_1' = 0, \quad \underline{\text{com}}\,(a_1) = a_1 \vee a_1' = 1,$$

$$\overline{\text{com}}\,(a_1,a_2) = (a_1 \vee a_2) \wedge (a_1 \vee a_2') \wedge (a_1' \vee a_2) \wedge (a_1' \vee a_2') = \text{com}\,(a_1,a_2),$$

$$\underline{\text{com}}\,(a_1,a_2) = (a_1 \wedge a_2) \vee (a_1 \wedge a_2') \vee (a_1' \wedge a_2) \vee (a_1' \wedge a_2').$$

We next prove the following purely technical results that will enable us to study the basic properties of the commutators of n elements.

Lemma 1.2. In any orthomodular lattice,

(i) $(\overline{\text{com}}\,(a_1,a_2,\ldots,a_n))' = \underline{\text{com}}\,(a_1,a_2,\ldots,a_n)$

and, more generally,

$$\overline{\text{com}}_{[p,q]}(a_1,a_2,\ldots,a_n) \vee \underline{\text{com}}_{[p,q]}(a_1,a_2,\ldots,a_n) = q,$$

$$\overline{\text{com}}_{[p,q]}(a_1,a_2,\ldots,a_n) \wedge \underline{\text{com}}_{[p,q]}(a_1,a_2,\ldots,a_n) = p;$$

(ii) $\overline{\text{com}}\,(a_1,a_2,\ldots,a_{n+1}) \geqslant \overline{\text{com}}\,(a_1,a_2,\ldots,a_n)$;

(iii) $\underline{\text{com}}\,(a_1,a_2,\ldots,a_{n+1}) \leqslant \underline{\text{com}}\,(a_1,a_2,\ldots,a_n)$;

(iv) $\overline{\text{com}}\,(a_1,a_2,\ldots,a_{n+1}) = \overline{\text{com}}\,(a_1,a_2,\ldots,a_n)$

whenever $a_n = a_{n+1}$;

(v) $\overline{\text{com}}\,(a_1^{i_1},a_2^{i_2},\ldots,a_n^{i_n}) = \overline{\text{com}}\,(a_1,a_2,\ldots,a_n) = \overline{\text{com}}\,(a_{j_1},a_{j_2},\ldots,a_{j_n})$

for any permutation of the subscripts $1,2,\ldots,n$ and for any choice of $i_k \in \{-1,1\}$;

(vi) $\overline{\text{com}}\,(a_1,b_1) \vee \ldots \vee \overline{\text{com}}\,(a_n,b_n) \leqslant \overline{\text{com}}\,(a_1,\ldots,a_n,b_1,\ldots,b_n)$;

(vii) $\overline{\text{com}}_{[p,q]}(a_1,a_2,\ldots,a_n) = [p \vee \overline{\text{com}}\,(a_1,a_2,\ldots,a_n)] \wedge q =$

$= p \vee [\overline{\text{com}}\,(a_1,a_2,\ldots,a_n) \wedge q]$; $\underline{\text{com}}_{[p,q]}(a_1,a_2,\ldots,a_n) =$

$= [p \vee \underline{\text{com}}\,(a_1,a_2,\ldots,a_n)] \wedge q = p \vee [\underline{\text{com}}\,(a_1,a_2,\ldots,a_n) \wedge q]$

provided $a_1,a_2,\ldots,a_n \in [p,q]$.

Proof. Ad (i) and (vii). Since $p \leqslant a_i \leqslant q$ for every $i = 1,2,\ldots,n$, we obtain from Theorem II.2.3 that pCa_i, qCa_i for every i. Clearly, $a_i = p \vee (a_i \wedge q)$, $a_i^{-1} = p \vee (a_i' \wedge q)$, and therefore, by Foulis-Holland Theorem II.3.10,

$$a_1^{i_1} \vee a_2^{i_2} \vee \ldots \vee a_n^{i_n} = p \vee [(a_1^{e_1} \vee a_2^{e_2} \vee \ldots \vee a_n^{e_n}) \wedge q]$$

for any choice of the superscripts $i_1 = e_1$, $i_2 = e_2$, ... , $i_n = e_n$ of the set $\{-1,1\}$. Here we put $a_i^{e_i} = a_i$ whenever $e_i = 1$ and $a_j^{e_j} = a_j'$ whenever $e_j = -1$. By Theorem II.4.2,

$pCa_1^{e_1} \vee \ldots \vee a_n^{e_n}$, $qCa_1^{e_1} \vee \ldots \vee a_n^{e_n}$, $pCa_1^{e_1} \wedge \ldots \wedge a_n^{e_n}$,

$qCa_1^{e_1} \wedge \ldots \wedge a_n^{e_n}$.

By using Theorem II.3.10 once again, we get

$$\overline{\text{com}}_{[p,q]}(a_1,a_2,\ldots,a_n) = p \vee [(\bigwedge(a_1^{e_1} \vee \ldots \vee a_n^{e_n})) \wedge q] =$$

$$= p \vee [\overline{\text{com}}\,(a_1,a_2,\ldots,a_n) \wedge q] = [p \vee \overline{\text{com}}\,(a_1,a_2,\ldots,a_n)] \wedge q.$$

The second part of (vii) can be proved similarly. To prove (i), note that $(\overline{\text{com}}\,(a_1,\ldots,a_n))' = \underline{\text{com}}\,(a_1,\ldots,a_n)$ is a simple consequence of De Morgan laws. For brevity write $\bar{c} = \overline{\text{com}}_{[p,q]}(\ldots)$, $\underline{c} = \underline{\text{com}}_{[p,q]}(\ldots)$. As we have already seen,

$$\bar{c} \vee \underline{c} = p \vee [\overline{\text{com}}\,(a_1,\ldots,a_n) \wedge q] \vee p \vee [\underline{\text{com}}\,(a_1,\ldots,a_n) \wedge q].$$

Since q commutes with the upper and lower commutator of the elements $a_1,a_2,\ldots,a_n$, Theorem II.3.10 yields

$$\bar{c} \vee \underline{c} = p \vee \{[\overline{\text{com}}\,(a_1,\ldots,a_n) \vee \underline{\text{com}}\,(a_1,\ldots,a_n)] \wedge q\} = p \vee q = q.$$

The assertion $\bar{c} \wedge \underline{c} = p$ follows by a similar argument.

Ad (ii). It is easily verified that

$$\overline{\mathrm{com}}\,(a_1,a_2,\ldots,a_{n+1}) = \bigwedge(a_1^{i_1} \vee a_2^{i_2} \vee \ldots \vee a_n^{i_n} \vee a_{n+1}^{i_{n+1}}) \geq$$
$$\geq \bigwedge(a_1^{i_1} \vee a_2^{i_2} \vee \ldots \vee a_n^{i_n}) = \overline{\mathrm{com}}\,(a_1,a_2,\ldots,a_n).$$

Ad (iii). This follows from (ii) and (i).

Ad (iv). If $a_n = a_{n+1}$ and $i_n \neq i_{n+1}$, then

$$a_1^{i_1} \vee a_2^{i_2} \vee \ldots \vee a_n^{i_n} \vee a_{n+1}^{i_{n+1}} = a_1^{i_1} \vee a_2^{i_2} \vee \ldots \vee a_n^{i_n} \vee (a_n^{i_n})' = 1.$$

If $a_n = a_{n+1}$ and $i_n = i_{n+1}$, then

$$a_1^{i_1} \vee a_2^{i_2} \vee \ldots \vee a_n^{i_n} \vee a_{n+1}^{i_{n+1}} = a_1^{i_1} \vee a_2^{i_2} \vee \ldots \vee a_n^{i_n}.$$

The equality of the considered upper commutators is now obvious.

Ad (v). The statement follows by definition.

Ad (vi). As a consequence of (ii) and (v) we find that

$$\overline{\mathrm{com}}\,(a_i,b_i) \leq \overline{\mathrm{com}}\,(a_1,\ldots,a_n,b_1,\ldots,b_n)$$

and this implies the validity of (vi). //

<u>Theorem 1.3</u>. If $a_1,a_2,\ldots,a_n \in [p,q]$ and m are elements of an orthomodular lattice L such that mCp, $m \leq q$ and mCa_i for every $i = 1,2,\ldots,n$, then

(i) $mC\overline{\mathrm{com}}_{[p,q]}(a_1,a_2,\ldots,a_n)$;

(ii) $m = (m \wedge \overline{\mathrm{com}}_{[p,q]}(a_1,\ldots,a_n)) \vee (m \wedge \underline{\mathrm{com}}_{[p,q]}(a_1,\ldots,a_n))$.

<u>Proof</u>. Ad (i). First observe that the element m commutes with p,q and with the elements a_i. Hence, by Remark II.4.3, m commutes also with $a_1^{i_1} \vee a_2^{i_2} \vee \ldots \vee a_n^{i_n}$ for every choice of $i_1,i_2,\ldots,i_n \in \{-1,1\}$. By the same remark we conclude that m commutes with the element $\bigwedge(a_1^{i_1} \vee a_2^{i_2} \vee \ldots \vee a_n^{i_n})$.

Ad (ii). Apply (i) with $m = \underline{\mathrm{com}}_{[p,q]}(a_1,a_2,\ldots,a_n)$. Then it is clear that the elements $\underline{c} = \underline{\mathrm{com}}_{[p,q]}(\ldots)$, $\bar{c} = \overline{\mathrm{com}}_{[p,q]}(\ldots)$

commute. Therefore, from (i), Lemma 2 (i) and Theorem II.3.10,

$$m = m \wedge q = m \wedge (\bar{c} \vee \underline{c}) = (m \wedge \bar{c}) \vee (m \wedge \underline{c}). \quad //$$

Theorem 1.4. Let $a_1, a_2, \ldots, a_n \in [p,q]$ be elements of an orthomodular lattice. Then

(1) $\overline{\text{com}}_{[p,q]}(a_1, a_2, \ldots, a_n) = p$

if and only if

(2) $\overline{\text{com}}\,(a_1, a_2, \ldots, a_n) = 0.$

Proof. 1. If (2) holds, then (1) is true by Lemma 2 (vii).

2. Now suppose (1) is valid. By Lemma 2 (vii),

$$\overline{\text{com}}\,(a_1, a_2, \ldots, a_n) \wedge q \leqslant p \vee [\overline{\text{com}}\,(a_1, a_2, \ldots, a_n) \wedge q] = p.$$

Since $a_i \leqslant q$, $\overline{\text{com}}\,(a_1, a_2, \ldots, a_n) = (a_1 \vee a_2 \vee \ldots \vee a_n) \wedge \ldots \leqslant q$ and, therefore,

$$\overline{\text{com}}\,(a_1, a_2, \ldots, a_n) \wedge q = \overline{\text{com}}\,(a_1, a_2, \ldots, a_n) \leqslant p \leqslant a_1 \wedge a_2 \wedge \ldots \wedge a_n.$$

But we also have $\overline{\text{com}}\,(a_1, a_2, \ldots, a_n) \leqslant a_1' \vee a_2' \vee \ldots \vee a_n'$ and so

$$\overline{\text{com}}\,(a_1, a_2, \ldots, a_n) \leqslant (a_1 \wedge a_2 \wedge \ldots \wedge a_n) \wedge (a_1' \vee a_2' \vee \ldots \vee a_n') = 0. \quad //$$

Corollary 1.5. Two elements a, b of an orthomodular lattice commute if and only if $\overline{\text{com}}_{[a \wedge b, a \vee b]}(a,b) = a \wedge b$.

Proof. By Theorem III.2.11, aCb if and only if $\overline{\text{com}}\,(a,b) = 0$. Setting $n = 2$, $a_1 = a$, $a_2 = b$, $p = a \wedge b$, $q = a \vee b$ and using Theorem 4, we conclude that a commutes with b if and only if the mentioned commutator is equal to the element $a \wedge b$. //

Theorem 1.6. Let s, t be elements of an orthomodular lattice L. Let

$$i = s \wedge (s' \vee t) \wedge (s' \vee t'), \quad j = t \wedge (s \vee t') \wedge (s' \vee t'),$$

$$i^+ = s' \wedge (s \vee t) \wedge (s \vee t'), \quad j^+ = t' \wedge (s \vee t) \wedge (s' \vee t).$$

Then $\{0, i, i^+, j, j^+, com\ (s,t)\}$ is the base set for a sublattice `M of the lattice `L and either `M is a one-element lattice or $`M \cong `M_{02}$.

Proof. Apply the last part of the proof given for Lemma III. 2.3 to the sublattice $`T = \langle s,t \rangle$ generated by the elements s,t in `L. //

Theorem 1.7. Let s,t be elements of an orthomodular lattice `L. Then the following conditions are equivalent:

(i) s does not commute with t;

(ii) `L contains a sublattice $`M \cong `M_{02}$ which has $(s \wedge t) \vee \vee (s \wedge t')$ and $(s \vee t) \wedge (s \vee t')$ as its least and greatest element, respectively, and which contains s as an inner element.

(iii) there is no distributive subalgebra of the orthomodular lattice `L containing s and t.

Proof. (i) ⇒ (ii). By Remark II.4.3 and Theorem II.4.4, the elements $a = com\ (s,t)$ and $b = (s \wedge t) \vee (s \wedge t')$ commute with every element of the subalgebra $`T = \langle s,t \rangle$. It follows from Theorem III.1.1 that $b \dot{\wedge} a = b \wedge a \leq \underline{com}\ (s,t) \wedge \overline{com}\ (s,t) = 0$ and $a \dot{\vee} b = a \vee b = (s \vee t) \wedge (s \vee t')$. Put $y = s \wedge (s' \vee t)$ and note that $y \vee b = [s \wedge (s' \vee t)] \vee (s \wedge t') \vee (s \wedge t) = s$ by orthomodularity. The rest follows from Theorems III.1.2, III.2.11 and 6.

(ii) ⇒ (iii). If s and t are elements of a subalgebra `D of `L, then `D contains as a sublattice the lattice `M constructed in the proof of Theorem 6. Hence `D is not distributive.

(iii) ⇒ (i). Suppose to the contrary that sCt. Consider the subalgebra $`T = \langle s,t \rangle$ of `L generated by s and t. From Lemma III.2.1 and Remark III.2.2 we see that $`T = (\overline{com}\ (s,t)] \oplus \oplus (\underline{com}\ (s,t)]$. However, by Theorem III.2.11, $`T = (\underline{com}\ (s,t)]$.

In view of Corollary III.2.6 we find that $`T$ is distributive, a contradiction. Thus s,t do not commute. //

Corollary 1.8. An element s of an orthomodular lattice $`L$ does not commute with an element of this lattice if and only if $s$ is an inner element of a sublattice $`M$ of $`L$ satisfying $`M \cong `M_{O2}$. In this case $s$ does not commute with an inner element of $`M$.

Proof. 1. If s has the described property, use Theorem 7.

2. Suppose s is an inner element of $`M$ where $M = \{o,s,x,z,u\}$ and $o < s < u$. If $sCx$ and $sCz$, then $s \wedge (x \vee z) = (s \wedge x) \vee (s \wedge z)$, by Theorem II.3.10. However, this is not true in $`M$. //

Corollary 1.9. In an orthomodular lattice $`L$, for all $s, t \in L$,

(i) the quotient $s/(s \wedge t) \vee (s \wedge t')$ is an allele;

(ii) $(s,t) \in \hat{C}$ if and only if

$$(*) \quad (0, s \wedge t') \in \hat{C} \ \& \ (0, s' \wedge t) \in \hat{C}$$

where $\hat{C}$ denotes the coreflective congruence of $`L$.

Proof. Ad (i). Use Theorem 7.

Ad (ii). 1. Suppose $(s,t) \in \hat{C}$. Then $(s \wedge t, s \vee t) \in \hat{C}$ and $s \vee t/s \wedge t \searrow r/0$ for the relative orthocomplement $r = (s \vee t) \wedge (s \wedge t)'$ of $s \wedge t$ in $[0, s \vee t]$. Hence $(0,r) \in \hat{C}$. Since $r = (s \vee t) \wedge (s' \vee t') \geqslant (s \wedge t') \vee (s' \wedge t)$, we find the claim of $(*)$.

2. Conversely, from $(*)$ it follows $(0, (s \wedge t') \vee (s' \wedge t)) \in \hat{C}$ and, by (i), $(s, (s \wedge t) \vee (s \wedge t')) \in \hat{C}$. Since $\hat{C}$ is a congruence, we have $(s, (s \wedge t) \vee (s \wedge t') \vee (s' \wedge t)) \in \hat{C}$. By symmetry, $(t, (s \wedge t) \vee (s \wedge t') \vee (s' \wedge t)) \in \hat{C}$ and so $(s,t) \in \hat{C}$. //

Theorem 1.10 (cf. [21]). Let $x_1, x_2, \ldots, x_n$ be elements of an orthomodular lattice and let $n \geq 2$. Then

$$\overline{\text{com}}\,(x_1,x_2,\dots,x_n) = \overline{\text{com}}\,(x_1,x_2,\dots,x_{n-1}) \vee$$
$$\vee \bigvee \overline{\text{com}}\,(x_1^{e_1} \vee x_2^{e_2} \vee \dots \vee x_{n-1}^{e_{n-1}}, x_n)$$

where the join is over all (n-1)-tuples $(e_1,e_2,\dots,e_{n-1})$ such that $e_i = \pm 1$.

<u>Proof</u>. Using Lemma 2 (ii) and (v), we get

$$\overline{\text{com}}\,(x_1,x_2,\dots,x_n) \geq \overline{\text{com}}\,(x_1,x_2,\dots,x_{n-1}),$$
$$\overline{\text{com}}\,(x_1,x_2,\dots,x_n) = \overline{\text{com}}\,(x_1^{e_1},x_2^{e_2},\dots,x_n^{e_n}).$$

By definition,

$$\overline{\text{com}}\,(x_1,x_2,\dots,x_n) \geq \overline{\text{com}}\,(x_1 \vee x_2,x_3,\dots,x_n) \geq \overline{\text{com}}\,(x_1 \vee x_2 \vee x_3, x_4,\dots,x_n) \geq \dots \geq \overline{\text{com}}\,(x_1 \vee x_2 \vee \dots \vee x_{n-1},x_n).$$

It follows that

$$s = \overline{\text{com}}\,(x_1,x_2,\dots,x_n) \geq t' = \overline{\text{com}}\,(x_1,x_2,\dots,x_{n-1}) \vee \bigvee \overline{\text{com}}\,(x_1^{e_1} \vee x_2^{e_2} \vee \dots \vee x_{n-1}^{e_{n-1}}, x_n).$$

Denote by $W(e_1,e_2,\dots,e_{n-1})$ the element

$$(x_1^{e_1} \vee x_2^{e_2} \vee \dots \vee x_{n-1}^{e_{n-1}} \vee x_n) \wedge (x_1^{e_1} \vee x_2^{e_2} \vee \dots \vee x_{n-1}^{e_{n-1}} \vee x_n') \wedge$$
$$\wedge\, \underline{\text{com}}\,(x_1^{e_1} \vee x_2^{e_2} \vee \dots \vee x_{n-1}^{e_{n-1}}, x_n) = A \wedge B \wedge (C \vee D \vee A' \vee B')$$

where

$$A = x_1^{e_1} \vee x_2^{e_2} \vee \dots \vee x_{n-1}^{e_{n-1}} \vee x_n, \quad B = x_1^{e_1} \vee x_2^{e_2} \vee \dots \vee x_{n-1}^{e_{n-1}} \vee x_n',$$
$$C = (x_1^{e_1} \vee x_2^{e_2} \vee \dots \vee x_{n-1}^{e_{n-1}}) \wedge x_n, \quad D = (x_1^{e_1} \vee x_2^{e_2} \vee \dots \vee x_{n-1}^{e_{n-1}}) \wedge x_n'.$$

Note that $A \wedge B \geq x_1^{e_1} \vee x_2^{e_2} \vee \dots \vee x_{n-1}^{e_{n-1}} \geq C \vee D$. Hence, by orthomodularity,

$$W(e_1,e_2,\dots,e_{n-1}) = C \vee D \leq x_1^{e_1} \vee x_2^{e_2} \vee \dots \vee x_{n-1}^{e_{n-1}}.$$

It is now evident that

$$s \wedge t \leq \overline{\text{com}}\,(x_1,x_2,\dots,x_n) \wedge \bigwedge \underline{\text{com}}\,(x_1^{-e_1} \vee x_2^{-e_2} \vee \dots \vee x_{n-1}^{-e_{n-1}}, x_n) =$$

$$= \bigwedge_{(e_1,\bullet)} W(-e_1,-e_2,\dots,-e_{n-1}) \leq \bigwedge_{(e_1,\bullet)} (x_1^{-e_1} \vee x_2^{-e_2} \vee \dots \vee x_{n-1}^{-e_{n-1}}) =$$

$$= \overline{\text{com}}\,(x_1,x_2,\dots,x_{n-1}).$$

On the other hand, $s \wedge t \leq t \leq \underline{\text{com}}\,(x_1,x_2,\dots,x_{n-1})$. By using Lemma 2 (i) we conclude that $s \wedge t = 0$. From orthomodularity we infer that $s = t'$. //

Corollary 1.11. Let $x_1,x_2,\dots,x_n$, $n \geq 2$, be elements of an orthomodular lattice. Then

$$\overline{\text{com}}\,(x_1,x_2,\dots,x_n) = \bigvee_{(e_1,e_2,\dots,e_{n-1})} \overline{\text{com}}\,(x_1^{e_1} \vee \dots \vee x_{n-1}^{e_{n-1}}, x_n) \vee$$

$$\vee \bigvee_{(e_1,e_2,\dots,e_{n-2})} \overline{\text{com}}\,(x_1^{e_1} \vee x_2^{e_2} \vee \dots \vee x_{n-2}^{e_{n-2}}, x_{n-1}) \vee \dots$$

$$\dots \vee \bigvee_{(e_1,e_2)} \overline{\text{com}}\,(x_1^{e_1} \vee x_2^{e_2}, x_3) \vee \overline{\text{com}}\,(x_1,x_2).$$

Proof. Immediate from Theorem 10. //

Remark 1.12. From Theorem 10 we conclude that the commutator $\overline{\text{com}}\,(x_1,x_2,\dots,x_n)$ belongs to the commutator sublattice $`L'$ of an orthomodular lattice $`L$, provided $x_1,x_2,\dots,x_n \in L$.

Theorem 1.13. Let $a_1,a_2,\dots,a_n,a_{n+1}$ be elements of an orthomodular lattice. Then

(3) $\overline{\text{com}}\,(a_1,a_2,\dots,a_n \wedge a_{n+1}) \leq \overline{\text{com}}\,(a_1,a_2,\dots,a_n,a_{n+1})$

and

(4) $\overline{\text{com}}\,(a_1,a_2,\dots,a_n \vee a_{n+1}) \leq \overline{\text{com}}\,(a_1,a_2,\dots,a_n,a_{n+1})$.

Proof. First observe that

$$a_n \wedge a_{n+1} \leq a_n \vee a_{n+1},\ a_n \wedge a_{n+1} \leq a_n \vee a_{n+1}',\ a_n \wedge a_{n+1} \leq a_n' \vee a_{n+1}$$

and $(a_n \wedge a_{n+1})' = a_n' \vee a_{n+1}'$. Hence, to any choice of e_n, e_{n+1} in

$\{-1,1\}$ there exists f_n such that $(a_n \wedge a_{n+1})^{f_n} \leqslant a_n^{e_n} \vee a_{n+1}^{e_{n+1}}$. In other words, to any choice of $e_1, e_2, \ldots, e_n, e_{n+1} \in \{-1,1\}$ there exist $f_1, f_2, \ldots, f_n$ such that

$$a_1^{f_1} \vee a_2^{f_2} \vee \ldots \vee a_{n-1}^{f_{n-1}} \vee a_n^{f_n} \leqslant a_1^{e_1} \vee a_2^{e_2} \vee \ldots \vee a_{n-1}^{e_{n-1}} \vee a_n^{e_n} \vee a_{n+1}^{e_{n+1}}.$$

(Of course, here $f_1 = e_1$, $f_2 = e_2$, ... , $f_{n-1} = e_{n-1}$.) Thus (3) is valid.

Furthermore, by Lemma 2 (v) and by (3), we have

$$\overline{\text{com}}\,(a_1, a_2, \ldots, a_n \vee a_{n+1}) = \overline{\text{com}}\,(a_1, a_2, \ldots, a_n' \wedge a_{n+1}') \leqslant$$

$$\leqslant \overline{\text{com}}\,(a_1, a_2, \ldots, a_n', a_{n+1}') = \overline{\text{com}}\,(a_1, a_2, \ldots, a_n, a_{n+1})$$

and we see that (4) is also true. //

<u>Corollary 1.14</u>. Let $x_1, x_2, \ldots, x_n$ be elements of an orthomodular lattice `L and let $p_1 = p_1(x_1, x_2, \ldots, x_n)$, $p_2 = p_2(x_1, x_2, \ldots, x_n), \ldots$, $p_k = p_k(x_1, x_2, \ldots, x_n)$ be polynomials in $\vee, \wedge, '$. Then

(5) $\overline{\text{com}}\,(p_1, p_2, \ldots, p_k) \leqslant \overline{\text{com}}\,(x_1, x_2, \ldots, x_n)$.

<u>Proof</u>. Let r_i denote the rank of p_i and let $m = \max\{r_1, r_2, \ldots, r_k\}$.

1. If $m = 1$, then the expression on the left-hand side gives

$$\overline{\text{com}}\,(x_{i_1}^{e_1}, x_{i_2}^{e_2}, \ldots, x_{i_k}^{e_k}) \qquad (e_i \in \{-1,1\})$$

and the validity of (5) for any k is a consequence of Lemma 2 (ii), (iv) and (v).

2. Suppose by induction that (5) is true for any k whenever $\max\{r_1, r_2, \ldots, r_k\} < q$ and consider the case $m = q$. By Lemma 2 (v), we may assume that $m = r_1 = r_2 = \ldots = r_j > \max\{r_{j+1}, \ldots \ldots, r_k\}$ $(1 \leqslant j \leqslant k)$ and that $p_i = s_i \wedge t_i$ where $s_i = s_i(x_1, x_2, \ldots, x_n)$ and $t_i = t_i(x_1, x_2, \ldots, x_n)$ $(i = 1, 2, \ldots, j)$ are polynomials in $\vee, \wedge, '$ having their ranks less than q. By Theorem

13, Lemma 2 (v) and by induction hypothesis,

$$\overline{\text{com}}\ (p_1,p_2,\ldots,p_k) = \overline{\text{com}}\ (s_1 \wedge t_1,p_2,\ldots,p_k) \leqslant$$
$$\leqslant \overline{\text{com}}\ (s_1,t_1,p_2,\ldots,p_k) = \overline{\text{com}}\ (s_1,t_1,s_2 \wedge t_2,p_3,\ldots,p_k) \leqslant \ldots$$
$$\ldots \leqslant \overline{\text{com}}\ (s_1,t_1,\ldots,s_j,t_j,p_{j+1},\ldots,p_k) \leqslant \overline{\text{com}}\ (x_1,x_2,\ldots,x_n).$$

This completes the proof. //

As has already been noted in Section V.4, two elements a,b of a lattice with 0 are said to be perspective (written $a \underset{p}{\sim} b$) if and only if there exists an element t such that $a \vee t = b \vee t$ and $a \wedge t = b \wedge t = 0$. By Exercise VII;5, two elements of an orthomodular lattice are perspective if and only if they have a common complement.

We shall say (see, e.g., [38]) that two elements p,q are <u>strongly perspective</u> and we shall write $p \underset{sp}{\sim} q$ if and only if there exists an element x such that $p \vee x = q \vee x = p \vee q$ and $p \wedge x = q \wedge x = 0$.

<u>Remark 1.15</u>. (A) Obviously, any two elements which are strongly perspective are also perspective.

(B) By Exercise VII;6, only in a nonmodular lattice it is possible to find two elements which are perspective but which are not strongly perspective. For an example, see the lattice of Figure 12. That $a \underset{p}{\sim} b$ is apparent. However, the elements a,b of this lattice are not strongly perspective.

<u>Corollary 1.16</u>. Let x,y be elements of an orthomodular lattice `L. Then there exist elements $c,d \in L$ such that $c \underset{sp}{\sim} d$, $c \wedge d = 0$ and $c \vee d = \text{com}\ (x,y)$.

<u>Proof</u>. This is a straightforward consequence of Theorem 6. //

<u>Theorem 1.17</u>. The smallest congruence of an orthomodular lattice which contains the relation of (strong) perspectivity

is equal to the coreflective congruence of this lattice.

Proof. Denote by F and T the smallest congruence containing the relation $\underset{p}{\sim}$ and $\underset{sp}{\sim}$, respectively. Clearly, $T \subset F$. Let $a \underset{p}{\sim} b$. Denote by c a common complement of the elements a,b. Then we have $a/a \wedge b \nearrow a \vee b/b \nearrow^{w} 1/b \searrow c/0 \nearrow 1/a$ and, therefore, $a/a \wedge b \in A(`L)$. By a dual argument we get $a \vee b/a \in A(`L)$. It follows that $(a \wedge b, a \vee b) \in \hat{C}$ and, a fortiori, $(a,b) \in \hat{C}$. Hence $T \subset F \subset \hat{C}$.

Let P be a congruence relation on a relatively complemented lattice. Then $(a,b) \in P$ if and only if $(0,r) \in P$ where r is a relative complement of the element $a \wedge b$ in the interval $[0, a \vee b]$.

Hence, to prove the converse inclusion $\hat{C} \subset T$, it suffices to show that $(0,z) \in \hat{C}$ implies $(0,z) \in T$.

Suppose that $(0,z) \in \hat{C}$. Referring to Theorem VI.3.2, we see that the element z must belong to the commutator sublattice $`L'$ of $`L$. By definition and by Lemma V.5.2, $`L'$ is the ideal of $`L$ which is generated by commutators. Hence $z \leq k_1 \vee k_2 \vee \ldots \vee k_t$ where $k_i = \mathrm{com}\,(x_i, y_i)$ for some $x_i, y_i \in L$. Let c_i, d_i denote the elements constructed similarly as the elements c,d in Corollary 16, $c_i \wedge d_i = 0$, $c_i \vee d_i = k_i$ and $c_i \underset{sp}{\sim} d_i$. Hence, in particular, $(c_i, d_i) \in T$ and so $(0, k_i) = (c_i \wedge d_i, c_i \vee d_i) \in T$. Consequently, $(0, k_1 \vee k_2 \vee \ldots \vee k_t) \in T$. However, this means that also $(0,z) \in T$. Thus, in summary, $\hat{C} = T$. //

2. FINITELY GENERATED ORTHOMODULAR LATTICES

An orthomodular lattice $`F$ is said to be finitely generated if and only if there exist elements $x_1, x_2, \ldots, x_n$ such that the smallest subalgebra of the algebra $`F = (F, \vee, \wedge, ', 0, 1)$ which contains $x_1, x_2, \dots, x_n$ is equal to $`F$. We write $`F = \langle x_1, x_2, \dots, x_n \rangle$. This notation will be kept throughout this section.

Theorem 2.1. Let a be an element of a finitely generated orthomodular lattice $`F = \langle x_1, x_2, \dots, x_n \rangle$. If $a \leq \overline{\text{com}}\,(x_1, x_2, \dots, x_n)$, then $(0, a) \in \hat{C}$.

Proof. Denote by $[x]$ the class of the coreflection $`F/\hat{C} = $ = Coref $`F$ which is determined by an element $x \in F$. By Lemma VI.3.1,

$$[0] = [\overline{\text{com}}\,(x_i, x_j)] = \overline{\text{com}}\,([x_i], [x_j]).$$

Hence, by Theorem III.2.11, the elements $[x_i], [x_j]$, $1 \leq i, j \leq n$, commute. In view of Theorem II.3.10 we can use the distributive identities. This, together with Exercise VII;7, implies that

$$[\overline{\text{com}}\,(x_1, x_2, \dots, x_n)] = \overline{\text{com}}\,([x_1], [x_2], \dots, [x_n]) = 0 .$$

Thus $(0, \overline{\text{com}}\,(x_1, x_2, \dots, x_n)) \in \hat{C}$. Because of Theorem VI.3.2, the element $\overline{\text{com}}\,(x_1, x_2, \dots, x_n)$ is an element of the commutator sublattice $`F'$ of the lattice $`F$. Since $`F'$ is an ideal of $`F$, it follows that $a \in F'$ and so $(0, a) \in \hat{C}$. //

Theorem 2.2. Let $q_1, q_2, \dots, q_n$ and x_0 be elements of an orthomodular lattice $`L$. Suppose x_0 commutes with every element q_i, $i = 1, 2, \dots, n$. Then

$$x_0 \wedge \overline{\text{com}}\,(q_1 \wedge x_0, q_2 \wedge x_0, \dots, q_n \wedge x_0) = \overline{\text{com}}\,(q_1, q_2, \dots, q_n) \wedge x_0.$$

Proof. Let $W = x_0 \wedge [(q_1 \wedge x_0)^{e_1} \vee \dots \vee (q_n \wedge x_0)^{e_n}]$ where $e_i \in \{-1, 1\}$. Since $q_i \wedge x_0 \leq x_0$, $x_0 C (q_i \wedge x_0)^{e_i}$ by Theorem II.2.3. Hence, by Foulis-Holland Theorem II.3.10,

$$W = [x_0 \wedge (q_1 \wedge x_0)^{e_1}] \vee \dots \vee [x_0 \wedge (q_n \wedge x_0)^{e_n}].$$

If $e_1 = -1$, then

$$w = x_0 \wedge (q_1 \wedge x_0)^{e_1} = x_0 \wedge (q_1' \vee x_0').$$

By assumption, x_0 commutes with q_1'. Since x_0 commutes also with x_0', we conclude that $w = x_0 \wedge q_1' = x_0 \wedge q_1^{e_1}$. If $e_1 = 1$, then

$$x_0 \wedge (q_1 \wedge x_0)^{e_1} = x_0 \wedge q_1 = x_0 \wedge q_1^{e_1}.$$

Therefore,

$$W = (x_0 \wedge q_1^{e_1}) \vee \ldots \vee (x_0 \wedge q_n^{e_n}).$$

Then, since x_0 commutes with every element $q_i^{e_i}$, we have

$$W = x_0 \wedge (q_1^{e_1} \vee \ldots \vee q_n^{e_n}).$$

The theorem now follows by the definition of an upper commutator. //

The next result concerns lattice polynomials.

Theorem 2.3. Let $p(y_1, y_2, \ldots, y_n, z_1, z_2, \ldots, z_n)$ be a lattice polynomial and let $a_1, a_2, \ldots, a_n$ be elements of an orthomodular lattice. Let $a_0 = a_1 \wedge a_2 \wedge \ldots \wedge a_n$. Then

$$p(a_1, a_2, \ldots, a_n, a_1', a_2', \ldots, a_n') \wedge a_0 = p(a_0, a_0, \ldots, a_0, 0, 0, \ldots, 0).$$

Proof. We shall prove the assertion by induction on the rank r of the polynomial p.

1. Suppose $r = 1$. Then $p = x_i$, $1 \leqslant i \leqslant 2n$. In the case $1 \leqslant i \leqslant n$ we have

$$p(a_1, \ldots, a_n, a_1', a_2', \ldots, a_n') \wedge a_0 = a_i \wedge a_0 = a_0$$

and $p(a_0, a_0, \ldots, a_0, 0, 0, \ldots, 0) = a_0$. If $n < i \leqslant 2n$, then

$$p(a_1, \ldots, a_n, a_1', a_2', \ldots, a_n') \wedge a_0 = a_{i-n}' \wedge a_0 =$$
$$= a_{i-n}' \wedge a_1 \wedge \ldots \wedge a_{i-n} \wedge \ldots \wedge a_n = 0$$

and $p(a_0, a_0, \ldots, a_0, 0, 0, \ldots, 0) = 0$.

2. Assume that the assertion is true whenever the rank r'

of a polynomial p' satisfies the inequalities $1 \leqslant r' < r$.

(A) Let $p = s \vee t$ where s,t are lattice polynomials with ranks $r_1 < r$ and $r_2 < r$, respectively. Then

$$u = p(a_1,\ldots,a_n,a_1',a_2',\ldots,a_n') \wedge a_0 = [s(a_1,\ldots,a_n,a_1',\ldots$$
$$\ldots,a_n') \vee t(a_1,\ldots,a_n,a_1',\ldots,a_n')] \wedge a_0.$$

Since $a_0 \leqslant a_i$ for every $i = 1,2,\ldots,n$, it is evident that a_0 commutes with every a_i and every a_i'. Hence, by Remark II.4.3, a_0 commutes with the elements

$$s(a_1,\ldots,a_n,a_1',\ldots,a_n'),\ t(a_1,\ldots,a_n,a_1',\ldots,a_n').$$

Therefore,

$$u = [s(a_1,\ldots,a_n,a_1',\ldots,a_n') \wedge a_0] \vee [t(a_1,\ldots,a_n,a_1',\ldots,a_n') \wedge a_0].$$

Because of the induction assumption, we have

$$u = s(a_0,\ldots,a_0,0,\ldots,0) \vee t(a_0,\ldots,a_0,0,\ldots,0).$$

This gives $u = p(a_0,\ldots,a_0,0,\ldots,0)$.

(B) Let $p = s \wedge t$. Then

$$v = p(a_1,\ldots,a_n,a_1',\ldots,a_n') \wedge a_0 = s(a_1,\ldots,a_n,a_1',\ldots,a_n') \wedge$$
$$\wedge a_0 \wedge t(a_1,\ldots,a_n,a_1',\ldots,a_n') \wedge a_0.$$

From our induction hypothesis, it then follows that

$$v = s(a_0,\ldots,a_0,0,\ldots,0) \wedge t(a_0,\ldots,a_0,0,\ldots,0) =$$
$$= p(a_0,\ldots,a_0,0,\ldots,0). \quad //$$

Theorem 2.4. Let x_0 be an element which commutes with every element of an orthomodular lattice `L.

(i) If

$$(1)\quad q/p = a_0^*/a_0 \sim a_1^*/a_1 \sim \ldots \sim a_n^*/a_n = d/c$$

and $q \leqslant x_0$, then

$$q/p = a_0^*/a_0 \sim a_1^* \wedge x_0/a_1 \wedge x_0 \sim \ldots \sim a_n^* \wedge x_0/a_n \wedge x_0 = d \wedge x_0/c \wedge x_0.$$

(ii) If $q/p \in A(`L)$, $p \neq q$ and $x_0 \geqslant q$, then there exist elements h,k such that $h \leqslant x_0$, $k \leqslant x_0$ and $0 \neq \overline{\mathrm{com}}\,(h,k) \leqslant x_0$.

Proof. Ad (i). First suppose that $a_j^*/a_j \searrow a_{j+1}^*/a_{j+1}$. Then $(a_j \wedge x_0) \wedge (a_{j+1}^* \wedge x_0) = a_{j+1} \wedge x_0$. Since $x_0 C a_j$ and $x_0 C a_{j+1}^*$, Theorem II.3.10 gives

$$(a_j \wedge x_0) \vee (a_{j+1}^* \wedge x_0) = (a_j \vee a_{j+1}^*) \wedge x_0 = a_j^* \wedge x_0.$$

This means that

$$a_j^* \wedge x_0/a_j \wedge x_0 \searrow a_{j+1}^* \wedge x_0/a_j \wedge x_0.$$

The proof of the dual case is essentially the same.

Ad (ii). Put $r = q \wedge p'$. Since $q/p \in A(`L)$ and $q/p \searrow r/0$, we have $0 \neq r$ and $r \in L'$, by Theorem VI.3.2. This and Lemma V.5.2 implies that there exist $x_i, y_i, b_i \in L$ such that

$$r \leqslant \bigvee_{i=1}^{n} \mathrm{com}_{[0,b_i]}(x_i, y_i) = \bigvee_{i=1}^{n} \mathrm{com}\,(x_i, y_i).$$

By Theorem II.3.10 and Theorem 2,

$$0 \neq r = r \wedge x_0 \leqslant x_0 \wedge [\bigvee_{i=1}^{n} \mathrm{com}\,(x_i, y_i)] =$$

$$= x_0 \wedge [\bigvee_{i=1}^{n} \mathrm{com}\,(x_i \wedge x_0, y_i \wedge x_0)].$$

Hence, at least one i is such that $0 \neq \mathrm{com}\,(x_i \wedge x_0, y_i \wedge x_0) \leqslant$ $\leqslant x_0$. //

A characterization of the reflective congruence $\hat{B}$ gives the following theorem.

Theorem 2.5. In any orthomodular lattice,

$$(0,z) \in \hat{B} \Leftrightarrow (\forall u,y \in L \;\; u \leqslant z \Rightarrow uCy).$$

Proof. 1. Suppose that uCy for every $u \leqslant z$ and any $y \in L$.

We claim that $(0,z) \in \hat{B}$. For, if this were false, then by Corollary VI.2.5 (i), there would exist an allele $q/p \in A(`L)$ such that $[p,q] \subset [0,z]$ and $p \neq q$. By assumption, $z$ commutes with every element $y$ of $`L$. From Theorem 4 (ii) we deduce that there exist elements $h,k \leq z$ such that $0 \neq \overline{\text{com}}\,(h,k)$. By assumption and Theorem III.2.11, $\overline{\text{com}}\,(h,k) = 0$, a contradiction. Thus $(0,z) \in \hat{B}$.

2. Conversely, let $(0,z) \in \hat{B}$ and let u be such that $u \leq z$. Choose $y \in L$. Since $\hat{B}$ is a congruence, $(0,u) \in \hat{B}$ and, a fortiori, $((u \wedge y) \vee (u \wedge y'),u) \in \hat{B}$. However, $((u \wedge y) \vee (u \wedge y'), u) \in \hat{C}$, by Corollary 1.9 (i). From Theorem VI.1.11 (ii) we conclude that $u = (u \wedge y) \vee (u \wedge y')$. //

Theorem 2.6. Let $`F = x_1, x_2, \ldots, x_n$ be a finitely generated orthomodular lattice. Then

$$`F = `F[0,\overline{\text{com}}\,(x_1,x_2,\ldots,x_n)] \oplus `F[0,\underline{\text{com}}\,(x_1,x_2,\ldots,x_n)].$$

Proof. By Theorem II.4.4, $\overline{\text{com}}\,(x_1,x_2,\ldots,x_n)$ commutes with every element of F. Hence, Lemma III.2.1 completes the proof. //

We wish to establish a general characterization for the coreflective congruence on finitely generated orthomodular lattices. To this end, we first prove a technical assertion on $\hat{C}$. For brevity we write $\bar{c}$ instead of $\overline{\text{com}}\,(x_1,x_2,\ldots,x_n)$ and $\underline{c}$ for the lower commutator $\underline{\text{com}}\,(x_1,x_2,\ldots,x_n)$.

Lemma 2.7. Let t be an element of an orthomodular lattice $`F = \langle x_1,x_2,\ldots,x_n \rangle$. Then

$$(t \geq \overline{\text{com}}\,(x_1,x_2,\ldots,x_n)\ \&\ (0,t) \in \hat{C}) \Rightarrow t = \overline{\text{com}}\,(x_1,x_2,\ldots,x_n).$$

Proof. Since $(0,t) \in \hat{C}$, there exist c_i, d_i $(i = 1,2,\ldots,k)$ such that $t \leq \overline{\text{com}}\,(c_1,d_1) \vee \ldots \vee \overline{\text{com}}\,(c_k,d_k)$. From Corollary 1.14 we see that $\overline{\text{com}}\,(c_i,d_i) \leq \bar{c}$. Thus $t \leq \bar{c}$ and so $t = \bar{c}$. //

The following theorem, from which we shall draw many useful corollaries, characterizes the coreflective congruence in the case of finitely generated orthomodular lattices.

<u>Theorem 2.8 (cf. [20, Thm 4.14])</u>. In any orthomodular lattice $`F = \langle x_1, x_2, \ldots, x_n \rangle$,

$$(0,a) \in \hat{C} \Leftrightarrow a \leqslant \overline{\text{com}}\,(x_1, x_2, \ldots, x_n).$$

<u>Proof</u>. By Theorem 1 it suffices to prove that $(0,a) \in \hat{C}$ implies $a \leqslant \bar{c}$. Let $t = a \vee \bar{c}$. Since $(0,a) \in \hat{C}$ and since, by Theorem 1, $(0,\bar{c}) \in \hat{C}$, we find that $(0,t) \in \hat{C}$. From Lemma 7 we infer that $\bar{c} = t = a \vee \bar{c} \geqslant a$. //

<u>Corollary 2.9</u>. The commutator sublattice $`F'$ of an orthomodular lattice $`F = \langle x_1, x_2, \ldots, x_n \rangle$ is given by

$$F' = \{a \in F;\ a \leqslant \overline{\text{com}}\,(x_1, x_2, \ldots, x_n)\}.$$

<u>Proof</u>. By Theorem VI.3.2, the element a belongs to F' if and only if $(0,a) \in \hat{C}$. //

<u>Corollary 2.10</u>. In any orthomodular lattice $`L$, $\overline{\text{com}}\,(a_1, a_2, \ldots, a_n) \in L'$ for every $n \in \underline{N}$ and any $a_1, a_2, \ldots, a_n \in L$.

<u>Proof</u>. Let $`F$ be the subalgebra of the algebra $(L, \vee, \wedge, ', 0, 1)$ generated by the elements $a_1, a_2, \ldots, a_n$. According to Corollary 9, $\overline{\text{com}}\,(a_1, a_2, \ldots, a_n) \in F'$. Since we obviously have $F' \subset L'$, the corollary is proved. //

<u>Corollary 2.11</u>. In any orthomodular lattice $`L$,

$$(0,a) \in \hat{C} \Leftrightarrow [\exists m \in \underline{N} \exists y_1, y_2, \ldots, y_m \in L :: a \leqslant \overline{\text{com}}\,(y_1, y_2, \ldots, y_m)].$$

<u>Proof</u>. 1. If $a \leqslant \overline{\text{com}}\,(y_1, \ldots, y_m)$, then $\overline{\text{com}}\,(y_1, \ldots, y_m) \in L'$ by Corollary 10 and so $a \in L'$, i.e., $(0,a) \in \hat{C}$.

2. Conversely, if $(0,a) \in \hat{C}$, then $a \in L'$ by Theorem VI.3.2. By the definition of the commutator sublattice,

$$a \leq \overline{\text{com}}\,(a_1,b_1) \vee \ldots \vee \overline{\text{com}}\,(a_n,b_n)$$

for suitable $a_i, b_i \in L$. Lemma 1.2 (vi) now implies that

$$a \leq \overline{\text{com}}\,(a_1,\ldots,a_n,b_1,\ldots,b_n). \quad //$$

Corollary 2.12. In any orthomodular lattice $`F = \langle x_1, x_2, \ldots$ $\ldots, x_n \rangle$,

$$(a,b) \in \hat{C} \Leftrightarrow (a \vee b) \wedge (a' \vee b') \leq \overline{\text{com}}\,(x_1,x_2,\ldots,x_n).$$

Proof. Since $\hat{C}$ is a congruence relation, $(a,b) \in \hat{C}$ if and only if $(a \wedge b, a \vee b) \in \hat{C}$. This holds if and only if the relative orthocomplement $(a \vee b) \wedge (a' \vee b')$ of the element $a \wedge b$ in the interval $[0, a \vee b]$ is an element of the kernel of $\hat{C}$. //

Corollary 2.13. The following conditions are equivalent in any orthomodular lattice $`F = \langle x_1, x_2, \ldots, x_n \rangle$:

(i) $(a,b) \in \hat{C}$;

(ii) $(a \wedge b') \vee (a' \wedge b) \leq \overline{\text{com}}\,(x_1,x_2,\ldots,x_n)$.

Proof. 1. In any ortholattice, for all a,b, we always have $(a \wedge b') \vee (a' \wedge b) \leq (a \vee b) \wedge (a' \vee b')$. Thus, from Corollary 12, we see that $(a,b) \in \hat{C}$ implies (ii).

2. Suppose (ii) is true. Let $u = (a \wedge b') \vee (a' \wedge b)$ so that $u \leq \overline{\text{com}}\,(x_1, \bullet)$. Then $(0,u) \in \hat{C}$ by Corollary 11. The rest follows from Corollary 1.9. //

Corollary 2.14. In any orthomodular lattice $`F = \langle x_1, x_2, \ldots$ $\ldots, x_n \rangle$, for every $m \in \underline{N}$ and every choice of the elements a_1, $a_2, \ldots, a_m$,

$$\overline{\text{com}}\,(a_1,a_2,\ldots,a_m) \leq \overline{\text{com}}\,(x_1,x_2,\ldots,x_n).$$

Proof. The corollary follows from Theorem 8 and Corollary 11.

Theorem 2.15. In any orthomodular lattice $`F = \langle x_1, \ldots, x_n \rangle$,

$$(0,b) \in \hat{B} \Leftrightarrow b \leq \underline{\text{com}}\,(x_1,x_2,\ldots,x_n).$$

Proof. Put $a = 1$, $e = c = \overline{\text{com}}\,(x_1, x_2, \ldots, x_n)$, $c^+ = \underline{\text{com}}\,(x_1, x_2, \ldots, x_n)$ and $`L = `F$. By Theorem 8, $(a] \cap \hat{C}(`L) = (c]$. Observe that the assumption on the ascending chain condition has been used only in the first two lines in the proof of Theorem VI.3.14. Hence, by this proof, $(0, c^+) \in \hat{B}$ and $(c^+, 1) \in \hat{C}$. Therefore, for any b such that $(0, b) \in \hat{B}$ we have $(c^+, b \vee c^+) \in \hat{B} \cap \hat{C} = \Delta$ and so $c^+ = b \vee c^+ \geqslant b$. //

Theorem 2.16. Let $`F = \langle x_1, x_2, \ldots, x_n \rangle$ be an orthomodular lattice. Then

$$\text{Ref } `F \cong (\overline{\text{com}}\,(x_1, x_2, \ldots, x_n)] \;\&\; \text{Coref } `F \cong (\underline{\text{com}}\,(x_1, x_2, \ldots, x_n)].$$

Proof. In view of Theorem 6 we have $`F = `F[0, \underline{c}] \oplus `F[0, \bar{c}]$. From Theorem 8 and Theorem 15 we can see that $(\underline{c}]$ is the kernel of the congruence $\hat{B}$ and that $(\bar{c}]$ is the kernel of $\hat{C}$. This together with I.24 yields

$$(\bar{c}] \cong F/(\underline{c}] = `F/\hat{B} = \text{Ref } `F$$

and

$$(\underline{c}] = `F/(\bar{c}] = `F/\hat{C} = \text{Coref } `F. \quad //$$

Proposition 2.17. For any orthomodular lattice $`F = \langle x_1, x_2, \ldots, x_n \rangle$,

(i) $`F[0, \underline{c}] = ((\underline{\text{com}}\,(x_1, x_2, \ldots, x_n)], \vee, \wedge)$ is a distributive lattice;

(ii) if an element a of F is an atom of the lattice $`F[0, \underline{c}]$, then $a = x_1^{e_1} \wedge x_2^{e_2} \wedge \ldots \wedge x_n^{e_n}$ for some $e_i \in \{-1, 1\}$;

(iii) $`F[0, \underline{c}]$ has at most 2^n atoms and it is a finite lattice.

Proof. Note that $(0, \underline{c}) \in \hat{B}$ by Theorem 15. Hence, by Theorem VI.1.11 (iv), the principal ideal $(\underline{c}] = (\underline{\text{com}}\,(x_1, x_2, \ldots, x_n)]$ is distributive. By Corollary II.6.3 it is also an orthomodular lattice. Thus it is a Boolean algebra.

Let a be an element of the lattice $(\underline{c}]$. Then

$$a = p(x_1, x_2, \ldots, x_n, x_1', x_2', \ldots, x_n')$$

where p is a lattice polynomial. The elements $x_1^{e_1} \wedge \ldots \wedge x_n^{e_n}$ commute with the elements $x_1, x_2, \ldots, x_n, x_1', \ldots, x_n'$. Hence, by Remark II.4.3, they commute with every polynomial expression in these elements. By Foulis-Holland Theorem II.3.10, therefore,

$$a = a \wedge \underline{c} = a \wedge (\bigvee (x_1^{e_1} \wedge \ldots \wedge x_n^{e_n})) = \bigvee (a \wedge x_1^{e_1} \wedge \ldots \wedge x_n^{e_n}).$$

Let

$$w = x_1^{e_1} \wedge \ldots \wedge x_n^{e_n}.$$

Then

$$a \wedge w = p(x_1, \ldots, x_n, x_1', \ldots, x_n') \wedge w.$$

Observe that the element w commutes with the elements $x_1, x_2, \ldots$ $\ldots, x_n'$. This implies that

$$a \wedge w = p(x_1 \wedge w, \ldots, x_n \wedge w, x_1' \wedge w, \ldots, x_n' \wedge w).$$

It is easily seen that exactly n elements of $x_1 \wedge w, \ldots, x_n' \wedge w$ are equal to 0 and exactly n other elements are equal to w. It follows that $a \wedge w = p(\ldots, 0, \ldots, 1, \ldots) \wedge w$. Clearly, $p(\ldots)$ is either 0 or 1. Hence, $a \wedge w = 0$ or $a \wedge w = w$. Putting these results together, we see that $a \neq 0$ is an atom of the lattice $(\underline{c}]$ if and only if $a = x_1^{e_1} \wedge x_2^{e_2} \wedge \ldots \wedge x_n^{e_n}$ for a suitable choice of $e_1, e_2, \ldots, e_n \in \{-1, 1\}$.

By (ii), the lattice $`F[0,\underline{c}]$ has at most $k = 2^n$ atoms $a_1, a_2, \ldots, a_k$. Since $\underline{c} = a_1 \vee a_2 \vee \ldots \vee a_k$ and since $`F[0,\underline{c}]$ is distributive, we have $d = d \wedge \underline{c} = (d \wedge a_1) \vee (d \wedge a_2) \vee \ldots \vee (d \wedge a_k)$ for any $d \in (\underline{c}]$. Now, $d \wedge a_i$ is either 0 or a_i. Hence every d of $(\underline{c}]$ is a join of some atoms of $`F[0,\underline{c}]$. Therefore, $`F[0,\underline{c}]$ is finite. //

Corollary 2.18. Every orthomodular lattice $`F = \langle x_1, x_2, \ldots, x_n \rangle$ is isomorphic to the direct product $\underline{2}^m \times \mathrm{Ref}\ `F$ where $m \leqslant 2^n$.

Proof. By Theorem 6, Corollary I.17 and Theorem 16, $`F \cong \mathrm{Ref}\ `F \times `F[0,\underline{c}]$. From Proposition 17 (iii) and Theorem I.20 we deduce that $`F[0,\underline{c}] \cong \underline{2}^m$ with $m \leqslant 2^n$. //

The following theorem shows that the choice of the polynomials defining the commutators is in a sense the best possible.

Theorem 2.19. Let $p = p(y_1, y_2, \ldots, y_n)$, $n \in \underline{N}$, be an element of the free orthomodular lattice F_n with n generators $y_1, y_2, \ldots, y_n$. Suppose p is such that $p(a_1, a_2, \ldots, a_n) = 0$ for any distributive orthomodular lattice $`L$ and any elements $a_1, a_2, \ldots, a_n$ of $`L$. Then

$$p(y_1, y_2, \ldots, y_n) \leqslant \overline{\mathrm{com}}\ (y_1, y_2, \ldots, y_n).$$

Proof. The proof is very simple. We use square brackets to denote elements of the lattice $F_n/\hat{C}$. Since, by Theorem VI.1.10, $F_n/\hat{C}$ is distributive, we have $[0] = p([y_1], \bullet\ ) = [p(y_1, \bullet\ )]$ by assumption and Theorem V.3.2. It follows from Theorem 8 that $p(y_1, \bullet\ ) \leqslant \overline{\mathrm{com}}\ (y_1, y_2, \ldots, y_n)$. //

Remark 2.20. Let F_n be a free orthomodular lattice with n generators $y_1, y_2, \ldots, y_n$. Since $F_n = (\bar{c}] \oplus (\underline{c}]$ where $\bar{c} = \mathrm{com}\ (y_1, y_2, \ldots, y_n)$ and $\underline{c} = \bar{c}'$, we have $y_i = c_i \vee d_i$ with $c_i \in (\bar{c}]$ and $d_i \in (\underline{c}]$ $(i = 1, 2, \ldots, n)$. From Proposition I.17 and Lemma III.2.1 (iv) we see that $\{d_1, d_2, \ldots, d_n\}$ is a generating set of $`F_n[0,\underline{c}]$. Now let $`B$ be a Boolean algebra and let $g: \{d_1, d_2, \ldots, d_n\} \to B$ be a mapping. Define $g^+: \{y_1, y_2, \ldots, y_n\} \to B$ by $g^+(y_i) = g(d_i)$ $(i = 1, 2, \ldots, n)$. Since F_n is free, there exists an orthohomomorphism h^+ of F_n into $`B$ extending g^+. Let

$h:(\underline{c}] \to B$ be the restriction of h^+ on $(\underline{c}]$, i.e., $h(x) = h^+(x)$ for any $x \in (\underline{c}]$. Then h is an orthohomomorphism of $`F_n[0,\underline{c}]$ into $`B$ and this orthohomomorphism extends the mapping g. Hence $F_n[0,\underline{c}]$ is a free Boolean algebra with n generators. By I.23, $`F_n[0,\underline{c}] \cong \underline{2}^t$ where $t = 2^n$. From Theorem I.20 we infer that $`F_n[0,\underline{c}]$ has exactly 2^n atoms. Hence, by Proposition 17, in this case the atoms are represented by all the elements $y_1^{e_1} \wedge y_2^{e_2} \wedge \ldots \wedge y_n^{e_n}$ where $e_i \in \{-1,1\}$.

3. FORMULAS FOR ORTHOMODULAR LATTICES

As noted in Section II.1, we write $a \perp b$ for two elements a,b of an ortholattice if and only if $a \leqslant b'$. Here we shall introduce a similar notation for ordered pairs of elements in an ortholattice. We write $(a,b) \perp (c,d)$ if and only if

$$a \perp b \ \& \ b \perp c \ \& \ c \perp d \ \& \ d \perp a.$$

Using this notation, we arrive at the following theorem:

<u>Theorem 3.1</u>. (i) In any ortholattice, $(a,b) \perp (c,d)$ and $(c,d) \perp (e,f)$ imply $(a \wedge c, b \wedge d) \perp (e,f)$.

(ii) If a_i, b_i, $i = 1,2,\ldots,n$, are elements of an ortholattice such that $(a_i,b_i) \perp (a_{i+1},b_{i+1})$ for every $i = 1,2,\ldots$ $\ldots,n-1$, and $(a_n,b_n) \perp (e,f)$, then

$$\left(\bigwedge_{i=1}^{n} a_i, \bigwedge_{i=1}^{n} b_i\right) \perp (e,f).$$

<u>Proof</u>. Ad (i). By assumption, $b' \geqslant a$, $d' \geqslant c$ and, a fortiori, $(b \wedge d)' = b' \vee d' \geqslant a \wedge c$, i.e., $a \wedge c \perp b \wedge d$. According to $e' \geqslant d$, we also have $e' \geqslant b \wedge d$ and so $b \wedge d \perp e$. Similarly, $f' \geqslant c$ implies $f' \geqslant a \wedge c$ and, therefore, $f \perp a \wedge c$.

Ad (ii). The assertion is trivial if $n = 1$. To continue induction on n, assume that the assertion is true when $1 \leq n = k$. Let $a_1, b_1, \ldots, a_{k+1}, b_{k+1}$ be corresponding elements. From our induction hypothesis, it then follows that

$$(a_1 \wedge a_2 \wedge \ldots \wedge a_k, b_1 \wedge b_2 \wedge \ldots \wedge b_k) \perp (a_{k+1}, b_{k+1}).$$

Since $(a_{k+1}, b_{k+1}) \perp (e,f)$, we can conclude from (i) that

$$(a_1 \wedge a_2 \wedge \ldots \wedge a_k \wedge a_{k+1}, b_1 \wedge b_2 \wedge \ldots \wedge b_k \wedge b_{k+1}) \perp (e,f). \quad //$$

Theorem 3.2. Let a_i, b_i, $i = 1,2,\ldots,n$, be elements of an orthomodular lattice satisfying $(a_i, b_i) \perp (a_{i+1}, b_{i+1})$ for every $i = 1,2,\ldots,n-1$. Then $\bigwedge_{i=1}^{n} (a_i \vee b_i) = (\bigwedge_{i=1}^{n} a_i) \vee (\bigwedge_{i=1}^{n} b_i)$.

Proof. We shall prove the theorem by induction on n.

1. The case $n = 1$ being trivial, we shall confine our attention to the case $n = 2$. Since $(a_1, b_1) \perp (a_2, b_2)$, $b_1 \leq a_1'$ and $b_1 \leq a_2'$. It follows that $b_1 \leq a_1' \wedge a_2'$. Similarly, $b_2 \leq a_1' \wedge a_2'$ and so $b_1 \vee b_2 \leq a_1' \wedge a_2'$. Then it is immediate that $(a_1 \vee a_2) \wedge (b_1 \vee b_2) \leq (a_1 \vee a_2) \wedge a_1' \wedge a_2' = 0$. From $b_2 \perp a_1$ and $b_2 \perp a_2$ we conclude that b_2 commutes with a_1 and a_2. By Theorem II.4.2 we see that $a_1 \vee a_2 C b_2$. Using Remark II.4.3 and the fact that $a_1 \leq a_1 \vee a_2$ we obtain $a_1 \vee a_2 C a_1$. Hence (taking Theorem II.4.2 again into account) $a_1 \vee a_2 C a_1 \vee b_2$. By the same argument we can establish that $b_1 \vee b_2$ commutes with $a_1 \vee b_2$. Applying Theorem II.3.10 we have

$$\begin{aligned} a_1 \vee b_2 &= (a_1 \vee b_2) \vee [(a_1 \vee a_2) \wedge (b_1 \vee b_2)] = \\ &= (a_1 \vee a_2 \vee b_2) \wedge (a_1 \vee b_1 \vee b_2) \geq (a_1 \vee b_1) \wedge (a_2 \vee b_2). \end{aligned}$$

By symmetry,

$$a_2 \vee b_1 \geq (a_1 \vee b_1) \wedge (a_2 \vee b_2).$$

Since $a_1 C b_i$, $a_2 C b_i$ and also $b_1 \wedge b_2 C a_i$, Theorem II.3.10 yields

$$(a_1 \wedge a_2) \vee (b_1 \wedge b_2) = [a_1 \vee (b_1 \wedge b_2)] \wedge [a_2 \vee (b_1 \wedge b_2)] =$$
$$= (a_1 \vee b_1) \wedge (a_1 \vee b_2) \wedge (a_2 \vee b_1) \wedge (a_2 \vee b_2) = (a_1 \vee b_1) \wedge (a_2 \vee b_2).$$

2. Next suppose that $n \geqslant 3$ and that the assertion is already true for every $k \leqslant n-1$. Then

$$v = \bigwedge_{i=1}^{n} (a_i \vee b_i) = \bigwedge_{i=1}^{n-1} (a_i \vee b_i) \wedge (a_n \vee b_n).$$

Let $A_{n-1} = \bigwedge_{i=1}^{n-1} a_i$ and $B_{n-1} = \bigwedge_{i=1}^{n} b_i$. Because of the induction assumption, we have

$$v = (A_{n-1} \vee B_{n-1}) \wedge (a_n \vee b_n).$$

By Theorem 1 (ii), $(A_{n-1}, B_{n-1}) \perp (a_n, b_n)$. The first part of our proof implies that

$$v = (A_{n-1} \wedge a_n) \vee (B_{n-1} \wedge b_n). \quad //$$

The proof of the following resultat may be useful for the light it sheds on the role of commutators in some proofs of formulas in orthomodular lattices.

Corollary 3.3. Let a,b,c,d be elements of an orthomodular lattice satisfying aCb, bCc, cCd and dCa. Then

$$(a \vee b) \wedge (c \vee d) = (a \wedge c) \vee (a \wedge d) \vee (b \wedge c) \vee (b \wedge d).$$

Proof. Let

$$L = (a \vee b) \wedge (c \vee d), \; R = (a \wedge c) \vee (a \wedge d) \vee (b \wedge c) \vee (b \wedge d).$$

Then

$$L \wedge R' = (a \vee b) \wedge (c \vee d) \wedge (a' \vee c') \wedge (a' \vee d') \wedge (b' \vee c') \wedge (b' \vee d') \leqslant$$
$$\leqslant [(a \vee b) \wedge (a' \vee d')] \wedge [(c \vee d) \wedge (b' \vee c')].$$

By Remark II.4.3, $a \vee bCa'$. Using $a'Cd'$, together with Foulis-Holland Theorem II.3.10, we get

$$B_1 = (a \vee b) \wedge (a' \vee d') = [a' \wedge (a \vee b)] \vee [d' \wedge (a \vee b)].$$

Since $a'Ca$, aCb and $d'Ca$, $B_1 = (a' \wedge b) \vee (a \wedge d') \vee (b \wedge d')$ by Theorem II.3.10. A similar reasoning, involving $c \vee dCc'$, $b'Cc'$, cCd and $c'Cc$, yields

$$B_2 = (c \vee d) \wedge (b' \vee c') = [b' \wedge (c \vee d)] \vee [c' \wedge (c \vee d)] = \\ = (b' \wedge c) \vee (c' \wedge d) \vee (b' \wedge d).$$

From $c' \wedge d \leqslant (b \wedge d')' = b' \vee d$ and $b' \wedge c \leqslant (b \wedge d')'$ it follows that $b \wedge d'$ commutes with $c' \wedge d$ and $b' \wedge c$. Therefore, by Theorem II.3.10,

$$[(b' \wedge c) \vee (c' \wedge d)] \wedge (b \wedge d') = (b \wedge b' \wedge c \wedge d') \vee (b \wedge c' \wedge d \wedge d') = 0.$$

Similarly, $[(a' \wedge b) \vee (a \wedge d')] \wedge (b' \wedge d) = 0$. From Theorem II.4.2 we see that $b' \wedge d$ and $b \wedge d'$ commute with a,b,c,d and so they commute with all the elements considered here. Hence, by Theorem II.3.10,

$$B_1 \wedge B_2 = [(a' \wedge b) \vee (a \wedge d') \vee (b \wedge d')] \wedge [(b' \wedge c) \vee (c' \wedge d) \vee \\ \vee (b' \wedge d)] = \{[(a' \wedge b) \vee (a \wedge d')] \wedge [(b' \wedge c) \vee (c' \wedge d)]\} \vee \\ \{[(a' \wedge b) \vee (a \wedge d')] \wedge (b' \wedge d)\} \vee \{[(b' \wedge c) \vee (c' \wedge d)] \wedge \\ \wedge (b \wedge d')\} \vee [(b \wedge d') \wedge (b' \wedge d)] = \\ = [(a' \wedge b) \vee (a \wedge d')] \wedge [(c' \wedge d) \vee (b' \wedge c)].$$

Let $a_1 = a' \wedge b$, $b_1 = a \wedge d'$, $a_2 = c' \wedge d$ and $b_2 = b' \wedge c$. Since $(a_1, b_1) \perp (a_2, b_2)$, Theorem 4 gives

$$B_1 \wedge B_2 = (a_1 \vee b_1) \wedge (a_2 \vee b_2) = (a_1 \wedge a_2) \vee (b_1 \wedge b_2) = \\ = (a' \wedge b \wedge c' \wedge d) \vee (a \wedge b' \wedge c \wedge d').$$

This means that $L \wedge R' \leqslant$ <u>com</u> (a,b,c,d). Let $`S$ denote the subalgebra generated in the considered orthomodular lattice by the elements $a,b,c,d$. Let $\hat{C}^+$ denote the coreflective congruence of the lattice $`S$. By Theorem VI.2.10, the coreflection $`S/\hat{C}^+$ is a distributive lattice. Hence $L \wedge R'$ belongs to the zero class of $`S/\hat{C}^+$, i.e., $(0, L\wedge R') \in \hat{C}^+$. It follows from Corollary 2.11 that $L\wedge R' \leqslant \overline{\text{com}}\,(a,b,c,d)$. In summary,

$$L\wedge R' \leqslant \underline{\text{com}}\,(a,b,c,d) \wedge \overline{\text{com}}\,(a,b,c,d) = 0.$$

Thus $L\wedge R' = 0$. By orthomodularity, $L \geqslant R$ implies that $L = R$.

Remark 3.4. The proof of Corollary 3 given here can be found in [21]. The corollary is also proved in [84] by a different approach (cf. the original statement in [148, p. A-51]).

We are now in a position to establish a generalization of Foulis-Holland Theorem II.3.10 and of the Marsden's result mentioned above.

Theorem 3.5. Let $`L$ be an orthomodular lattice, $R = \{x_{r_1}, x_{r_2}, \dots, x_{r_h}, y_{u_1}, y_{u_2}, \dots, y_{u_m}\}$, $S = \{x_{s_1}, x_{s_2}, \dots, x_{s_k}, y_{v_1}, y_{v_2}, \dots, y_{v_n}\}$, $h \geqslant 0$, $k \geqslant 0$, $m \geqslant 0$, $n \geqslant 0$, $h+m > 0$, $k+n > 0$, be subsets of L such that $R\cup S = \{x_1, x_2, \dots, x_p, y_1, y_2, \dots, y_q\}$ and aCb for each $a\in R, b\in S$. If $M = (x_1 \vee x_2 \vee \dots \vee x_p) \wedge (y_1 \vee y_2 \vee \dots \vee y_q)$, then

$$\begin{aligned} M = {} & [(x_{r_1} \vee x_{r_2} \vee \dots \vee x_{r_h}) \wedge (y_{u_1} \vee y_{u_2} \vee \dots \vee y_{u_m})] \vee \\ & \vee [(x_{s_1} \vee x_{s_2} \vee \dots \vee x_{s_k}) \wedge (y_{v_1} \vee y_{v_2} \vee \dots \vee y_{v_n})] \vee \\ & \vee \bigvee (x_{r_i} \wedge y_{v_j};\ 1 \leqslant i \leqslant h,\ 1 \leqslant j \leqslant n) \vee \\ & \vee \bigvee (x_{s_i} \wedge y_{u_j};\ 1 \leqslant i \leqslant k,\ 1 \leqslant j \leqslant m). \end{aligned}$$

Proof. Let

$$a = x_{r_1} \vee x_{r_2} \vee \dots \vee x_{r_h},\quad b = x_{s_1} \vee x_{s_2} \vee \dots \vee x_{s_k},$$
$$c = y_{u_1} \vee y_{u_2} \vee \dots \vee y_{u_m},\quad d = y_{v_1} \vee y_{v_2} \vee \dots \vee y_{v_n}.$$

By Theorem II.4.2 we have aCb, bCc, cCd and dCa. Hence, by Corollary 5, $M = (a\vee b) \wedge (c\vee d) = (a\wedge c) \vee (a\wedge d) \vee (b\wedge c) \vee (b\wedge d)$. Since aCy_{v_j} by Theorem II.4.2, $a\wedge d = \bigvee (a\wedge y_{v_j};\ 1 \leqslant j \leqslant n)$. Moreover,

$x_{r_i} C y_{v_j}$ implies $a \wedge y_{v_j} = \bigvee(x_{r_i} \wedge y_{v_j};\ 1 \leqslant i \leqslant h)$. This gives

$$a \wedge d = \bigvee(x_{r_i} \wedge y_{v_j};\ 1 \leqslant i \leqslant h,\ 1 \leqslant j \leqslant n)$$

and, similarly,

$$b \wedge c = \bigvee(x_{s_i} \wedge y_{u_j};\ 1 \leqslant i \leqslant k,\ 1 \leqslant j \leqslant m). \quad //$$

4. EXCHANGE THEOREMS

We begin our considerations with the following theorem which is a slight modification of a result due to Gudder and Schelp (cf. [87]).

Gudder-Schelp's Theorem 4.1. Let a, b, c be elements of an orthomodular lattice such that aCb, bCc and $aCb \wedge c$. Then $a \wedge bCc$.

Proof. Write $d = [(a \wedge b) \vee c] \wedge [(a \wedge b) \vee c']$. Since bCc and bCa, $(a \wedge b) \vee c = (a \vee c) \wedge (b \vee c)$ follows from Foulis-Holland Theorem II.3.10. But we also have bCa and bCc'. Using Theorem II.3.10 once again, we see that $d = (a \vee c) \wedge (b \vee c) \wedge (a \vee c') \wedge (b \vee c')$. From bCc we conclude that $b'Cc'$ and so, by definition, $b' = (b' \wedge c') \vee (b' \wedge c)$. By De Morgan laws we therefore find $b = (b \vee c) \wedge (b \vee c')$, i.e., $d = (a \vee c) \wedge b \wedge (a \vee c')$. Furthermore, because bCa and bCc, we obtain $d = [(a \wedge b) \vee (b \wedge c)] \wedge (a \vee c')$. We now arrive at the key step in our proof. First we note that $b \wedge c$ commutes with b and c. By assumption, $b \wedge cCa$. Hence, by Theorem II.4.2, the element $b \wedge c$ commutes with $a \wedge b$ and $a \vee c'$. Then, by Theorem II.3.10, we see that

$$d = [a \wedge b \wedge (a \vee c')] \vee [(b \wedge c) \wedge (a \vee c')] =$$

$$= (a \wedge b) \vee (b \wedge c \wedge a) \vee (b \wedge c \wedge c') = a \wedge b.$$

By what was done above we get that

$$a \wedge b = [(a \wedge b) \vee c] \wedge [(a \wedge b) \vee c'].$$

Thus, by Theorem II.3.6, $a \wedge bCc$. //

A general investigation of the question as to when we can use an analogue of Gudder-Schelp's Theorem shows that the hypotheses of Theorem 1 can be weakened. In fact, we can delete the first hypothesis. This omission makes the theorem more symmetric. The strengthened version can be formulated as follows (cf. [23]):

Theorem 4.2. Let x, y, z be elements of an orthomodular lattice such that yCz and $xCy \wedge z$. Then:

(i) $x \wedge yCz$; (ii) $x \wedge zCy$;

(iii) $x' \wedge yCz$; (iv) $x' \wedge zCy$;

(v) $x' \vee y'Cz$; (vi) $x' \vee z'Cy$;

(vii) $x \vee y'Cz$; (viii) $x \vee z'Cy$.

Proof. Since $xCy \wedge z$, we also have $x'Cy \wedge z$. Clearly, $y'Cy \wedge z$. Hence, by Theorem II.4.2, we get $x' \vee y'Cy \wedge z$. Let $a = x' \vee y'$, $b = y$, $c = z$. From $x' \vee y'Cy$ and from Theorem 1, we find $(x' \vee y') \wedge yCz$. This implies that the element z commutes with the orthocomplement of the element $(x' \vee y') \wedge y$, i.e., z commutes with $y' \vee (x \wedge y)$. This together with the assumption zCy and Theorem II.4.2 shows that $zC\{y \wedge [y' \vee (x \wedge y)]\}$. In addition, yCy' and $yCx \wedge y$. Therefore, by orthomodularity,

$$y \wedge [y' \vee (x \wedge y)] = (y \wedge y') \vee (x \wedge y) = x \wedge y.$$

In other words, we have just deduced that $zCx \wedge y$. By the symmetry in y and z we conclude that we have proved both (i) and (ii). However, by Theorem II.3.5, $xCy \wedge z$ is equivalent to

$x'Cy \wedge z$. It therefore follows that also the assertions (iii) and (iv) are true. The remaining statements follow by Theorem II.3.5 and De Morgan laws. //

At this point, it is worth mentioning that Theorem 2 can motivate some interesting investigations. Indeed, write Theorem 2 in a concise form as follows:

$$\text{If } \ xCy \wedge z \ \text{ and } \ yCz, \ \text{ then } \ yCx \wedge z.$$

This suggests the natural question: "Is this a special case of a more general result?" By inspecting the theorems we have already proved we note that there is another situation of this type. In fact, Theorem II.3.4 can be rewritten in a shortened formulation in the following way:

$$\text{If } \ xCy, \ \text{ then } \ yCx.$$

Hence it is reasonable to conjecture that there exists a positive answer to our question. We shall now furnish a solution of this problem and settle the above-mentioned question. Before formulating the theorem which is crucial for our purposes we need the following three results which are useful in attacking the problem.

In Theorems 3, 5 and in Remark 4 below, `L is an orthomodular lattice and $\{b, c_2, \ldots, c_n\}$ denotes a subset of L where any two elements commute.

<u>Theorem 4.3</u>. Let p be an n-ary lattice polynomial. Then

$$(1) \quad p(b, c_2, \ldots, c_n) = B_1(b, c_2, \ldots, c_n) \wedge B_2(b, c_2, \ldots, c_n)$$

where

$$(2) \quad B_1(y_1, x_2, \ldots, x_n) = y_1 \vee [\bigwedge(\bigvee(x_i;\ i \in I_j);\ j \in J)],$$

$$(3) \quad B_2(y_1, x_2, \ldots, x_n) = \bigwedge(\bigvee(x_k;\ k \in K_t);\ t \in T)$$

and where I_j, J, K_t, T are finite sets.

Proof. By Theorem II.4.5 the lattice generated by $b, c_2, \ldots, c_n$ is distributive. Hence (taking the distributive identities into account) the polynomial expression $p(b, c_2, \ldots, c_n)$ can be written as a meet of joins formed from some elements of $\{b, c_2, \ldots, c_n\}$. Collecting all the terms containing b yields the polynomial B_1. The remaining terms give the polynomial B_2. For further details the reader is referred to the Birkhoff's monograph [27, Thm 12, p. 145]. //

Remark 4.4. Notice the following consequence of the preceding proof: If $e_1 = d, e_2, \ldots, e_n$ are any elements of L such that $e_i C e_j$ for every $1 \leqslant i, j \leqslant n$ and if p is the polynomial of Theorem 3, then

$$p(d, e_2, \ldots, e_n) = B_1(d, e_2, \ldots, e_n) \wedge B_2(d, e_2, \ldots, e_n)$$

where B_1 and B_2 are the same polynomials as in (2) and (3).

Theorem 4.5. Let p be an n-ary lattice polynomial and let (1) be valid. Then, for every $a \in L$,

$$D_1 \wedge D_2 \leqslant p(a, c_2, \ldots, c_n) \leqslant D_2$$

where

$$D_1 = \bigwedge(\bigvee(c_i;\ i \in I_j);\ j \in J),\ D_2 = \bigwedge(\bigvee(c_k;\ k \in K_t); t \in T).$$

Proof. Since p is a lattice polynomial, we have (cf. [74, Lemma 6, p. 33])

$$p(0, c_2, \ldots, c_n) \leqslant p(a, c_2, \ldots, c_n) \leqslant p(1, c_2, \ldots, c_n).$$

By Remark 4,

$$p(0, c_2, \ldots, c_n) = B_1(0, c_2, \ldots, c_n) \wedge B_2(0, c_2, \ldots, c_n) = D_1 \wedge D_2.$$

Moreover,

$$p(1, c_2, \ldots, c_n) = B_1(1, c_2, \ldots, c_n) \wedge B_2(1, c_2, \ldots, c_n) = D_2. \quad //$$

Theorem 4.6 (cf. [23]). Let $c_1, c_2, \ldots, c_n$ and a be elements of an orthomodular lattice. If $c_i C c_j$ for every $1 \leqslant i,j \leqslant n$ and if $aCp(c_1, c_2, \ldots, c_n)$ where p is an n-ary lattice polynomial, then $c_k Cp(c_1, c_2, \ldots, c_{k-1}, a, c_{k+1}, \ldots, c_n)$ for every $k = 1,2,\ldots,n$.

Proof. Since $q(x_1, \ldots, x_n) = p(x_k, x_2, \ldots, x_{k-1}, x_1, x_{k+1}, \ldots, x_n)$ is also a lattice polynomial, we may assume that $k = 1$. Let $b = c_1$. Then $aCp(b, c_2, \ldots, c_n)$. From Remark II.4.3 we conclude that $c_i Cp(b, c_2, \ldots, c_n)$ for every $i = 2, \ldots, n$. Hence, by the same remark, we see that $p(a, c_2, \ldots, c_n)$ commutes with $p(b, c_2, \ldots, c_n)$. Using the notation of Theorems 3 and 5, we get

$$B_1(b, c_2, \ldots, c_n) \wedge B_2(b, c_2, \ldots, c_n) Cp(a, c_2, \ldots, c_n).$$

Now

$$B_1(b, c_2, \ldots, c_n) = b \vee (\bigwedge(\bigvee(c_i;\ i \in I_j);\ j \in J)) = b \vee D_1.$$

Replacing $B_2(b, c_2, \ldots, c_n)$ by D_2, we infer that

$$(b \vee D_1) \wedge D_2 Cp(a, c_2, \ldots, c_n).$$

In view of Theorem II.4.4 we have bCD_1 and bCD_2. Then, by Foulis-Holland Theorem II.3.10,

$$(b \wedge D_2) \vee (D_1 \wedge D_2) Cp(a, c_2, \ldots, c_n).$$

According to Theorem II.4.2, $b \wedge D_2 C D_1 \wedge D_2$. By the assertion dual to Theorem 2,

$$b \wedge D_2 Cp(a, c_2, \ldots, c_n) \vee (D_1 \wedge D_2).$$

This together with Theorem 5 implies that $b \wedge D_2 Cp(a, c_2, \ldots, c_n)$. Thus, by Theorem 2, $bCp(a, c_2, \ldots, c_n) \wedge D_2$. Finally Theorem 5 shows that $bCp(a, c_2, \ldots, c_n)$. //

Remark 4.7. We end this Section with the following gloss on the "best possible" aspect of Theorem 6. It is not difficult to show that the conclusion of Theorem 6 does not follow if we assume that p is a polynomial in $\vee, \wedge$ and $'$. For consider the orthomodular lattice of Figure 45 and put $p(x,y) = (x \wedge y) \vee x'$. Then $p(d',e) = (d' \wedge e) \vee d = e \vee d = c'$ and so $aCp(d',e)$, while $p(a,e) = (a \wedge e) \vee a' = 0 \vee a' = a'$. Now it is an easy matter to verify that d' and $p(a,e)$ do not commute.

5. CENTER OF AN ORTHOMODULAR LATTICE

As usual, let $`L$ denote an orthomodular lattice. By Theorems II.4.2, II.3.5 and II.3.4 it is clear that the elements c such that cCd for every element d of $`L$ form a subalgebra of the orthomodular lattice $`L$. This subalgebra will be denoted by $C(`L)$ and we shall call it the <u>center</u> of the orthomodular lattice $`L$.

<u>Remark 5.1</u>. According to [179], the center of a lattice $`V$ with 0 and 1 is usually defined as the set $Z(`V)$ consisting of the elements $e \in V$ which have a complement in $`V$ and which are neutral, i.e., the sublattice of $`V$ generated by the elements x,y,e is distributive for every $x,y \in V$. Thus, if c is an element of the center $C(`L)$ of an orthomodular lattice $`L$, then by Theorem II.3.10, $c \in Z(`L)$ and so $C(`L) \subset Z(`L)$. Next suppose that $e \in Z(`L)$ where $`L$ is again orthomodular. Then $(e \wedge d) \vee (e \wedge d') = e \wedge (d \vee d') = e$ so that eCd. Hence, $e \in C(`L)$ and it easily follows that $C(`L) = Z(`L)$ for every orthomodular lattice $`L$. Notice that $C(`L)$ is a Boolean subalgebra of $`L$ which is an important by-product of our considerations.

<u>Theorem 5.2</u>. An element of an orthomodular lattice belongs

to its center if and only if it has exactly one complement.

Proof. Let a and $`L$ be the mentioned element and the corresponding orthomodular lattice, respectively.

1. Suppose a^* is a complement of $a \in C(`L)$. Then, by Foulis-Holland Theorem II.3.10,

$$a' = a' \vee (a^* \wedge a) = (a' \vee a^*) \wedge (a' \vee a) = (a' \vee a^*) \wedge 1 =$$
$$= (a' \vee a^*) \wedge (a^* \vee a) = a^* \vee (a' \wedge a) = a^*.$$

2. Conversely, let a be an element which has exactly one complement and let $y \in L$. Hence, by Theorem II.6.6, a has not more than one relative complement in the interval $[(a \wedge y) \vee (a \wedge y'), (a \vee y) \wedge (a \vee y')]$. Combining this with Theorem 1.7 we can derive easily that aCy. //

Theorem 5.3. If an orthomodular lattice $`L$ is the direct product $`L_1 \times `L_2 \times \ldots \times `L_n$ of orthomodular lattices $`L_1, `L_2, \ldots, `L_n$, then $C(`L) = C(`L_1) \times C(`L_2) \times \ldots \times C(`L_n)$.

Proof. The proof is an easy exercise left to the reader. //

Theorem 5.4. The center of an orthomodular lattice $`F = \langle x_1, x_2, \ldots, x_n \rangle$ is isomorphic to the direct product $\underline{2}^m \times C(\mathrm{Ref}\ `F)$ where $m \leqslant 2^n$.

Proof. By Theorems 2, 3 and Corollary 2.16,

$$C(`F) \cong C(\mathrm{Coref}\ `F) \times C(\mathrm{Ref}\ `F).$$

Since, by Proposition 2.17, Coref $`F$ is a Boolean algebra having at most $m \leqslant 2^n$ atoms, $C(\mathrm{Coref}\ `F) \cong \underline{2}^m$. //

Remark 5.5. (A) An easy argument shows that an orthomodular lattice $`F = \langle x_1, x_2, \ldots, x_n \rangle$ which is simple and nondistributive is necessarily such that $`F = (\overline{\mathrm{com}}\ (x_1, \ldots, x_n)]$. Indeed, since the lattice is simple, its reflective congruence $\hat{B}$ is equal either to $F \times F$ or to the diagonal Δ. Now if $\hat{B} = F \times F$, then $`F$ would be distributive, a contradiction. Hence $\hat{B} = \Delta$. Applying Theorem 2.15 we have that $\underline{\text{com}}\,(x_1, x_2, \ldots, x_n) = 0$. Then, by Lemma 1.2 (i), $\overline{\text{com}}\,(x_1, x_2, \ldots, x_n) = 1$.

(B) One should point out that the condition $`F = (\text{com}\,(a_1, a_2, \ldots, a_n)]$ is not enough to force the simplicity of the lattice $`F$. A simple counter-example - $`F = `M_{02} \times `M_{02}$ (see Figure 9d) - shows this to be the case. In fact, suppose $a, b$ and $c, d$ are generators of these two copies of $`M_{02}$, respectively. Let $a_1 = (a, c)$, $a_2 = (b, d)$. Then $(\overline{\text{com}}\,(a_1, a_2)] = `F$.

(C) Recall that $c \in C(`L)$ if and only if $`L = (c] \oplus (c']$ (see Theorem III.2.13). In this case $`L \cong (c] \times (c']$ by I.17. It is well-known that $`L \cong (c] \times (c']$ is also sufficient for c to belong to $Z(`L) = C(`L)$ (cf. [29, Lemma 1, p. 67]).

Theorem 5.6 (cf. [23]). An element e of an orthomodular lattice $`L$ belongs to the center of $`L$ if and only if the following is true:

Whenever $a < e < b$ and $e/a \in A(`L)$, $b/e \in A(`L)$, then b/e and e/a are not projective.

Proof. 1. Let e be an element of $C(`L)$, $a < e < b$ and $e/a \S b/e$. Using Theorem 2.4 with $x_0 = e$, $q/p = e/a$ and $d/c = b/e$, we get

$$e/a = a_0^*/a_0 \sim a_1^* \wedge e/a_1 \wedge e \sim \ldots \sim a_n^* \wedge e/a_n \wedge e = b \wedge e/e \wedge e = e/e.$$

By Lemma VI.2.1 (iv), with $T = \Delta$, we have $e = a$, a contradiction. Thus e/a and b/e are not projective.

2. Suppose s does not belong to $C(`L)$. Consider the sublattice $`M$ of Theorem 1.7 (ii) and write $M = \{o, s, t, u, i\}$ with $o < s < i$. Then $i/s \S s/o$. //

<u>Theorem 5.7</u>. An element c belongs to the center of an orthomodular lattice `L if and only if it is not an inner element of a sublattice isomorphic to $`M_5$.

<u>Proof</u>. 1. If c is not an element of the center, use Corollary 1.8.

2. Suppose $M = \{o,c,d,e,i\}$ determines such a sublattice `M with $o < c < i$. Then i/c§c/o and c does not belong to C(`L) by Theorem 6. //

<u>Theorem 5.8</u>. In any orthomodular lattice $`F = \langle x_1,x_2,\dots,x_n\rangle$,

$$C(`F) \supset (\underline{\mathrm{com}}\ (x_1,x_2,\dots,x_n)] \cup [\overline{\mathrm{com}}\ (x_1,x_2,\dots,x_n)).$$

<u>Proof</u>. From Theorems 2.5 and 2.15 it follows that uCy for every $y \in F$ and every $u \leqslant \underline{\mathrm{com}}\ (x_1,x_2,\dots,x_n)$.

On the other hand, $v \geqslant \overline{\mathrm{com}}\ (x_1,x_2,\dots,x_n)$ if and only if $v' \leqslant \underline{\mathrm{com}}\ (x_1,x_2,\dots,x_n)$. Hence v'Cy for every such v and every $y \in L$. Thus, by Theorem II.3.5, vCy. //

<u>Remark 5.9</u>. The following counter-example shows that the inclusion of Theorem 8 cannot be in general replaced by the sign of equality. Let `F be the lattice of Remark 5.B. Then $(1,0) \in C(`F)$, while $(\underline{c}] \cup [\bar{c}) = \{(0,0),(1,1)\}$.

6. IDENTITIES AND OPERATIONS

We have already studied some relations between the operations defined in a Boolean skew lattice (see Section III.1). Now we shall investigate further properties of skew operations.

H. Kröger (cf. [127]) has recently obtained the following necessary and sufficient condition in order that the skew operations should be associative.

Theorem 6.1. Let $`L$ be an orthomodular lattice and let $L^{\cdot} = (L, \dot\vee, \dot\wedge, ', 0, 1)$ be the Boolean skew lattice associated with $`L$. If bCc, then $a \dot\wedge (b \dot\wedge c) = (a \dot\wedge b) \dot\wedge c$ and $a \dot\vee (b \dot\vee c) = (a \dot\vee \dot\vee b) \dot\vee c$.

Proof. Using Theorem III.1.1 (iv), together with bCc, we get $b \dot\wedge c = b \wedge c$. Hence

$$f = a \dot\wedge (b \dot\wedge c) = a \dot\wedge (b \wedge c) = [a \vee (b \wedge c)'] \wedge b \wedge c = (a \vee b' \vee c') \wedge b \wedge c.$$

By Remark II.4.3, b commutes with $a \vee b'$ and c'. We therefore have from Foulis-Holland Theorem II.3.10 that

$$h = (a \dot\wedge b) \dot\wedge c = \{[(a \vee b') \wedge b] \vee c'\} \wedge c = (a \vee b' \vee c') \wedge (b \vee c') \wedge c.$$

However, $(b \vee c') \wedge c = b \dot\wedge c = b \wedge c$ and the first assertion follows. The second one can be proved by duality. //

In analogy with the symmetric difference $(M \setminus N) \cup (N \setminus M)$ of two sets M, N, we define the lower symmetric difference $a\underline{D}b$ of two elements a, b of an ortholattice $`L$ as the element $a\underline{D}b = (a \wedge b') \vee (a' \wedge b)$. The upper symmetric difference $a\bar{D}b$ is defined by $a\bar{D}b = (a' \vee b') \wedge (a \vee b)$. For convenience, we shall use the same symbols $\underline{D}, \bar{D}$ to denote the corresponding operations on the ortholattice $`L$.

Clearly, whatever a, b are, we have $a\underline{D}b \leqslant a\bar{D}b$ and $(a\bar{D}b)' = a'\underline{D}b = a\underline{D}b'$.

The following theorem deals with the question of equality in $a\underline{D}b \leqslant a\bar{D}b$.

Theorem 6.2. The following conditions on elements a, b of an orthomodular lattice are equivalent:

(i) aCb;

(ii) $a\underline{D}b = a\bar{D}b$;

(iii) $a\underline{D}b \geqslant a\bar{D}b$.

Proof. If aCb, then, by Foulis-Holland Theorem II.3.10,

$$a\underline{D}b = (a' \wedge b) \vee (a \wedge b') = [a' \vee (a \wedge b')] \wedge [b \vee (a \wedge b')] =$$
$$= (a' \vee a) \wedge (a' \vee b') \wedge (a \vee b) \wedge (b \vee b') = a\bar{D}b.$$

Evidently, (ii) implies (iii).

Now, let $(a' \wedge b) \vee (a \wedge b') \geqslant (a' \vee b') \wedge (a \vee b)$. Then we have $\bar{c}' = \underline{\text{com}}\,(a,b) \geqslant \overline{\text{com}}\,(a,b) = \bar{c}$, so $\bar{c} = 0$, from which aCb follows by Theorem III.2.11. //

Theorem 6.3. Let $L^{\bullet}$ be the Boolean skew lattice associated with an orthomodular lattice $`L$. Then

$$a\bar{D}b = (a' \dot{\wedge} b) \dot{\vee} (a \dot{\wedge} b'), \quad a\underline{D}b = (a' \dot{\vee} b') \dot{\wedge} (a \dot{\vee} b).$$

Proof. Let $V = (a' \dot{\wedge} b) \dot{\vee} (a \dot{\wedge} b')$. Then

$$V = [(a' \vee b') \wedge b] \dot{\vee} [(a \vee b) \wedge b'].$$

Now $(a' \vee b') \wedge b \leqslant b \leqslant [(a \vee b) \wedge b']'$. Hence these elements commute, and, since $a\bar{D}b$ commutes with b and b',

$$V = [(a' \vee b') \wedge (a \vee b) \wedge b] \vee [(a' \vee b') \wedge (a \vee b) \wedge b'] =$$
$$= (a' \vee b') \wedge (a \vee b) \wedge (b \vee b') = a\bar{D}b,$$

by Theorem III.1.1 and Theorem II.3.10.

The second assertion follows similarly. //

Theorem 6.4. Let $`L$ be an orthomodular lattice. Then the following conditions are equivalent:

(i) $`L$ is a Boolean lattice;

(ii) the operation $\underline{D}$ is associative on L;

(iii) the operation $\bar{D}$ is associative on L.

Proof. 1. If $`L$ is a Boolean lattice, then it is well-known (cf. [179, Thm 57, p. 127]) that $\underline{D}$ and $\bar{D}$ are associative.

2. Suppose that $\underline{D}$ is associative. Consider the element $k = (a\underline{D}b)\underline{D}b$. Since b commutes with $a' \wedge b$ and $a \wedge b'$, we see

from Foulis-Holland Theorem II.3.10 that

$$k = \{[(a' \wedge b) \vee (a \wedge b')]' \wedge b\} \vee \{[(a' \wedge b) \vee (a \wedge b')] \wedge b'\} =$$
$$= [(a \vee b') \wedge (a' \vee b) \wedge b] \vee (a \wedge b') = [(a \vee b') \wedge b] \vee (a \wedge b').$$

By assumption, $k = a\underline{D}(b\underline{D}b) = a$. Hence $a = [(a \vee b') \wedge b] \vee (a \wedge b')$. Now $a \wedge b' \leqslant [(a \vee b') \wedge b]' = (a' \wedge b) \vee b'$ and so $a \wedge b' C (a \vee b') \wedge b$. In addition, $aCa \wedge b'$. It follows that

$$a = a \wedge a = \{[(a \vee b') \wedge b] \vee (a \wedge b')\} \wedge a =$$
$$= [(a \vee b') \wedge b \wedge a] \vee (a \wedge b') = (a \wedge b) \vee (a \wedge b').$$

Thus, by definition, aCb.

3. Evidently,

$$[(a\bar{D}b)\bar{D}c]' = (a\bar{D}b)'\underline{D}c = (a\underline{D}b')\underline{D}c$$

and

$$[a\bar{D}(b\bar{D}c)]' = a\underline{D}(b\bar{D}c)' = a\underline{D}(b'\underline{D}c).$$

Hence $\bar{D}$ is associative if and only if $\underline{D}$ is associative. //

<u>Theorem 6.5</u>. Let x,y be elements of an orthomodular lattice. Then x and y do not commute if and only if the interval $[x\underline{D}y, x\bar{D}y]$ contains a sublattice $`M$ isomorphic to $`M_{O2}$. In this case it is possible to suppose that the greatest element of $`M$ is equal to $x\bar{D}y$ and that the least one is equal to $x\underline{D}y$.

<u>Proof</u>. Let $a = \overline{\text{com}}\,(x,y)$ and $b = (x' \wedge y) \vee (x \wedge y')$. Then aCb and, by Theorem III.1.1, $b \dot{\wedge} a = a \wedge b \leqslant \overline{\text{com}}\,(x,y) \wedge \underline{\text{com}}\,(x,y) = 0$. Furthermore, $a \dot{\vee} b = a \vee b = [(x \vee y) \wedge (x \vee y') \wedge (x' \vee y) \wedge (x' \vee y')] \vee (x' \wedge y) \vee (x \wedge y') = (x \vee y) \wedge (x' \vee y')$, by orthomodularity. From Theorem III.1.2 we conclude that we have $`L[0, \overline{\text{com}}\,(x,y)] \cong `L[(x \wedge y') \vee (x' \wedge y), (x' \vee y') \wedge (x \vee y)]$. Theorem III.2.11 and Corollary III.2.4 yield the remainder of the proof. //

The Marsden's result stated in Corollary 3.3 has the following analogue for Boolean skew lattices (cf. [22]):

Theorem 6.6. Let $L^{\bullet}$ be the Boolean skew lattice associated with an orthomodular lattice $`L$. Suppose $a,b,c,d \in L$ are such that aCb, bCc, cCd and dCa. Then

$$(a \dot\vee b) \dot\wedge (c \dot\vee d) = [(a \dot\wedge d) \dot\vee (b \dot\wedge c)] \dot\vee [(a \dot\wedge c) \dot\vee (b \dot\wedge d)]$$

if and only if $a \vee b \vee c' Cd$ and $a \vee b \vee d' Cc$.

Proof. 1. By assumption and by Theorem III.1.1 (vii), $a \dot\vee b = a \vee b$ and $c \dot\vee d = c \vee d$. Hence

$$k = (a \dot\vee b) \dot\wedge (c \dot\vee d) = (a \vee b) \dot\wedge (c \vee d) = [a \vee b \vee (c' \wedge d')] \wedge (c \vee d).$$

Since bCc' and $c'Cd'$, Foulis-Holland Theorem II.3.10 yields $k = \{a \vee [(b \vee c') \wedge (b \vee d')]\} \wedge (c \vee d)$. Referring to Remark II.4.3, we see that the element $b \vee d'$ commutes with every element of the subalgebra $S = \langle a,b,c,d \rangle$ of $`L$ generated by the elements a, b,c,d. Therefore, by Theorem II.3.10, $k = (a \vee b \vee c') \wedge (a \vee b \vee d') \wedge (c \vee d)$. It is evident from Theorem III.1.1 (iv) that

$$r = [(a \dot\wedge d) \dot\vee (b \dot\wedge c)] \dot\vee [(a \dot\wedge c) \dot\vee (b \dot\wedge d)] =$$
$$= [(a \wedge d) \dot\vee (b \wedge c)] \dot\vee \{[(a \vee c') \wedge c] \dot\vee [(b \vee d') \wedge d]\}.$$

Furthermore, by Theorem II.4.2, we conclude that the elements $(a \vee c') \wedge c$, $(b \vee d') \wedge d$ commute. Hence, the third symbol $\dot\vee$ on the right-hand side can be replaced by $\vee$. This gives

$$r = (\{[a \wedge d \wedge (b' \vee c')] \vee (b \wedge c)\} \wedge [c' \vee (a' \wedge c)] \wedge [(b' \wedge d) \vee d']) \vee \{[(a \vee c') \wedge c] \vee [(b \vee d') \wedge d]\}.$$

From Remark II.4.3 we infer that $a \wedge d \wedge (b' \vee c') Cb \wedge c$, $b \wedge cCc'$, $c'Ca' \wedge c$ and $a' \wedge cCa \wedge d \wedge (b' \vee c')$. By Corollary 3.3,

$$r = (\{[a \wedge d \wedge (b' \vee c') \wedge c'] \vee (b \wedge c \wedge c') \vee [a \wedge d \wedge (b' \vee c') \wedge a' \wedge c] \vee [b \wedge c \wedge a' \wedge c]\} \wedge [(b' \wedge d) \vee d']) \vee \{..\} =$$
$$= ([(a' \wedge b \wedge c) \vee (a \wedge c' \wedge d)] \wedge [d' \vee (b' \wedge d)]) \vee \{...\}.$$

It follows from Remark II.4.3 that $a'\wedge b\wedge c \mathrm{C} a\wedge c'\wedge d$, $a\wedge c'\wedge d \mathrm{C} d'$, $d' \mathrm{C} b'\wedge d$ and $b'\wedge d \mathrm{C} a'\wedge b\wedge c$. Again using Corollary 3.3, we find

$r = (a\wedge b'\wedge c'\wedge d)\vee(a'\wedge b\wedge c\wedge d')\vee[(a\vee c')\wedge c]\vee[(b\vee d')\wedge d]$.

2. Suppose that $k = r$. Then

(1) $(b\vee d')\wedge d \leqslant a\vee b\vee c'$

and

(2) $(a\vee c')\wedge c \leqslant a\vee b\vee d'$.

Because of (1), $b\vee[(b\vee d')\wedge d] \leqslant a\vee b\vee c'$. Now, it is easily shown, using $b \mathrm{C} b\vee d'$, $b\vee d' \mathrm{C} d$, and Theorem II.3.10, that

(3) $(b\vee d')\wedge(b\vee d) \leqslant a\vee b\vee c'$.

By assumption and by Remark II.4.3, the element $a\vee c'$ commutes with every element of the subalgebra S. This together with (3) and Theorem II.3.10 implies that

$$(1') \quad \begin{aligned} a\vee b\vee c' &\geqslant [(b\vee d')\wedge(b\vee d)]\vee(a\vee c') = \\ &= (a\vee b\vee c'\vee d')\wedge(a\vee b\vee c'\vee d). \end{aligned}$$

This result yields $a\vee b\vee c' = (a\vee b\vee c'\vee d')\wedge(a\vee b\vee c'\vee d)$. By Theorem II.3.6, this is equivalent to

(1'') $a\vee b\vee c' \mathrm{C} d$.

In other words,

$$(1) \Rightarrow (1') \Leftrightarrow (1'').$$

Similarly,

$$(2) \Rightarrow (2') \Leftrightarrow (2'')$$

where

(2'') $a\vee b\vee d' \mathrm{C} c$.

3. We shall show that also (1) and (1'') are equivalent.

Indeed, if $(1'')$ is valid, then $(1')$ implies that $(b\vee d')\wedge d \leqslant \leqslant a\vee b\vee c'$. The same remark applies to the conditions $(2),(2'')$.

4. Finally, suppose the conditions $(1'')$ and $(2'')$ are valid. Then, by the part 3, the conditions (1) and (2) are true. Clearly, this means that

$$r = (a\wedge b'\wedge c'\wedge d)\vee(a'\wedge b\wedge c\wedge d')\vee[(a\vee c')\wedge c]\vee[(b\vee d')\wedge d] \leqslant$$
$$\leqslant (a\vee b\vee c')\wedge(a\vee b\vee d')\wedge(c\vee d) = k.$$

We shall now show that also the converse inequality holds. By Theorem II.4.4 it is obvious that the elements $a'\wedge b\wedge c\wedge d'$, $a\wedge b'\wedge \wedge c'\wedge d$ commute with every element of the subalgebra S. Moreover, we have $a\vee c'\mathrm{C}c$, $c\mathrm{C}d$, $d\mathrm{C}b\vee d'$ and $b\vee d'\mathrm{C}a\vee c'$. Hence by the dual of Corollary 3.3, $[(a\vee c')\wedge c]\vee[d\wedge(b\vee d')]$ is equal to

$$(a\vee c'\vee d)\wedge(c\vee d)\wedge(a\vee b\vee c'\vee d')\wedge(b\vee c\vee d').$$

It easily follows that

$$r = [a\vee c'\vee d\vee(a'\wedge b\wedge c\wedge d')]\wedge(c\vee d)\wedge(a\vee b\vee c'\vee d')\wedge$$
$$\wedge[b\vee c\vee d'\vee(a\wedge b'\wedge c'\wedge d)].$$

Since $a\vee c'$ commutes with every element of the subalgebra S, the first expression on the right side becomes

$$[\ldots] = d\vee(a\vee c')\vee[(a'\wedge c)\wedge(b\wedge d')] =$$
$$= d\vee a\vee c'\vee(b\wedge d') = d\vee[(a\vee b\vee c')\wedge(a\vee c'\vee d')].$$

By $(1'')$, the element $a\vee b\vee c'$ commutes with every element of the subalgebra S. Hence, by Theorem II.3.10, $[\ldots] = a\vee b\vee c'\vee \vee d$. Similarly, the last expression $[\ldots]$ is equal to $a\vee b\vee \vee c\vee d'$. This means that

$$r = (a\vee b\vee c'\vee d)\wedge(c\vee d)\wedge(a\vee b\vee c'\vee d')\wedge(a\vee b\vee c\vee d')\geqslant$$
$$\geqslant (a\vee b\vee c')\wedge(c\vee d)\wedge(a\vee b\vee d') = k.$$

Thus, $r = k$ and the theorem is proved. //

7. ANALOGUES OF FOULIS–HOLLAND THEOREM

In studying orthomodularity we have seen the central role played here by Foulis-Holland Theorem II.3.10. This suggests that there could be some interesting analogues in the skew operations $\dot\wedge, \dot\vee$.

In the sequel $L^\cdot$ denotes the Boolean skew lattice associated with an orthomodular lattice $`L$.

We begin with the following translation of such results for orthomodular lattices into the language of Boolean skew lattices.

Theorem 7.1. Let a,b,c be elements of a Boolean skew lattice $L^\cdot$. If aCb and aCc, then

(i) $aCb\dot\vee c$; (i′) $aCb\dot\wedge c$;

(ii) $a\dot\wedge(b\dot\vee c)=(a\dot\wedge b)\dot\vee(a\dot\wedge c)$; (ii′) $a\dot\vee(b\dot\wedge c)=(a\dot\vee b)\dot\wedge(a\dot\vee c)$;

(iii) $(b\dot\vee c)\dot\wedge a=(b\dot\wedge a)\dot\vee(c\dot\wedge a)$; (iii′) $(b\dot\wedge c)\dot\vee a=(b\dot\vee a)\dot\wedge(c\dot\vee a)$.

Proof. Ad (i). By Theorem II.4.2, $aCb\wedge c'$ and, again by II.4.2, a commutes with $(b\wedge c')\vee c = b\dot\vee c$.

Ad (ii). We have just seen that $aCb\dot\vee c$. Hence, by Theorem III.1.1 (iv),

$$v = a\dot\wedge(b\dot\vee c) = a\wedge(b\dot\vee c) = a\wedge[(b\wedge c')\vee c].$$

Now aCc and, according to Theorem II.4.2, $aCb\wedge c'$. Thus, by Theorem II.3.10, $v = (a\wedge b\wedge c')\vee(a\wedge c)$. From the same theorem, recalling that a commutes with b and c, we have $a\dot\wedge b = a\wedge b$, $a\dot\wedge c = a\wedge c$. Observe that $a\wedge(a'\vee c') = a\wedge c'$ by Theorem II.3.10. Therefore,

$$w = (a\dot\wedge b)\dot\vee(a\dot\wedge c) = (a\wedge b)\dot\vee(a\wedge c) = [(a\wedge b)\wedge(a\wedge c)']\vee(a\wedge c) =$$
$$= [(a\wedge b)\wedge(a'\vee c')]\vee(a\wedge c) = (a\wedge b\wedge c')\vee(a\wedge c) = v.$$

Ad (iii). One observes from (i) that $aCb\dot{\vee}c$. Hence, by Theorem III.1.1 (viii) and by what we have already proved,

$$z = (b\dot{\vee}c)\dot{\wedge}a = a\dot{\wedge}(b\dot{\vee}c) = (a\dot{\wedge}b)\dot{\vee}(a\dot{\wedge}c).$$

Using aCb, aCc and the same theorem, we conclude that $z = (b\dot{\wedge}a)\dot{\vee}(c\dot{\wedge}a)$.

Ad (i′). As a consequence of (i), we find that $a'Cb'\dot{\vee}c'$. From De Morgan laws applied to $L^{\bullet}$ and from Remark II.4.3 we infer that $aCb\dot{\wedge}c$.

The assertions (ii′) and (iii′) can be established by a similar argument as (i′). //

Corollary 7.2. An element a of a Boolean skew lattice $L^{\bullet}$ commutes with an element b if and only if $a = (a\dot{\wedge}b)\dot{\vee}(a\dot{\wedge}b')$.

Proof. First suppose that aCb. Then by Theorem II.2.3, we see that aCb'. We therefore have from Theorem 1 (ii) that $a = a\dot{\wedge}(b\dot{\vee}b') = (a\dot{\wedge}b)\dot{\vee}(a\dot{\wedge}b')$.

Conversely, let $a = (a\dot{\wedge}b)\dot{\vee}(a\dot{\wedge}b')$. Using Exercise III;3, we obtain $a\dot{\wedge}b' \leqslant a$. Hence, by Theorem III.1.1, aCb. //

Theorem 7.3. Let a,b,c be elements of a Boolean skew lattice $L^{\bullet}$ such that aCb and bCc. Then

(i) $(b\dot{\vee}c)\dot{\wedge}a = (b\dot{\wedge}a)\dot{\vee}(c\dot{\wedge}a)$;

(ii) $(b\dot{\wedge}c)\dot{\vee}a = (b\dot{\vee}a)\dot{\wedge}(c\dot{\vee}a)$.

Proof. It follows from bCc and Theorem III.1.1 (vii) that $b\dot{\vee}c = b\vee c$. Since bCa and $bCc\vee a'$, Theorem II.3.10 and Theorem III.1.1 yield

$$u = (b\dot{\vee}c)\dot{\wedge}a = (b\vee c)\dot{\wedge}a = (b\vee c\vee a')\wedge a = $$
$$= (b\wedge a)\vee[(c\vee a')\wedge a] = (b\dot{\wedge}a)\vee(c\dot{\wedge}a).$$

However, $bCc\dot{\wedge}a$ and $aCc\dot{\wedge}a$ by Remark II.4.3. Hence $b\dot{\wedge}aCc\dot{\wedge}a$. Thus, by Theorem III.1.1, $u = (b\dot{\wedge}a)\dot{\vee}(c\dot{\wedge}a)$.

The remaining assertion follows dually. //

Remark 7.4. Unlike the situations discussed above, an analogue of Theorems 1 and 3 does not hold for the expression $a \dot{\wedge} (b \dot{\vee} c)$ in the case where aCb and bCc, as the following example shows: Consider the lattice of Figure 45. Let $x = a'$, $y = c'$, $z = e'$. Then xCy and yCz. Therefore, $x \dot{\wedge} (y \dot{\vee} z) = x \dot{\wedge} (y \vee z) = x \dot{\wedge} 1 = x = a'$, while

$$(x \dot{\wedge} y) \dot{\vee} (x \dot{\wedge} z) = (x \wedge y) \dot{\vee} (x \dot{\wedge} z) = b \dot{\vee} (a' \dot{\wedge} e') =$$
$$= b \dot{\vee} [(a' \vee e) \wedge e'] = b \dot{\vee} e' = (o \wedge e) \vee e' = 0 \vee e' = e'.$$

Hence, in this orthomodular lattice, $x \dot{\wedge} (y \dot{\vee} z)$ and $(x \dot{\wedge} y) \dot{\vee} (x \dot{\wedge} z)$ are two distinct elements.

We shall now consider conditions under which one can assert that these two elements are equal. We shall investigate what can be salvaged from Theorems 1 and 3 under the following convention which we shall adopt for the remainder of this section:

(*) By a, b, c we denote elements of a Boolean skew lattice $L^{\bullet}$ such that aCb and bCc. We write

$$L = a \dot{\wedge} (b \dot{\vee} c), \quad R = (a \dot{\wedge} b) \dot{\vee} (a \dot{\wedge} c).$$

We now turn our attention to some theorems about the expressions L and R.

Theorem 7.5. Under the assumption (*),

(i) $L = (a \vee b') \wedge (a \vee c') \wedge (b \vee c)$;

(ii) $L = (a \wedge b) \vee \{[a \vee (b' \wedge c')] \wedge c\}$;

(iii) $L = (a \vee b') \wedge (a \vee c') \wedge (b \vee c) \wedge (a \vee c)$;

(iv) $R = (a \vee c') \wedge (b \vee c) \wedge [c \vee (a \wedge c')]$.

Proof. Ad (i). By Theorem III.1.1 (vii), $b \dot{\vee} c = b \vee c$. From aCb', $b'Cc'$ and Foulis-Holland Theorem II.3.10 it is seen that

$$L = a \dot{\wedge} (b \vee c) = [a \vee (b' \wedge c')] \wedge (b \vee c) = (a \vee b') \wedge (a \vee c') \wedge (b \vee c).$$

Ad (ii) and (iii). Since b commutes with a and $b' \wedge c'$, Theorem II.4.2 gives $bCa \vee (b' \wedge c')$. We know that bCc, aCb and $b' \wedge c'Cb$. Hence, by Theorem II.3.10,

$$L = [a \vee (b' \wedge c')] \wedge (b \vee c) = \{[a \vee (b' \wedge c')] \wedge b\} \vee$$
$$\vee \{[a \vee (b' \wedge c')] \wedge c\} = (a \wedge b) \vee \{[a \vee (b' \wedge c')] \wedge c\} \leqslant a \vee c.$$

Ad (iv). By Theorem III.1.1 (iv), $a \dot{\wedge} b = a \wedge b$. Consequently,

$$R = (a \wedge b) \dot{\vee} (a \dot{\wedge} c) = \{(a \wedge b) \wedge [(a' \wedge c) \vee c']\} \vee [(a \vee c') \wedge c].$$

From $a \wedge b \leqslant (a' \wedge c)' = a \vee c'$ it follows that $a \wedge b$ commutes with $a' \wedge c$. Moreover, $a' \wedge cCc'$. Therefore, by Foulis-Holland Theorem II.3.10, $R = (a \wedge b \wedge c') \vee [(a \vee c') \wedge c]$. One has also $a \vee c'Cc$, $a \vee c'Ca \wedge b \wedge c'$, cCb and $cCa \wedge c'$. Hence, again by Theorem II.3.10,

$$R = (a \vee c') \wedge [c \vee (b \wedge a \wedge c')] = (a \vee c') \wedge (c \vee b) \wedge [c \vee (a \wedge c')]. \ //$$

The theorem 6 below is basic for the equality $L = R$ (cf. [19]).

Theorem 7.6. Under the assumption (*), the following conditions are equivalent:

(1) $L \geqslant R$; (1′) $L \leqslant R$;

(2) $a \dot{\wedge} c \leqslant a \vee b'$; (2′) $a \wedge b \leqslant a \dot{\vee} c$;

(3) $L = R$;

(4) $cCa \wedge b$; (4′) $cCa \vee b'$

(5) $b \wedge a' \wedge (a \vee c) \wedge (a \vee c') = 0$.

Proof. The diagram of Figure 55 reveals the structure of our proof.

Ad (1) ⇒ (2). Since $c \wedge L \geqslant c \wedge R$, we have, by Theorem 5 (i) and (iv),

$$c \wedge (a \vee b') \wedge (a \vee c') \wedge (b \vee c) \geqslant c \wedge (a \vee c') \wedge (b \vee c) \wedge [c \vee (a \wedge c')].$$

Thus $a \vee b' \geqslant (a \vee c') \wedge c = a \dot{\wedge} c$.

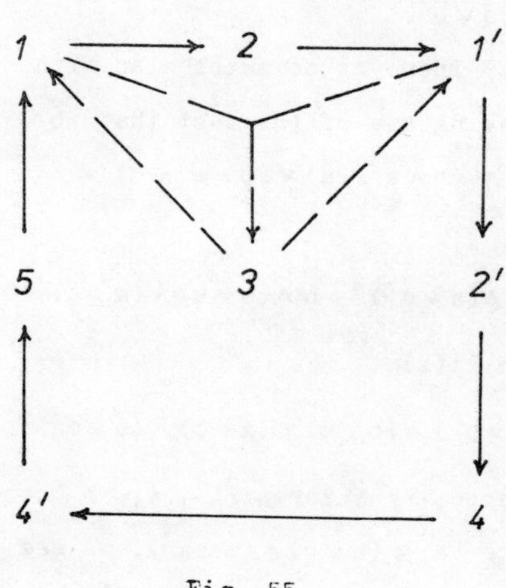

Fig. 55

Ad (2) $\Rightarrow$ (1′). If $a \dot{\wedge} c \leqslant a \vee b'$, then $a \dot{\wedge} c C a \vee b'$. Since aCb and cCb, Theorems II.4.2 and III.1.1 (iv) imply that $a \dot{\wedge} c$ commutes with b and with $(a \vee b') \wedge b = a \dot{\wedge} b = a \wedge b$. Therefore, by Theorem III.1.1 (iv) and (vii), $R = (a \wedge b) \vee (a \dot{\wedge} c) = (a \wedge b) \vee [(a \vee c') \wedge c]$. From $a \wedge b \leqslant a \leqslant a \vee c'$ we infer that $a \wedge b C a \vee c'$. Clearly, $a \vee c' C c$. Hence, by Foulis-Holland Theorem II.3.10, $R = (a \vee c') \wedge [(a \wedge b) \vee c]$. Using Theorems II.3.10 and 5 (iii), together with bCc and aCb, we get $R = (a \vee c') \wedge (a \vee c) \wedge (b \vee c) \geqslant L$.

Ad (1′) $\Rightarrow$ (2′). By Theorem 5 (i) and (iv), $a \wedge L \leqslant a \wedge R$ can be written in the form

$$a \wedge (a \vee b') \wedge (a \vee c') \wedge (b \vee c) \leqslant a \wedge (a \vee c') \wedge (b \vee c) \wedge [(a \wedge c') \vee c].$$

Thus $a \wedge b \leqslant a \wedge (b \vee c) \leqslant (a \wedge c') \vee c = a \dot{\vee} c$.

Ad (2′) $\Rightarrow$ (4). If $a \wedge b \leqslant (a \wedge c') \vee c$, then $a \wedge b = a \wedge b \wedge [(a \wedge c') \vee c]$. Evidently, $aCa \wedge c'$. Since bCa and bCc, we see, by Theorem II.4.2, that $bCa \wedge c'$. By the same theorem, $a \wedge bCa \wedge c'$. From $cCa \wedge c'$ and Theorem II.3.10, it follows that $a \wedge b = (a \wedge b \wedge c') \vee (a \wedge b \wedge c)$. This means that $a \wedge bCc$ and the assertion follows by Theorem II.3.4.

Ad (4) ⇒ (4′). We note that $cCa \wedge b$, cCb' and it follows as a consequence of Theorem II.4.2 that c commutes with the element $(a \wedge b) \vee b' = a \dot{\vee} b'$. Combining aCb' with Theorem III.1.1 (vii) we can derive easily $cCa \vee b'$.

Ad (4′) ⇒ (5). Let $cCa \vee b'$. Then c commutes also with the element $(a \vee b')' = a' \wedge b$. Making use of the fact that $b \wedge a'Ca$, we obtain $b \wedge a' \wedge (a \vee c) = (b \wedge a' \wedge a) \vee (b \wedge a' \wedge c) = b \wedge a' \wedge c$. Hence

$$b \wedge a' \wedge (a \vee c) \wedge (a \vee c') = b \wedge a' \wedge c \wedge (a \vee c') = b \wedge (a' \wedge c) \wedge (a' \wedge c)' = 0.$$

Ad (5) ⇒ (1). From Theorem 5 (iii),

$$L = [b \wedge a' \wedge (a \vee c) \wedge (a \vee c')] \vee [(a \vee b') \wedge (b \vee c) \wedge (a \vee c) \wedge (a \vee c')].$$

By assumption, bCa and bCc. Hence, by Theorem II.4.4, b commutes with $(a \vee c) \wedge (a \vee c')$. Since $a \leqslant (a \vee c) \wedge (a \vee c')$, we see that $a'C(a \vee c) \wedge (a \vee c')$. Because of Theorem II.4.2, the element $u = (a \vee c) \wedge (a \vee c')$ commutes with $b \wedge a'$ and so $uCa \vee b'$. Moreover, $c \leqslant a \vee c$, $c' \leqslant a \vee c'$ and this shows that uCc. Using Theorem II.4.2 and the fact that uCb, we get $uCb \vee c$. In summary, $uC(a \vee b') \wedge (b \vee c)$. This, together with $uCb \wedge a'$ and Theorem II.3.10 implies that $L = \{(b \wedge a') \vee [(a \vee b') \wedge (b \vee c)]\} \wedge (a \vee c) \wedge (a \vee c')$. Note that $b \wedge a' = (a \vee b')'$ and $b \wedge a' \leqslant b \vee c$. Consequently, by orthomodularity, we find $(b \wedge a') \vee [(a \vee b') \wedge (b \vee c)] = b \vee c$, i.e., $L = (b \vee c) \wedge (a \vee c) \wedge (a \vee c')$. Thus, by Theorem 5 (iv), $L \geqslant R$. //

<u>Corollary 7.7</u>. Under the assumption (*), the following conditions are equivalent:

(6) $a \wedge (b \vee c) \leqslant a \dot{\vee} c$;

(6′) $a \dot{\wedge} c \leqslant a \vee (b' \wedge c)$;

(6″) $L = R$.

Proof. The assertion (2′) follows trivially from (6). Conversely, suppose (2′) is valid. Then $a \wedge c \leqslant c \leqslant (a \wedge c') \vee c = a \dot\vee c$. This and $a \wedge b \leqslant a \dot\vee c$ imply that $(a \wedge b) \vee (a \wedge c) \leqslant a \dot\vee c$. We therefore have from Theorem II.3.10 that $a \wedge (b \vee c) \leqslant a \dot\vee c$.

In a similar manner we can see that (6′) is equivalent to the assertion (2). //

Let $M = c \dot\wedge (b \dot\vee a)$ and $P = (c \dot\wedge b) \dot\vee (c \dot\wedge a)$.

Corollary 7.8. Under the assumption (∗), $L = R$ if and only if one of the following conditions is true:

(7) $M \geqslant P$; (7′) $M \leqslant P$;

(8) $c \dot\wedge a \leqslant b' \vee c$; (8′) $b \wedge c \leqslant c \dot\vee a$;

(9) $M = P$;

(10) $aCb \wedge c$; (10′) $aCc \vee b'$;

(11) $b \wedge c' \wedge (a \vee c) \wedge (a' \vee c) = 0$.

Proof. By Theorem 4.1, $cCa \wedge b$ is equivalent to $aCb \wedge c$. In other words, the conditions (4) and (10) are equivalent and the corollary follows by Theorem 6. //

Remark 7.9. (A) One observes from Theorem II.3.5 and De Morgan laws that (2) is equivalent to $a' \wedge b \leqslant a' \dot\vee c'$. A similar argument shows that (4′) is equivalent to $cCa' \wedge b$.

(B) By duality we get from (4),(4′) and (10),(10′), respectively, the following statement: If a, b, c are elements of a Boolean skew lattice such that aCb and bCc, then $a \dot\vee (b \dot\wedge c) = (a \dot\vee b) \dot\wedge (a \dot\vee c)$ if and only if one of the following conditions holds:

(12) $cCa \vee b$; (12′) $cCa \wedge b'$;

(13) $aCb \vee c$; (13′) $aCb' \wedge c$.

Theorem 7.10. Under the assumption (*),

(i) $L = [a \wedge (b \vee c)] \vee [(a \dot{\wedge} c) \wedge b']$;

(ii) $R = (a \wedge b \wedge c') \vee (a \dot{\wedge} c)$.

Proof. We have aCb' and $b'Cc'$ which, owing to Theorem II.3.10 and Theorem 5 (ii), gives $L = (a \wedge b) \vee [(a \vee b') \wedge (a \vee c') \wedge c]$. Since aCb' and $b'Cc$, we find by using Theorem II.3.10 that

$$L = (a \wedge b) \vee \{[(a \wedge c) \vee (b' \wedge c)] \wedge (a \vee c')\}.$$

Note that $a \wedge c C a \vee c'$. Moreover, by Theorem II.4.2, $b' C a \vee c'$. In addition, $c C a \vee c'$. Hence, again by Theorem II.4.2, $b' \wedge c C a \vee c'$. Consequently, Foulis-Holland Theorem II.3.10 shows that

$$L = (a \wedge b) \vee (a \wedge c) \vee [b' \wedge c \wedge (a \vee c')] = [a \wedge (b \vee c)] \vee [(a \dot{\wedge} c) \wedge b'].$$

Ad (ii). From $a \vee c' C c$, $a \vee c' C a \wedge c'$ and Theorem II.3.10 we find that $(a \vee c') \wedge [c \vee (a \wedge c')] = [c \wedge (a \vee c')] \vee (a \wedge c')$. Now, by Theorem 5 (iv), we get $R = (b \vee c) \wedge \{[c \wedge (a \vee c')] \vee (a \wedge c')\}$. The element $a \wedge c'$ commutes with a, b and c. By Theorem II.4.4, $a \wedge c'$ commutes with $c \wedge (a \vee c')$ and $b \vee c$. Therefore, by Theorem II.3.10, $R = [c \wedge (a \vee c')] \vee [a \wedge c' \wedge (b \vee c)]$. Since $c'Cc$ and $c'Cb$, we obtain $a \wedge c' \wedge (b \vee c) = a \wedge b \wedge c'$. In summary, $R = (a \dot{\wedge} c) \vee (a \wedge b \wedge c')$. //

Corollary 7.11. Let $S = (a \wedge b) \vee (a \dot{\wedge} c)$. Under the assumption (*), $L = R$ if and only if either $L \geqslant S$ or $R \geqslant S$.

Proof. From aCb, bCc, Theorems 10 and II.3.10 we conclude that $L \vee R = [a \wedge (b \vee c)] \vee (a \wedge b \wedge c') \vee (a \dot{\wedge} c) = (a \wedge b) \vee (a \wedge c) \vee (a \dot{\wedge} c)$. According to $a \wedge c \leqslant a \dot{\wedge} c = (a \vee c') \wedge c$, we have $L \vee R = (a \wedge b) \vee (a \dot{\wedge} c) = S$.

Thus $L = R$ implies that $S = L = R$.

If $L \geqslant S = L \vee R \geqslant R$, then $L = R$ by Theorem 6. An entirely analogous argument leads to the conclusion in the case $R \geqslant S$. //

Theorem 7.12. Under the assumption (ж), the following equivalences hold:

(14) $a \dot{\wedge} (b \dot{\vee} c) = a \wedge (b \vee c) \Leftrightarrow aCb \vee c$;

(15) $a \dot{\vee} (b \dot{\wedge} c) = a \vee (b \wedge c) \Leftrightarrow aCb \wedge c$.

Proof. Let $aCb \vee c$. Then, by Theorem III.1.1 (iv) and (vii), $a \dot{\wedge} (b \dot{\vee} c) = a \dot{\wedge} (b \vee c) = a \wedge (b \vee c)$.

Conversely, suppose $a \dot{\wedge} (b \dot{\vee} c) = a \wedge (b \vee c)$. Then $a \dot{\wedge} (b \vee c) \leqslant a$ so that $aCb \vee c$ by Theorem III.1.1 (ii).

The statement (15) follows by duality. //

Theorem 7.13. Let $f = a \dot{\vee} (b \dot{\wedge} c)$, $h = (a \dot{\vee} b) \dot{\wedge} (a \dot{\vee} c)$. Under the assumption (ж), the following conditions are equivalent:

(i) aCc; (ii) $L = R$ & $f = h$;

(iii) $L = a \wedge (b \vee c)$ & $f = a \vee (b \wedge c)$;

(iv) $R = (a \wedge b) \vee (a \wedge c)$; (v) $h = (a \vee b) \wedge (a \vee c)$.

Proof. Indeed, making use of Theorem II.4.5 and of Theorem III.1.1, it is clear that (i) implies the assertions (ii)-(v).

Ad (ii) ⇒ (i). Using (4) of Theorem 6, we see that $cCa \wedge b$. From (12´) we get $cCa \wedge b'$. Hence, by Theorem II.4.2, c commutes with $(a \wedge b) \vee (a \wedge b')$. However, aCb and so $(a \wedge b) \vee (a \wedge b') = a$. Thus aCc.

Ad (iii) ⇒ (i). Here we can use (14) and (15). Then $aCb \vee c$ and $aCb \wedge c$. Since (13) and (13´) are two equivalent conditions, we have $aCb' \wedge c$. Consequently, a commutes with $(b \wedge c) \vee (b' \wedge c) = c$.

Ad (iv) ⇒ (i). If $R = (a \dot{\wedge} b) \dot{\vee} (a \dot{\wedge} c) = (a \wedge b) \vee (a \wedge c)$, then $(a \dot{\wedge} b) \dot{\vee} (a \dot{\wedge} c) \leqslant a$. Therefore, by Exercise III;3 (i),

$a \dot{\wedge} c \leqslant a$. Now, referring to Theorem III.1.1 (ii), we conclude that aCc.

Ad (v) $\Rightarrow$ (i). The proof is analogous to that of the implication (iv) $\Rightarrow$ (i). //

EXERCISES

VII;1. By means of Theorem 5.7 find the center $C(`L)$ of the lattice $`L$ shown in Figure 45.

VII;2. Suppose a_{n+1} commutes with the elements $a_1, a_2, \ldots, a_n$ of an orthomodular lattice. Prove that

$$\overline{\text{com}}\ (a_1, a_2, \ldots, a_n, a_{n+1}) = \overline{\text{com}}\ (a_1, a_2, \ldots, a_n).$$

VII;3. Prove that two elements x, y of an orthomodular lattice commute if and only if $((x \wedge y) \vee (x \wedge y'), x)$ belongs to the corresponding reflective congruence $\hat{B}$.

VII;4. Decide, whether the following is true: Two elements $x \neq y$ of an orthomodular lattice do not commute if and only if they are inner elements of a sublattice isomorphic to $`M_{02}$ (see Figure 8) and x is not the relative orthocomplement of y in the interval $[x \wedge y, x \vee y]$.

VII;5. Prove that two elements of a relatively complemented lattice with 0 and 1 are perspective if and only if they have a common complement.

VII;6. Show that in any modular lattice $a \underset{sp}{\sim} b$ if and only if $a \underset{p}{\sim} b$.

VII;7. Prove that an orthomodular lattice $`L$ is distributive if and only if $\overline{\text{com}}\ (a_1, a_2, \ldots, a_n) = 0$ for every $a_1, a_2, \ldots, a_n \in L$ and every $n \in \underline{N}$.

VII;8. Let $`F = \langle x_1, x_2, \ldots, x_n \rangle$ be an orthomodular lattice. Decide, whether an analogue of Corollary 2.13 holds also for the reflective congruence $\hat{B}$, i.e., verify whether the following two conditions are equivalent:

(i) $(a,b) \in \hat{B}$;

(ii) $(a \wedge b') \vee (a' \wedge b) \leqslant \underline{\mathrm{com}}\, (x_1, x_2, \ldots, x_n)$.

VII;9. Decide whether the conditions (i) and (ii) on two elements a,b of an orthomodular lattice are equivalent:

(i) $(a,b) \in \hat{B}$;

(ii) $\overline{\mathrm{com}}\, (a,b) = 0$.

VII;10. Prove Theorem 4.2 directly, without assuming Theorem 4.1.

VII;11. Let c be an element of an orthomodular lattice. Suppose that c commutes with every a_i, $i \in I$. Prove: If there exists the supremum $\bigvee(a_i;\ i \in I)$, then

(a) there exists also $\bigvee(a_i \wedge c;\ i \in I)$ and this element is equal to $[\bigvee(a_i;\ i \in I)] \wedge c$;

(b) the element c commutes with $\bigvee(a_i;\ i \in I)$.

State and prove similar assertions for the infimum $\bigwedge(a_i;\ i \in I)$.

VII;12. Let a,b,c be elements of an orthomodular lattice such that aCc and bCc. Show that

(i) $a \dot{\wedge} (b \dot{\vee} c) = (a \dot{\wedge} b) \dot{\vee} (a \dot{\wedge} c)$;

(ii) $a \dot{\vee} (b \dot{\wedge} c) = (a \dot{\vee} b) \dot{\wedge} (a \dot{\vee} c)$;

(iii) $(b \dot{\vee} c) \dot{\wedge} a = (b \dot{\wedge} a) \dot{\vee} (c \dot{\wedge} a)$;

(iv) $(b \dot{\wedge} c) \dot{\vee} a = (b \dot{\vee} a) \dot{\wedge} (c \dot{\vee} a)$.

VII;13. Let a,b,c be elements of an orthomodular lattice such that one of them commutes with the other two. Prove that

(i) $a \dot{\wedge} (b \dot{\vee} c) = (a \dot{\wedge} b) \dot{\vee} (a \dot{\wedge} c)$ if and only if $a \wedge b\, C\, b \wedge c$;

(ii) $(b \dot{\vee} c) \dot{\wedge} a = (b \dot{\wedge} a) \dot{\vee} (c \dot{\wedge} a)$ in every case.

VII;14. Let x,y be elements of an orthomodular lattice which commute. Prove that the intervals

$$I_1 = [(x \wedge y) \vee (x \wedge y'),(x \vee y) \wedge (x \vee y')];$$
$$I_2 = [(x \wedge y) \vee (x' \wedge y),(x \vee y) \wedge (x' \vee y)];$$
$$I_3 = [(x' \wedge y') \vee (x \wedge y'),(x' \vee y') \wedge (x \vee y')];$$
$$I_4 = [(x' \wedge y') \vee (x' \wedge y),(x' \vee y') \wedge (x' \vee y)]$$

determine distributive lattices.

Suppose I_j, $1 \leqslant j \leqslant 4$, determines a distributive lattice. Show that x,y commute.

VII;15. An ortholattice is orthomodular if and only if it satisfies the implication

$$(aCb \wedge c \ \& \ bCc) \Rightarrow bCa \wedge c$$

for every a,b,c.

VII;16. Let x,y be elements of an orthomodular lattice. Prove that the following conditions are equivalent:

(i) $(x' \vee y') \wedge [x \vee (x' \wedge y)]C(x' \vee y') \wedge [y \vee (x \wedge y')]$;

(ii) $(x' \vee y') \wedge [x \vee (x' \wedge y)]Cy \vee (x \wedge y')$;

(iii) $(x' \vee y') \wedge [x \vee (x' \wedge y)]Cy$;

(iv) $x \vee (x' \wedge y)Cy'$;

(v) xCy.

VII;17. An element $s \in L$ is said to be <u>standard element</u> (cf. [75]) of a lattice `L if and only if $a \wedge (s \vee b) = (a \wedge s) \vee (a \wedge b)$ for any $a,b \in L$. Prove that an element $s \in L$ of an orthomodular lattice `L is standard if and only if $s \in C(\text{`}L)$.

VII;18. Let s,t be elements of an orthomodular lattice `L. Prove that $(s \vee t) \wedge (s \vee t')/s$ is an allele of `L.

Chapter

APPLICATION VIII

1. ORTHOMODULARITY AND EXPERIMENTAL PROPOSITIONS

The present section consists of two parts, the first of which contains an explanation of the motivation and phenomenological background. From the mathematical point of view our following remark can be regarded as a practical, but nonrigorous contribution to the formulation of the picture which clarifies the relationship of the theory of orthomodularity to physical application.

Let us consider a physical system at some instant.

It can be a mechanical system S obeying classical mechanics, e.g. point particles. To describe the temporal evolution of S, a set of real-valued functions characterizing its positions and momenta is considered. One can suppose that these functions form a mapping in a "phase space" with a selected canonical coordinate system. If all these functions are known at a fixed point of time, then the state of the system at some later time (its trajectory in the phase space) is fully characterized by fundamental laws of classical mechanics.

For a single point particle we usually consider six functions $q_i(t), p_i(t)$ $(i = 1,2,3)$. At time t we then have a point $(q_1(t),$

$q_2(t), q_3(t), p_1(t), p_2(t), p_3(t))$ of the corresponding phase space. As customary, such a point is called a "state" of the mechanical system S.

However, for a system T obeying quantum mechanics, e.g. a moving electron, we have another situation. From Heisenberg uncertainty principle it follows that the concept of a point in a phase space loses here its meaning.

Basic physical magnitudes (e.g. energies, momenta, etc.) are called "observables" of the considered system T. In quantum mechanics not all the observables of T can be measured simultaneously. This is the case for the position and the momentum of a quantum particle. On the other hand, there are observables such as the energy and the total angular momentum of a particle in a central field which can be measured simultaneously. If we want to obtain a description of a quantum mechanical object at a single instant, then we can formulate physical statements about it such as "The quantum particle has its x coordinate $q_1(t)$ in the interval $[a,b]$ and its x momentum coordinate $p_1(t)$ in $[c,d]$". Statements of this type are usually regarded as "propositions" ("events", "questions"). They are characterized by the fact that there is an experiment which allows to decide whether the statement is true or not.

It has been noted above that the classical concept of a state as a point in a phase space is incompatible with fundamental ideas of quantum mechanics. A natural initial reaction to this situation is to form a suitable generalization of the classical concept, e.g. to replace points of the phase space by appropriate sets or functions.

There are various ways of defining correctly the notion of

a state. In some approaches the events and the states are treated as primitive entities. However, some authors regard the events as primitive entities and the states are derived from them. Of these, the approach due to Jauch and Piron [104] is closest both to classical mechanics and to quantum mechanics.

Properties of a physical system T under investigation can be described by being correlated with measurements. For example, if we know that a quantum particle is in a certain subset of a phase space at time t_1, then we can make some statistical predictions about the subset where it will be at time $t_1 + \Delta t$ ($\Delta t > 0$). So we are intuitively led to a hypothetical statistical theory where states are representable as probability functions called "probability gages" or "probability measures".

Let us consider a quantum-mechanical system T at a fixed point of time. Let $L = \{p,q,r,...\}$ be the set of propositions concerning T. We shall suppose that these propositions describe only those properties of T which can be completely characterized by a "yes-no" answer. The set L is called the "logic" of T. By a "state" ("probability gage") of T is then meant any mapping f of L into the unit interval $[0,1] \subset \underline{R}$ which satisfies some postulated properties (cf. the three conditions specified in the sequel for a state on `L). If f is a state and p an event, then f(p) expresses the probability of occurence of the event p in the state f. In other words, given a probability gage f and a question p, the number f(p) represents the probability that we obtain a "yes" answer observing experimentally the question p, provided the state of the system is described by f.

The starting point for the next step is a typical mathematical abstraction which provides a link between the theory of orthomodularity and physical experiments. Our following exposition is essentially based on the elegant treatment employed by Pool ([160]).

We shall say that a pair (L,F) is a logical structure if L is a nonvoid set and F is a set of functions from L into $[0,1] \subset \underline{R}$, and if the axioms (I)-(IV) below are satisfied:

(I) If $p,q \in L$ and $f(p) = f(q)$ for every $f \in F$, then $p = q$.

(II) There exists an element $u \in L$ such that $f(u) = 1$ for every $f \in F$.

(III) For each $p \in L$, there exists an element $p' \in L$ such that $f(p) + f(p') = 1$ for every $f \in F$.

Before formulating the fourth axiom we shall introduce some notions which we shall need later.

In the considered definition, the set L is said to be a logic and its elements are called questions (propositions, events). The set F is called a set of states (gage set) and each element of F is said to be a state (probability measure, probability gage) on L.

The axiom (I) identifies two experimental propositions which have the same probability for a "yes" answer in every state of the physical system.

The axiom (II) guarantees the existence of an experimental proposition which is valid in every state of the system. In the following we shall write 1 instead of u and 0 instead of u'.

Clearly, by (I), the element p' of the axiom (III) is uniquely determined for any question p. It represents the logical negation of the experimental proposition p. The question p' is said to be the negation of the question p.

Let $\leq$ be the relation defined on L in such a way that

$p \leq q$ if and only if $f(p) \leq f(q)$ for every $f \in F$.

Theorem 1.1. The system $(L, \leq, ', 0, 1)$ is a poset satisfying the following conditions for every $p, q \in L$:

(i) $p = p''$; (ii) $p \leq q \Rightarrow p' \geq q'$; (iii) $0 = 1'$.

Proof. Evidently, $(L, \leq)$ is a poset. By (III), $f(p) + f(p') = 1$ and $f(p') + f(p'') = 1$. Hence $f(p) = f(p'')$ for every $f \in F$ and, therefore, $p = p''$. If $p \leq q$, then $f(p) \leq f(q)$ for every $f \in F$. Consequently, $f(p') = 1 - f(p) \geq 1 - f(q) = f(q')$ and this yields $p' \geq q'$. By definition, $0=1'$. //

As is customary, we let $p \perp q$ denote the fact that $p \leq q'$. In this case we say that the propositions p, q are disjoint.

If $f(p) = 1$ for a state f, then p is said to be true in the state f. If $f(q) = 0$, then q is called false in the state f.

We are now able to give a concise formulation of the fourth axiom:

(IV) For any $p, q \in L$ with $p \perp q$ there exists $s = \sup_{(L, \leq)}\{p, q\}$ and $f(s) = f(p) + f(q)$ for every $f \in F$.

Clearly, if p and q are two disjoint propositions and p is true in a state f, then q is false in f. Hence it is physically permissible to postulate that the supremum s exists. In other words, the axiom (IV) has been stated in a form appropriate to the description of the properties that one usually postulates in the mentioned physical model and it is crucial for the probabilistic interpretation. The element s is uniquely determined by p and q and it is denoted by $p \vee q$ or $p + q$. In this case we can write $f(p + q) = f(p) + f(q)$.

Theorem 1.2. The system $(L, \leq, ', 0, 1)$ is an orthomodular poset.

Proof. Since $p \perp p'$, the supremum $s = p \vee p'$ exists and $f(s) = f(p) + f(p') = 1$. Hence $p \vee p' = 1$ and it follows that the poset is orthocomplemented.

Let $p \leqslant q$. Then $p \vee q'$ exists, $p \perp (p \vee q')'$ and we have

$$f(p \vee (p \vee q')') = f(p + (p + q')') = f(p) + f((p + q')') =$$
$$= f(p) + 1 - f(p + q').$$

Now, $f(p + q') = f(p) + f(q') = f(p) + 1 - f(q)$. In summary,

$$f(p \vee (p \vee q')') = f(p) + 1 - [f(p) + 1 - f(q)] = f(q).$$

Thus by (I), $p \vee (p \vee q')' = q$. //

The orthomodular poset of Theorem 2 is said to be associated with the logical structure (L,F).

Example 1.3. Let $A = \{a_1, a_2, \ldots, a_n\}$ be a set of possible outcomes $a_1, a_2, \ldots, a_n$ of an experiment. Assume that exactly one outcome occurs as a result of the experiment. Let $Q_1 = \emptyset, Q_2, \ldots$ $\ldots, Q_m = A$ $(m = 2^n)$ be all the subsets of the set A and let $L_1 = \{Q_1, Q_2, \ldots, Q_m\}$. Define $f_k : L_1 \to [0,1]$ $(k = 1,2,..,n)$ by

$$f_k(Q) = 1 \quad \text{if} \quad a_k \in Q;$$
$$f_k(Q) = 0 \quad \text{otherwise.}$$

Put $F_0 = \{f_1, f_2, \ldots, f_n\}$.

Then $f_k(Y) \leqslant f_k(Z)$ for all $k = 1,2,\ldots,n$ if and only if $Y \subset Z$.

One easily shows that (L_1, F_0) is a logical structure and that the associated orthomodular poset $(L_1, \leqslant, ', \emptyset, A)$ is a Boolean algebra having 2^n elements.

Notice that it is not hard to enlarge the set F_0. Indeed, if $0 \leqslant r_i$ $(i = 1,2,\ldots,n)$ are real numbers such that $r_1 + r_2 + \ldots + r_n = 1$ and $f_1, f_2, \ldots, f_n$ are the states defined above, then $f = r_1 f_1 + r_2 f_2 + \ldots + r_n f_n$ satisfies also (I)-(IV).

For example, if $a_1, a_2, \ldots, a_6$ are the possible outcomes in the experiment of tossing a die, then $f = \frac{1}{6}(f_1 + f_2 + \ldots + f_6)$ has the properties (I)-(IV) of a state. Here $f(Q)$ expresses the classical probability of an event $Q \subset \{a_1, a_2, \ldots, a_6\}$.

Example 1.4. Let H be a Hilbert space. Let $L(H)$ denote the set formed by all the closed subspaces of H (cf. Sections III.3 and III.4). If K is a closed subspace of H, then, by Theorem III.3.7 and Remark III.3.12, $H = K + K^{\perp}$. Therefore, any vector $v \in H$ can be uniquely written in the form $v = u + w$ where $u \in K$ and $w \in K^{\perp}$. Let $e_K : H \to K$ be the projection mapping defined by $e_K(v) = u$. Take a unit vector $v \in H$ (i.e., $\|v\| = 1$) and define $f_v : L(H) \to [0,1] \subset \underline{R}$ by $f_v(K) = \langle v, e_K(v) \rangle = \langle v, u \rangle$. Let F_1 be the set consisting of all the mappings f_v where v runs over all the unit vectors of H. Then $(L(H), F_1)$ is a logical structure.

Indeed, let $E \neq G$ be two closed subspaces of H and let v be a unit vector of E which does not belong to G. Clearly, $f_v(E) = \langle v, e_E(v) \rangle = \langle v, v \rangle = 1$. Since $H = G + G^{\perp}$, $v = u_1 + w_1$ where $u_1 \in G$ and $o \neq w_1 \in G^{\perp}$. Now $f_v(G) = \langle v, e_G(v) \rangle = \langle u_1 + w_1, u_1 \rangle = \langle u_1, u_1 \rangle + \langle w_1, u_1 \rangle = \|u_1\|^2$. However,

$$1 = \langle v, v \rangle = \langle u_1 + w_1, u_1 + w_1 \rangle = \|u_1\|^2 + \|w_1\|^2 > \|u_1\|^2 = f_v(G).$$

Therefore, the axiom (I) holds.

As a consequence, we have the following statement:

If E, G are two different closed subspaces of H such that the difference $E \setminus G$ is a nonvoid set, then there is a vector $v \in E \setminus G$ with $f_v(G) < f_v(E)$.

Consequently, if $f_v(E) \leq f_v(G)$ holds for every unit vector v of H, then $E \setminus G = \emptyset$, i.e., $E \subset G$.

Conversely, suppose that $E \subset G$ are two closed subspaces of H.

Let v be a unit vector of H and let $v = u + w = u_1 + w_1$ where $u \in E$, $u_1 \in G$, $w \in E^\perp$ and $w_1 \in G^\perp \subset E^\perp$. Then

$$f_v(E) = \langle v, e_E(v) \rangle = \langle v, u \rangle = \|u\|^2, \quad f_v(G) = \|u_1\|^2.$$

But $u_1 = u_2 + w_2$ where $u_2 \in E$ and $w_2 \in E^\perp$. We have $u + w = u_2 + (w_1 + w_2)$ with $u, u_2 \in E$ and $w, w_1 + w_2 \in E^\perp$. Hence $u = u_2$ and so, by Remark III.3.4.C,

$$f_v(G) = \|u_1\|^2 = \|u + w_2\|^2 = \|u\|^2 + \|w_2\|^2 \geq \|u\|^2 = f_v(E).$$

In summary, $E \leq G$ in the ordering induced by $(L(H), F_1)$ if and only if $E \subset G$.

Evidently, for any unit vector v of H,

$$f_v(H) = \langle v, e_H(v) \rangle = \langle v, v \rangle = \|v\|^2 = 1.$$

Therefore, the axiom (II) is verified.

If K is a closed subspace of H and v is a unit vector of H, then

$$f_v(K) + f_v(K^\perp) = \langle v, e_K(v) \rangle + \langle v, e_{K^\perp}(v) \rangle =$$
$$= \langle v, e_K(v) + e_{K^\perp}(v) \rangle = \langle v, v \rangle = 1.$$

Thus (III) is satisfied.

Let K, G be two closed subspaces of H such that $K \subset G^\perp$. Then $S = K + G$ is a closed subspace of H.

Indeed, let $\{z_n\}$ be a convergent sequence of vectors in $K + G$. Write $z_n = x_n + y_n$ where $x_n \in K$ and $y_n \in G \subset K^\perp$. Suppose that $z_n \to a = b + c$ where $b \in K$ and $c \in K^\perp$. By Remark III.3.4.C,

$$\|x_n + y_n - (b + c)\|^2 = \|x_n - b + y_n - c\|^2 = \|x_n - b\|^2 + \|y_n - c\|^2.$$

Hence $x_n \to b \in K$. Now let $a = d + e$ where $d \in G$ and $e \in G^\perp$. Similarly,

$$\|x_n + y_n - (d + e)\|^2 = \|x_n - e\|^2 + \|y_n - d\|^2.$$

Therefore, $y_n \to d \in G$ and, consequently, $x_n + y_n \to b + d \in K + G$.

By the remark above, S is the supremum of $\{K,G\}$ in $(L(H), \leq)$. At the same time, $f_v(S) = \langle v, e_S(v)\rangle = \langle v, e_{K+G}(v)\rangle$. Putting $v = u + w$ where $u \in K + G$ and $w \in (K + G)^{\perp}$, we obtain $v = u_1 + v_1 + w$ with $u_1 \in K$ and $v_1 \in G$. Therefore, $f_v(S) = $ $= \langle v,u\rangle = \langle v, u_1 + v_1\rangle = \langle v,u_1\rangle + \langle v,v_1\rangle$. Since $u_1 \in K$ and $v_1 +$ $+ w \in K^{\perp}$, $e_K(v) = u_1$ and, similarly, $e_G(v) = v_1$. Thus

$$f_v(S) = \langle v, e_K(v)\rangle + \langle v, e_G(v)\rangle = f_v(K) + f_v(G).$$

It follows that the axiom (IV) is also satisfied.

Notice that this example is closely related to the von Neumann's formulation of quantum mechanics constructed over a complex Hilbert space. According to this formulation (cf. [185]), the "states" of a physical system are described by nonzero vectors v of H, taken to within a complex factor, the "observables" are self-adjoint operators P (i.e., $P = P^{*}$) on H and "propositions" are represented by closed subspaces of H or, equivalently, by projection operators.

Let $`L = (L, \leq, ', 0, 1)$ be an orthomodular poset. A mapping $f: L \to [0,1] \subset \underline{R}$ is said to be a <u>state</u> on $`L$ if and only if it satisfies the conditions (1),(2) and (3):

(1) $f(0) = 0$;

(2) $f(1) = 1$;

(3) $f(a \vee b) = f(a) + f(b)$ whenever $a \leq b'$.

A set S of states on $`L$ is said to be <u>full</u> if and only if it has the following property: If a,b are elements of L such that $f(a) \leq f(b)$ for every $f \in S$, then $a \leq b$.

<u>Lemma 1.5</u>. If m is a state on $`L$ and $a \leq b$, then $m(a) \leq m(b)$.

Proof. By assumption, $a \leqslant (b')'$ and so $m(a \vee b') = m(a) + m(b')$. However, $1 = m(1) = m(b \vee b') = m(b) + m(b')$ and, therefore,

$$0 \geqslant m(a \vee b') - 1 = m(a) + m(b') - [m(b) + m(b')] = m(a) - m(b).$$

This yields $m(a) \leqslant m(b)$. //

Proposition 1.6. An orthomodular poset $`L = (L, \leqslant, ', 0, 1)$ is associated with a logical structure $(L,F)$ if and only if the set $F$ is a full set of states on $`L$.

Proof. 1. If $`L$ is associated with $(L,F)$, then the axioms (I)-(IV) show that $F$ is a full set of states on $`L$.

2. Conversely, let us suppose that F is a full set of states on $`L$. We have to prove that $(L,F)$ is a logical structure and that $`L$ is associated with (L,F).

If p,q of L are such that $f(p) = f(q)$ for every $f \in F$, then $f(p) \leqslant f(q)$ and $f(q) \leqslant f(p)$ in $\underline{R}$. Since F is full, we have $p \leqslant q$ and $q \leqslant p$ in $`L$ and, consequently, $p = q$.

From (2) and (3) we conclude that the axioms (II) and (III) are satisfied.

Using Lemma 5 and the fact that F is full we see that $a \leqslant b$ in $`L$ if and only if $f(a) \leqslant f(b)$ for any $f \in F$. Hence the axiom (IV) holds and $`L$ is associated with (L,F). //

Now we can formulate the following question: Does a full set of states exist for any orthomodular poset?

Unfortunately, the answer is no. M. K. Bennett [8] has shown that the lattice G_{32} (cf. IV.4) represents the corresponding counter-example. Below we shall offer a proof of her result.

Some preliminary results given in the following remark will be useful. We use m to denote a state on an orthomodular lattice $`L$.

Remark 1.7. (A) If the elements $a_1, a_2, \dots, a_n$ of `L are mutually orthogonal, that is, $a_i \leqslant a_j'$ whenever $1 \leqslant i \neq j \leqslant n$, then

$$m(a_1 \vee a_2 \vee \dots \vee a_n) = m(a_1) + m(a_2) + \dots + m(a_n).$$

Indeed, if $a_1 \leqslant a_i'$ for $i = 2,3,\dots,n$, then, a fortiori, $a_1 \leqslant \leqslant a_2' \wedge a_3' \wedge \dots \wedge a_n'$ and so $a_1 \perp a_2 \vee a_3 \vee \dots \vee a_n$. Hence $m(a_1 \vee \vee a_2 \vee \dots \vee a_n) = m(a_1) + m(a_2 \vee \dots \vee a_n)$. The remainder follows by induction.

(B) Let $a_1, a_2, \dots, a_n$ be all the distinct atoms of a finite Boolean algebra. Then

(i) $a_1 \vee a_2 \vee \dots \vee a_n = 1$;

(ii) for any $1 \leqslant i \neq j \leqslant n$, $a_i \perp a_j$;

(iii) given any state m on this algebra, $m(a_1) + m(a_2) + \dots \dots + m(a_n) = 1$.

For the sake of completeness, we shall prove these statements. Let $h = a_1 \vee a_2 \vee \dots \vee a_n$. Suppose that $h \neq 1$. Then $h' \neq \neq 0$ and there exists an atom b with $0 \prec b \leqslant h'$. Since b is one of the atoms a_i, $b \leqslant a_1 \vee a_2 \vee \dots \vee a_n = h$. Hence, $b \leqslant h' \wedge \wedge h = 0$, a contradiction.

To prove (ii), observe that

$$a_1 \wedge (a_2 \vee \dots \vee a_n) = (a_1 \wedge a_2) \vee \dots \vee (a_1 \wedge a_n) = 0,$$
$$a_1 \vee (a_2 \vee \dots \vee a_n) = 1,$$

i.e., $a_1' = a_2 \vee \dots \vee a_n \geqslant a_i$ for $i = 2,\dots,n$. In other words, $a_1 \perp a_i$ for every $i = 2,\dots,n$. Thus any two distinct atoms of our Boolean algebra are orthogonal.

The assertion (iii) follows from (ii), the part A and (2).

Theorem 1.8 (cf. M. K. Bennett, [8]). No set of states on G_{32} is full.

Proof. By Remark 7.B (iii) and Corollary IV.4.8, we have $m(a) + m(b) + m(c) = 1$ for the atoms $a,b,c \in G_{32}$ and for any state m on G_{32}. To simplify the notation we make the following convention: In such cases we shall simply write $a + b + c = 1$. Hence, we find

$$a + b + c = 1, \qquad g + f + e = 1, \qquad r + j + d = 1,$$
$$r + s + t = 1, \qquad a + l + k = 1, \qquad g + h + i = 1.$$

Adding and rearranging, we get

$$2(a + r + g) + (b + s + h) + (c + d + e) + (k + j + i) + \\ + (l + t + f) = 6.$$

Therefore, $2(a + r + g) + 4 = 6$ and this yields $a + r + g = 1$. Using the standard notation, we obtain $m(a) + m(r) + m(g) = 1$. Consequently, $m(a) = 1 - m(r) - m(g) \leq 1 - m(g)$. Since $g \perp g'$, $1 = m(1) = m(g \vee g') = m(g) + m(g')$. Thus $m(a) \leq m(g')$ for any state m of the considered set S of states. If S were full, we would have $a \leq g'$. Taking into account the construction of G_{32} we see that this is in G_{32} impossible. //

Theorem 8 gives us information on some sets of states in the Greechie's lattice G_{32}. The following remarkable example due to Greechie shows that even the existence of a state on an orthomodular lattice cannot be guaranteed.

Example 1.9 (cf. [82]). Let $`B_{ij} \cong \underline{2}^3$ be a Boolean algebra for any $1 \leq i \leq 3$ and $1 \leq j \leq 4$. Suppose that the algebra $`B_{ij}$ is generated by the atoms x_{ij}, y_{ij}, z_{ij}, i.e., $`B_{ij} = \langle x_{ij}, y_{ij}, z_{ij} \rangle$. Moreover, suppose that these twelve Boolean algebras $`B_{ij}$ have only 0 and 1 in common (see Figure 56).

In the first step take the elements x_{1i}, x_{2i}, x_{3i} and consider them as atoms of a Boolean algebra $`X_i \cong \underline{2}^3$ $(i = 1,2,3,4)$.

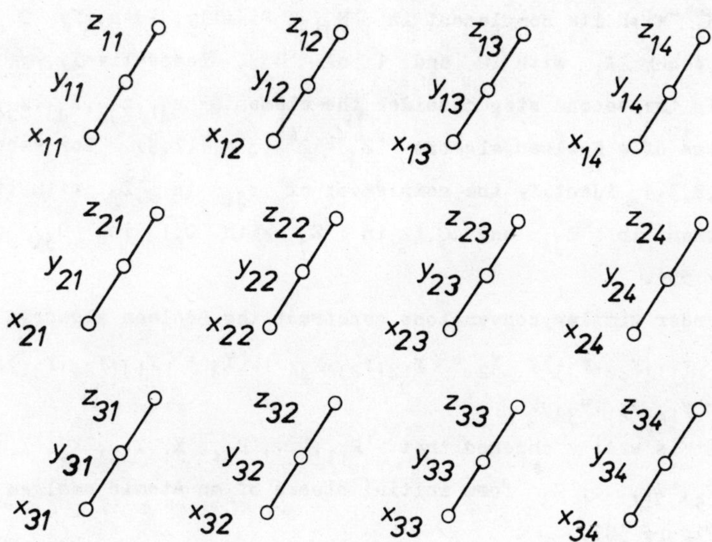

z11
y11
x11
z12
y12
x12
z13
y13
x13
z14
y14
x14
z21
y21
x21
z22
y22
x22
z23
y23
x23
z24
y24
x24
z31
y31
x31
z32
y32
x32
z33
y33
x33
z34
y34
x34

Fig. 56

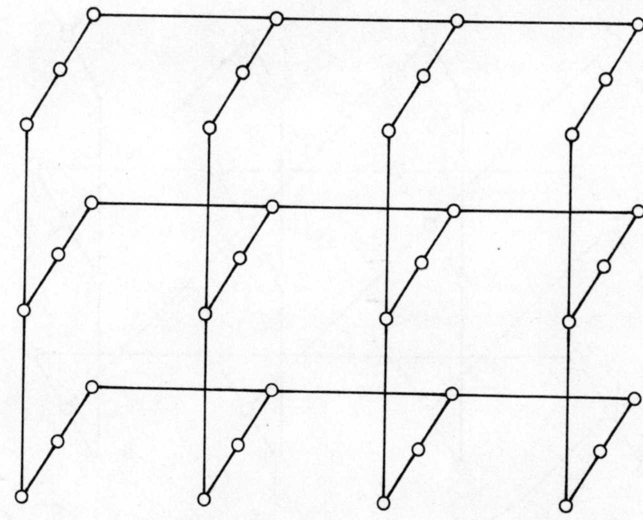

Fig. 57

Furthermore, for any $j = 1,2,3$ identify the complement of x_{ji} in $`X_i$ with its complement in $`B_{ji}$. Finally, identify 0 and 1 of each $`X_i$ with 0 and 1 of $`B_{ji}$, respectively.

In the second step consider the elements $z_{j1}, z_{j2}, z_{j3}, z_{j4}$ as atoms of a Boolean algebra $`Z_j \cong \underline{2}^4$ $(j = 1,2,3)$. For each $i = 1,2,3,4$ identify the complement of $z_{ji}$ in $`Z_j$ with its complement in $`B_{ji}$ and 0,1 in $`Z_j$ with 0,1 in $`B_{ji}$ (see Figure 57).

Under similar conventions construct the Boolean algebras $`Y_1 = \langle y_{11}, y_{22}, y_{33}\rangle$, $`Y_2 = \langle y_{12}, y_{23}, y_{34}\rangle$, $`Y_3 = \langle y_{13}, y_{24}, y_{31}\rangle$, $`Y_4 = \langle y_{14}, y_{21}, y_{32}\rangle$.

It is easily checked that $`B_{11}, \ldots, `B_{34}, `X_1, \ldots, `X_4, `Y_1, \ldots$ $\ldots, `Y_4, `Z_1, \ldots, `Z_3$ form initial blocks of an atomic amalgam (see **Figure 58**).

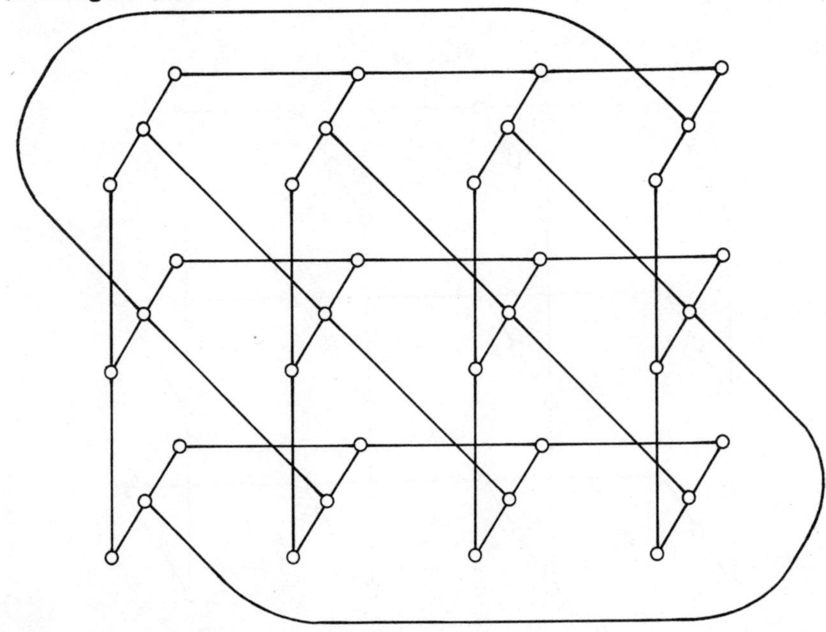

Fig. 58

By Greechie's Second Theorem IV.4.10, the resulting amalgam represents an orthomodular lattice. We shall show that it admits no state.

Suppose m is a state on this amalgam and use the convention made in the proof of Theorem 8. For instance, write $x_{11} + y_{11} + z_{11}$ for $m(x_{11}) + m(y_{11}) + m(z_{11})$. Find now $S = \sum m(x)$ where the sum runs over all the atoms of the amalgam. By Remark 7.B (iii) and Corollary IV.4.8,

$$x_{11} + y_{11} + z_{11} = 1, \ldots , x_{34} + y_{34} + z_{34} = 1$$

and so $S = 12$. On the other hand, by the same Remark,

$$x_{11} + x_{21} + x_{31} = 1, x_{12} + x_{22} + x_{32} = 1, x_{13} + x_{23} + x_{33} = 1,$$
$$x_{14} + x_{24} + x_{34} = 1, y_{11} + y_{22} + y_{33} = 1, y_{12} + y_{23} + y_{34} = 1,$$
$$y_{13} + y_{24} + y_{31} = 1, y_{14} + y_{21} + y_{32} = 1, z_{11} + z_{12} + z_{13} +$$
$$+ z_{14} = 1, z_{21} + z_{22} + z_{23} + z_{24} = 1, z_{31} + z_{32} + z_{33} + z_{34} = 1$$

which yields $S = 11$, a contradiction. This contradiction shows that the orthomodular lattice G_{32} has no state.

A sequence $\{e_n\}$ of vectors e_n in a Hilbert space H is called orthonormal if and only if $\langle e_i, e_j \rangle = 0$ for every $i \neq j$ and if $\| e_i \| = 1$ for all i. It is said to be an orthonormal basis for H if and only if every vector v of H is the limit of a sequence $\{s_n\}$ of partial sums $s_n = \sum_{i=1}^{n} c_i e_i$ $(c_i \in \underline{C})$.

A Hilbert space H is said to be separable if and only if it has a countable (i.e., finite or infinite) orthonormal basis.

By the logic of quantum mechanics is usually meant (cf. [160, p. 28], [31], [28]) the projection lattice (see III.4) constructed for a separable complex Hilbert space of infinite dimension.

However, there is a more general approach (cf. [161]) which defines such a logic as the collection of events pertaining to a physical system.

From this standpoint it is natural to propose other structures as models for the logic of quantum mechanics. In this connection we record the words of B. Jeffcott (see [105]): "Most contemporary practitioners of quantum logic seem to agree that a quantum logic is (at least) an orthomodular poset ... or some variation thereof ... ." If this belief is correct, the search for new approaches to the orthomodularity constitutes a worthwhile effort.

Before concluding this section, we shall add a notice of a little peculiar situation concerning terminology. The words "logic of quantum mechanics" have caused many misunderstandings. Of course, the calculus studied in a logical structure in the sense defined above has a certain similarity to the calculus of classical logic but its meaning is completely different. The former calculus represents a formalization of a set consisting of experimentally obtained results by means of techniques used in theory of orthomodular posets, whereas the latter one can be roughly sketched as a structural description and/or an analysis of propositions.

2. COMPATIBILITY

The purpose of this section is to present some fundamental properties of orthomodular posets which will be used in the following sections.

Lemma 2.1. Let a_i, $i \in I \neq \emptyset$, be elements of an orthocomplemented poset. Suppose the element $j = \bigvee\{a_i;\ i \in I\}$ exists. Then the element $\bigwedge\{a_i';\ i \in I\}$ exists and it is equal to j'.

Proof. The proof is straightforward and left to the reader.

Remark 2.2. Let $`P = (P, \leqslant, ', 0, 1)$ be an orthocomplemented poset. Assume that $`P$ has the following property:

(*) The supremum $a \vee b$ exists in $`P$ for all $a, b \in P$ such that $a \perp b$.

Let $x \leqslant y$ be elements of P. Then $x \leqslant (y')'$ and so $x \perp y'$. By hypothesis, the supremum $x \vee y'$ exists in $`P$. From this and Lemma 1 we see that also the infimum $x' \wedge y$ exists in $`P$.

The conditions (i)-(iv) below together with the condition (*) represent different possibilities how to characterize an orthomodular poset.

Theorem 2.3. The following conditions are equivalent in every orthocomplemented poset $`P = (P, \leqslant, ', 0, 1)$ satisfying the condition (*):

(i) For every $x, y \in P$

$$x \leqslant y \Rightarrow x \vee (x \vee y')' = y.$$

(ii) For all $a \leqslant b$ of P

$$b = a \vee (b \wedge a') \ \&\ (a \vee b') \wedge b = a.$$

(iii) For every $a \leqslant b$ of P there exists $c \in P$ such that $c \leqslant a'$ and $a \vee c = b$.

(iv) For all $s, t \in P$

$$(s \geqslant t' \ \&\ s \wedge t = 0) \Rightarrow s = t'.$$

Proof. (ii) $\Rightarrow$ (iii). Put $c = b \wedge a'$.

(iii) $\Rightarrow$ (iv). Let $s \geqslant t'$, $s \wedge t = 0$ and let c be such that $c \vee t' = s$ and $c \leqslant t$. Then $c \leqslant s \wedge t = 0$ and therefore $s = t'$.

The rest follows from Exercise IV;8 including the claim (ii) which represents a compendium of IV;8 (4) and (5) useful in calculations. //

Remark 2.4. If $`P$ is an orthomodular poset, then the element c of (iii) is uniquely determined. Indeed, $a' \wedge b = a' \wedge \wedge (a \vee c) = c$ by (ii). We denote this element c by $b - a$. In what follows, we shall, wherever this will lead to no essential confusion, denote it also a^+.

A set $\{a_i;\ i \in I\}$ of elements of an orthocomplemented poset is said to be orthogonal if and only if $a_i \perp a_j$ for any two different indices i,j of I. If the supremum $\bigvee\{a_i;\ i \in I\}$ of the orthogonal set $\{a_i;\ i \in I\}$ exists, it is denoted by $\sum a_i$. Similarly, we write $s = a_1 + a_2$ to denote the fact that the supremum $s = a_1 \vee a_2$ exists and that $a_1 \leqslant a_2'$.

Let a,b be elements of an orthocomplemented poset $`P$. We say that a commutes with b and write aCb if and only if $a \wedge b$, $a \wedge b'$ and $(a \wedge b) + (a \wedge b')$ exist and $a = (a \wedge b) + (a \wedge \wedge b')$.

Following the usage of [64] for c and d of $`P$, we shall say that c is compatible with d if and only if there exist mutually orthogonal elements $c_1, d_1, g \in P$ such that $c = = c_1 + g$ and $d = d_1 + g$. In this case we write cMd and say that the elements c,d ortogonally commute (cf. II.2). This notation, i.e. c_1, d_1, g, as well as the parallel terminology will be kept throughout the section.

Finally, if $e \vee f, e \vee f'$ and $(e \vee f) \wedge (e \vee f')$ exist and $e = = (e \vee f) \wedge (e \vee f')$, we say that e dually commutes with f and write eDf.

Remark 2.5. If $c \leqslant d$ are elements of an orthomodular poset, then cMd. Indeed, $c = c + 0$ and $d = c + (d \wedge c')$.

Theorem 2.6. Let `P be an orthocomplemented poset satisfying the condition (⁎) of Remark 2. If c orthogonally commutes with d, then $c \vee d$ exists.

Proof. Using the notation of the definition, we get $g \leqslant c_1'$ and $d_1 \leqslant c_1'$. Hence $d = d_1 + g \leqslant c_1'$. By hypothesis, the element $s = c_1 \vee d$ exists. We shall now show that $s = c \vee d$. From $s \geqslant$ $\geqslant d \geqslant g$ and $s \geqslant c_1$ one obtains $s \geqslant g + c_1 = c$. If $h \geqslant c$ and $h \geqslant d$, then $h \geqslant c \geqslant c_1$ and so $h \geqslant c_1 \vee d = s$. //

Lemma 2.7. Let `P be an orthomodular poset. If cMd, then cMd′.

Proof. With c_1, g as defined in the proof of Theorem 6, we have $c \vee d = c_1 \vee d$. By orthomodularity and Theorem 3, $c_1 \vee (c \vee$ $\vee d)' = c_1 \vee (c_1 \vee d)' = d'$. Hence $c = g + c_1$ and also $d' =$ $= (c \vee d)' + c_1$. //

Theorem 2.8. Let `P be an orthomodular poset and let cMd. Then:

(i) cMd′, c′Md, c′Md′.

(ii) The elements $c \vee d$ and $c \wedge d$ exist.

(iii) The elements $c \vee d'$, $c' \vee d$, $c' \vee d'$, $c \wedge d'$, $c' \wedge d$, $c' \wedge d'$ exist.

Proof. To justify the assertions of the theorem we need only refer to Theorem 6, Lemmas 7, 1 and to the symmetry of the relation M. //

Remark 2.9. We have shown that logical structures consisting of propositions give rise to orthomodular posets. We may ask what pairs of propositions correspond physically to compatibles ones.

We shall show that compatible propositions are exactly those which physically may be verified simultaneously. It comes to the same thing as to say that the experiments which correspond to the considered propositions do not interfere with each other. This is clear in the special case of two disjoint propositions p,q. As explained above, in this case p is true if and only if q is false. Hence only one experiment is needed if we want to verify p and q at the same time. Suppose now that c and d are two propositions of a logical structure associated with experiments which are compatible, i.e., $c = c_1 + g$ and $d = d_1 + g$ where c_1, d_1 and g are mutually orthogonal. Then c_1, d_1, g - as disjoint propositions - can be verified simultaneously and, a fortiori, $c = c_1 + g$, $d = d_1 + g$ are also simultaneously verifiable.

To show the converse, assume that c and d correspond to noninterfering experiments. Then it is physically meaningful to suppose that $g = c \wedge d$ exists. Since $(c \wedge d) \perp c'$, the physical interpretation of the axiom (IV) in Section 1 shows that the supremum $s = (c \wedge d) \vee c'$ exists. From Lemma 1 it follows that the infimum $c_1 = c \wedge (c \wedge d)' = s'$ exists. Similarly we can see that the infimum $d_1 = d \wedge (c \wedge d)'$ exists. By Theorem 1.2 and Theorem 3 (ii) we then have $c = c_1 + g$ and $d = d_1 + g$. Thus c and d are compatible.

At this point it is evident that the seemingly purely mathematical construction of the compatibility relation M has a direct physically intuitive meaning and, therefore, becomes relevant for physics.

The following theorem makes clear the fact that instead of dealing with the relation M we can also work with C.

<u>Theorem 2.10</u>. Let `P be an orthomodular poset. Then cMd if and only if cCd. In other words,

$$cMd \Leftrightarrow cCd \Leftrightarrow dCc.$$

<u>Proof</u>. 1. Suppose cMd. Then, by Theorem 8, the elements $c\wedge d'$, $c\wedge d$ and $c'\wedge d$ exist. Evidently, $g \leqslant c\wedge d$. Moreover, since $c_1 \leqslant d_1'$ and $c_1 \leqslant g'$, we have $c_1 \leqslant d_1'\wedge g' = (d_1 \vee g)' = d'$. Hence $c_1 \leqslant c\wedge d'$ and, similarly, $d_1 \leqslant c'\wedge d$. From

$$c \geqslant (c\wedge d) + (c\wedge d') \geqslant g + c_1 = c$$

we get $c = (c\wedge d) + (c\wedge d')$.

2. Suppose now that $c = (c\wedge d)\vee(c\wedge d')$. We shall show that the element $d\wedge c'$ exists and that $d\wedge c' = (c'\vee d')\wedge d$. Since $c\wedge d$ exists, it is clear that $c'\vee d'$ exists. Using Theorem 8 together with $d' \leqslant c'\vee d'$ we see that $(c'\vee d')\wedge d$ exists. Now $(c'\vee d')\wedge d = (c'\vee d')\wedge[(c'\vee d)\wedge d]$. Let $x = c'\vee d'$, $y = c'\vee d$ and $z = d$. From $c' = (c'\vee d')\wedge(c'\vee d)$ it is seen that $x\wedge y$ exists. We remark that $c\wedge d \leqslant d = (c'\vee d)\wedge d$. Therefore by Remark 2 $(c\wedge d)'\wedge[(c'\vee d)\wedge d] = x\wedge(y\wedge z)$ exists. By the dual of Lemma I.2,

$$(c'\vee d')\wedge d = x\wedge(y\wedge z) = (x\wedge y)\wedge z = [(c'\vee d')\wedge(c'\vee d)]\wedge d = c'\wedge d.$$

Thus $w = (d\wedge c) + (d\wedge c')$ exists and, by orthomodularity,

$$w = (d\wedge c)\vee[(c'\vee d')\wedge d] = (c\wedge d)\vee[(c\wedge d)\vee d']' = d.$$

Let $g = c\wedge d$, $d_1 = d\wedge c'$, $c_1 = c\wedge d'$. Then $d = (d\wedge c) + (d\wedge c') = g + d_1$, $c = (c\wedge d) + (c\wedge d') = g + c_1$, that is cMd.

The remaining assertion follows e.g. by the symmetry of M. //

<u>Theorem 2.11</u>. The following conditions are equivalent for any two elements c,d of an orthomodular poset:

(i) dCc; (ii) $(c\vee d')\wedge d = c\wedge d$; (iii) $(c\vee d')\wedge d \leqslant c$.

Proof. (i) ⇒ (ii). From Theorem 10 and 8 we see that dCc implies $c'Cd$. Hence, by the second part in the proof of Theorem 10, $(c \vee d') \wedge d = (c'' \vee d') \wedge d = c'' \wedge d = c \wedge d$.

(ii) ⇒ (iii). This is obvious.

(iii) ⇒ (i). Suppose $c \geqslant (c \vee d') \wedge d = s$. We shall prove that $c \wedge d$ exists and that $c \wedge d = s$. Evidently, $s \leqslant d$ and, by hypothesis, $s \leqslant c$. If $t \leqslant c$ and $t \leqslant d$, then $t \leqslant c \vee d'$. Thus $t \leqslant (c \vee d') \wedge d$. By Lemma 1, the element $d \wedge c'$ exists. Therefore,

$$(d \wedge c) + (d \wedge c') = (d \wedge c') + [(c \vee d') \wedge d] =$$
$$= (d \wedge c') + [(d \wedge c') \vee d']' = d$$

by orthomodularity. //

Theorem 2.12. Let `P be an orthomodular poset and let $c, d \in P$. Then the following conditions are equivalent:

(i) cMd; (ii) cCd;

(iii) $(c \vee d') \wedge d = c \wedge d$; (iv) $(c' \vee d) \wedge c = c \wedge d$;

(v) $(c \vee d') \wedge d \leqslant c$; (vi) $(c' \vee d) \wedge c \leqslant d$;

(vii) $c' \vee (c \wedge d) = c' \vee d$; (viii) cDd.

Proof. Using Theorem 8 and 10, we can see that cMd if and only if $c'Cd'$. Hence, cMd is equivalent to $c' = (c' \wedge d') \vee \vee (c' \wedge d)$ which means that $c = (c \vee d) \wedge (c \vee d')$. It follows that cMd if and only if cDd. Note that $c' \vee (c \wedge d) = c' \vee d$ is equivalent to $(c' \vee d') \wedge c = c \wedge d'$.

The rest follows from the same theorems. //

Theorem 2.13. Let `P be an orthomodular poset and let $c \in P$, $\{a_i;\ i \in I\} \subset P$ be such that c is compatible with a_i for every $i \in I$. Suppose that the elements $\bigvee\{a_i;\ i \in I\}$, $\bigvee\{c \wedge a_i;\ i \in I\}$ exist. Then $c \wedge (\bigvee\{a_i;\ i \in I\}) = d$ exists and $d = \bigvee\{c \wedge a_i;\ i \in I\}$.

Proof. It is convenient to use an abbreviation for the symbols considered here. Our aim is to prove that $s = \bigvee(c \wedge a_i)$ is the infimum of $\{c, \bigvee a_i\}$. Evidently, $s \leq c$ and $s \leq \bigvee a_i$.

Let $t \leq c$ and $t \leq \bigvee a_i$. We shall show that $t \leq s$. Now $a_i = (a_i \wedge c) \vee (a_i \wedge c')$ and it follows that

$$t \leq \bigvee[(a_i \wedge c) \vee (a_i \wedge c')] \leq [\bigvee(a_i \wedge c)] \vee c'.$$

Hence

$$t = t \wedge c \leq \{[\bigvee(a_i \wedge c)] \vee c'\} \wedge c = u.$$

Since $\bigvee(a_i \wedge c) \leq c$, we can conclude from Theorem 3 that $u = \bigvee(a_i \wedge c)$. Therefore, $t \leq \bigvee(a_i \wedge c)$. //

Theorem 2.14. Assume that P is an orthomodular poset, $c \in P$, $\{a_i;\ i \in I\} \subset P$, cMa_i for every $i \in I$ and that the supremum $\bigvee\{a_i;\ i \in I\}$ exists. If $\bigvee\{c \wedge a_i;\ i \in I\}$ exists, then $cM\bigvee\{a_i;\ i \in I\}$.

Proof. By Theorem 13, the element $c \wedge \bigvee a_i$ exists. Since $c \wedge (\bigvee a_i) \leq c$, the element $w = [c \wedge (\bigvee a_i)] \vee c'$ exists. From Theorem 13 we conclude that $v = c' \vee \bigvee(c \wedge a_i)$ also exists and that $v = w$. Because of $c \wedge a_i \leq c$, the element $c' \vee (c \wedge a_i)$ exists. It then follows from Lemma I.2 that the element $u = \bigvee[c' \vee (c \wedge a_i)]$ exists and that $u = v$. It is evident from Theorem 12 that $c' \vee (c \wedge a_i) = c' \vee a_i$. Thus $w = \bigvee(c' \vee a_i)$. Applying Lemma I.2 we see that the element $t = c' \vee \bigvee a_i$ exists and that $t = w$. In summary, $c' \vee [c \wedge (\bigvee a_i)] = c' \vee (\bigvee a_i)$. Therefore, by Theorem 12, $cM\bigvee a_i$. //

Remark 2.15. In Theorems 13 and 14, the symbols $\wedge, \bigvee$ can be replaced by the symbols $\vee$ and $\bigwedge$, respectively. The proof is word-for-word the same as the proof of those theorems, except that where a_iCc is used, one has to use the fact that a_iDc.

Let $`P$ be an orthomodular poset. The set $Z(`P)$ of all the elements $c \in P$ which are compatible with every element of P is called the <u>center</u> of $`P$.

We refer the reader to Exercises VIII;2 and VIII;3 for two basic properties of the center.

3. DIMENSION THEORY

In this section we derive some basic results on the dimension theory. The main tool in deriving these results are dimension posets in the sense of Šerstnev (cf. [176] and [110]).

The first standard definition here is the following convention: An orthomodular poset $`P = (P, \leqslant, ', 0, 1)$ is called a <u>dimension poset</u> if and only if there is an equivalence relation $\sim$ on P which satisfies the following axioms:

(D 1) If $a \sim 0$, then $a = 0$.

(D 2) If $\{a_i;\ i \in I\}$ and $\{b_i;\ i \in I\}$ are two orthogonal subsets of P and if $a_i \sim b_i$ for every $i \in I$, then the elements $a = \bigvee\{a_i;\ i \in I\}$ and $b = \bigvee\{b_i;\ i \in I\}$ exist and $a \sim b$.

(D 3) If $a + b \sim c + d$, then there exist elements $e, f, E, F \in P$ such that $a = e + f$, $b = E + F$, $e + E \sim c$ and $f + F \sim d$.

(D 4) For any two elements $a, b \in P$ which are not orthogonal there exist nonzero elements c, d such that $c \leqslant a$, $d \leqslant b$ and $c \sim d$.

Such a poset is called also a <u>dimension poset in the sense of Šerstnev</u>.

An alternative scheme for a dimension theory for complete orthomodular lattices has been proposed by Loomis in his classical work (cf. [130] and [167]). The results of his paper, especially of its second part, are related to the axioms (D 1), (D 2) and to the following two axioms:

(B) If $x \sim y + Y$, then there exist elements z, Z such that $x = z + Z$, $z \sim y$ and $Z \sim Y$.

(D′) If two elements x, y have a common complement, then $x \sim y$.

For convenience we say that a complete orthomodular lattice with $\sim$ satisfying (D 1), (D 2), (B) and (D′) is a dimension lattice in the sense of Loomis.

Remark 3.1. The axiom (B) is a consequence of the axiom (D 3). Indeed, if $x \sim y + Y$, then $x + 0 \sim y + Y$ and, by (D 3), there exist elements e, f, E, F such that $x = e + f$, $0 = E + F$ (hence $E = F = 0$) and such that $e \sim y$, $f \sim Y$.

Theorem 3.2. Every dimension lattice in the sense of Loomis is a dimension poset in the sense of Šerstnev.

Proof. The assertion of the theorem follows from Exercises VIII;9 and VIII;11. //

Remark 3.3. Note that there are finite orthomodular lattices which are dimension posets in the sense of Šerstnev but which are not dimension lattices in the sense of Loomis (cf. Exercise VIII; 13).

Convention 3.4. In what follows we suppose that `P is a dimension poset in the sense of Šerstnev.

If a, b are elements of `P, we shall write $a \precsim b$ if and only if there exists an element c such that $a \sim c$ and $c \leqslant b$.

An element $f \in P$ is said to be finite if and only if the relations $g \leqslant f$ and $g \sim f$ always imply $g = f$. If an element does not have this property, it is called <u>infinite</u>.

<u>Theorem 3.5</u>. If f is a finite element and $g \lesssim f$, then g is also finite.

<u>Proof</u>. 1. Suppose that f is a finite element and let $h \leqslant f$. We shall now show that h is finite. Let k be such that $k \leqslant h$ and $k \sim h$. Since $h \leqslant f$, the element $h' \wedge f$ exists. We have $h' \wedge f \sim h' \wedge f$, $k \sim h$, $k \leqslant h \vee f'$ and $h \leqslant h \vee f'$. Hence, by Remark 2.2, the elements $k \vee (h' \wedge f)$, $h \vee (h' \wedge f)$ exist and, by (D 2), $k \vee (h' \wedge f) \sim h \vee (h' \wedge f)$. But $h \vee (h' \wedge f) = h \vee (h \vee f')' = f$ by orthomodularity. Hence $k \vee (h' \wedge f) \sim f$ and $k \vee (h' \wedge f) \leqslant f$. By hypothesis, $k \vee (h' \wedge f) = f$. Therefore, $h = h \wedge f = h \wedge [k \vee (h' \wedge f)]$. Since $h \wedge (h' \wedge f) = 0$, the supremum $(h \wedge k) \vee [h \wedge (h' \wedge f)]$ exists. By Theorem 2.13, $h = h \wedge k = k$.

2. Let h be a finite element and let $g \sim h$. Assume that $c \leqslant g$ and $c \sim g$. Let $c^{+} = g - c$. Now $h \sim g = c + c^{+}$. By Remark 1, there exist elements z, Z such that $h = z + Z$, $z \sim c$ and $Z \sim c^{+}$. Since $z \sim c \sim g \sim h$, we have $z \sim h$ and $z \leqslant h$. Using the fact that h is finite, we see that $z = h$. It is clear that $h \geqslant Z$ and $h' = z' \geqslant Z$. This yields $c^{+} \sim Z = 0$ and, therefore, $c^{+} = 0$. It follows that $g = c + c^{+} = c$. Thus g is a finite element. //

Two elements a, b of P are called <u>related</u> if and only if there exist nonzero elements c, d such that $c \sim d$, $a \geqslant c$ and $d \leqslant b$.

An element i is said to be <u>invariant</u> if and only if the elements i and i' are not related. The set of all invariant elements of $`P$ will be denoted by $I(`P)$.

Clearly, if $i \in I(`P)$, then $i' \in I(`P)$.

<u>Example 3.6</u>. Consider the lattice sketched in Figure 31. For

the sake of brevity, we shall here write x instead of $[x]$. It is a dimension lattice in the sense of Šerstnev. The equivalence relation $\sim$ is determined by its classes $\{0\}$, $\{a,b,e,f\}$, $\{c\}$, $\{a',b',e',f'\}$, $\{c'\}$, $\{1\}$. The elements $0,1,c,c'$ are invariant.

<u>Theorem 3.7</u>. An element i is invariant in $`P$ if and only if it satisfies the following implication for every $a \in P$:

(+) $a \lesssim i \Rightarrow a \leqslant i$.

<u>Proof</u>. 1. Let $i \in I(`P)$ and let $a \sim j \leqslant i$. Our aim is to show that $a \leqslant i$ which is equivalent to $a \perp i'$. Suppose the contrary. Then, by (D 4), there exist nonzero elements c,d such that $c \leqslant a$, $d \leqslant i'$ and $c \sim d$. Let $c^+ = a - c$. Hence $c + c^+ = a \sim j$ and, therefore, by Remark 1, there exist elements z,Z with $i \geqslant j = z + Z$, $c \sim z$ and $c^+ \sim Z$. In particular, $i' \geqslant d \sim c \sim z \leqslant i$. Since $i \in I(`P)$, $d = 0$, a contradiction. Thus $a \leqslant i$.

2. Let i be an element which satisfies (+). If $i \geqslant c \sim d \leqslant i'$, then $d \lesssim i$. By hypothesis, $d \leqslant i$. But $d \leqslant i'$ and so $d = 0$. //

<u>Theorem 3.8</u>. Let $i \in I(`P)$. Then iMx for all $x \in P$.

<u>Proof</u>. Since $x + x' = 1 \sim 1 = i + i'$, there exist elements e,f,E,F such that $x = e + f$, $x' = E + F$, $e + E \sim i$ and $f + F \sim i'$. By Theorem 7, $e + E \leqslant i$ and $f + F \leqslant i'$. A fortiori $e \leqslant i$ and $f \leqslant i'$. We therefore have from Remark 2.5 and Theorem 2.8 that eMi and fMi. Now $i \wedge e = e$ and $i \wedge f \leqslant i \wedge i' = 0$. Hence, the supremum $(i \wedge e) \vee (i \wedge f)$ exists. Thus, from Theorem 2.14, we have $iMe + f$ and so iMx. //

A subset H of P is said to be <u>hereditary</u> if and only if the following implication holds for all $k \in P$ and all $h \in H$:

$$k \lesssim h \Rightarrow k \in H.$$

Observe that if $i \in H$ and $j \leq i$, then $j \lesssim i$ so that j belongs to the hereditary set H.

<u>Theorem 3.9</u>. Let H be a hereditary subset of $`P$. Then the supremum $s = \sup_{`P} H$ exists and $s \in I(`P)$.

<u>Proof</u>. No generality is lost in assuming that $H \neq \emptyset$. By Zorn's lemma there exists an orthogonal set $G = \{g_i;\ i \in I\}$ which is maximal with respect to $G \subset H$. By (D 2) there exists the supremum $s = \sup_{`P} G$. We now claim that $s = \sup_{`P} H$. To this end it is enough to show that $h \leq s$ for all $h \in H$. Suppose the contrary. Then there exists an element h such that h and s' are not orthogonal. Therefore, by (D 4), there exist nonzero elements c,d such that $h \geq c \sim d \leq s'$. Since $h \in H$ and H is hereditary, $d \in H$. But $g \leq s \leq d'$ for all $g \in G$ and it follows that the set $G \cup \{d\} \subset H$ is orthogonal. By the choice of G, $d \in G$. Hence $d \leq d'$ so that $d = 0$, a contradiction. Thus s is the supremum of H.

It remains to be shown that $s \in I(`P)$. To see it, we shall prove first that for every nonzero element c satisfying $c \leq s$ there exists a nonzero element $h \in H$ such that $h \leq c$. Really, suppose firstly that $k \in H$ is such that c and k are not orthogonal. Then c and k are related. Hence, there exist nonzero elements h,f such that $c \geq h$, $h \sim f$ and $f \leq k$. Since $h \lesssim k \in H$ and H is hereditary, $c \geq h \in H$ and our statement regarding c is evidently true. Therefore, from now on we may suppose that $c \perp k$ for every $k \in H$. It follows that $c' \geq k$ for every $k \in H$ and so $c' \geq s$ which means that $c \leq s'$. From $c \leq s$ we infer that $c = 0$, a contradiction. Thus the existence of the above mentioned h is guaranteed.

Finally, we shall prove that s and s' are not related.

Suppose this assertion were false. Then there exist nonzero elements c,d such that $s \geqslant c$, $c \sim d$ and $d \leqslant s'$. We can conclude from the above that there is a nonzero element $h \in H$ such that $h \leqslant c$. Let $h^+ = c - h$. Since $d \sim c = h + h^+$, there exist elements e,f such that $d = e + f$, $e \sim h$ and $f \sim h^+$. Furthermore, because H is hereditary and $e \sim h \in H$, we have $e \in H$ so that $e \leqslant s$. However, $e \leqslant s'$ and it follows that $e = 0$. Now $h \sim e = 0$ yields $h = 0$. This contradicts our assumption above and completes the proof. //

<u>Theorem 3.10</u>. For every set $E = \{e_i;\ i \in I\}$ of invariant elements the infimum $\inf_{`P} E$ exists and it is an invariant element.

<u>Proof</u>. We may assume, without any loss of generality that $E \neq \emptyset$. Let D be the set of all lower bounds of E. We shall prove that D is hereditary. Let $k \in P$ and $d \in D$ be such that $k \lesssim d$ so that $k \lesssim d \leqslant e_i$ for every $i \in I$. Since $k \lesssim e_i$ and e_i is invariant, $k \leqslant e_i$. Thus $k \in D$ and we see that D is hereditary. By Theorem 9, the supremum $s = \sup_{`P} D$ exists and $s \in I(`P)$. Clearly, $s = \inf_{`P} E$. //

<u>Theorem 3.11</u>. The set $I(`P)$ of all invariant elements of $`P$ determines a complete Boolean algebra $(I(`P), \leqslant, ', 0, 1)$.

<u>Proof</u>. Since $1 \in I(`P)$, I.13 and Theorem 10 imply that it is a complete lattice.

If $i \in I(`P)$, then $i' \in I(`P)$ by definition. Hence it is a complemented lattice.

By Theorem 8, $I(`P) \subset Z(`P)$. Exercise VIII;2 implies that it is a distributive lattice. //

An element $s \in P$ is called <u>simple</u> if and only if it has the following property:

For every $b \in P$ satisfying $b \leqslant s$ the elements b and $b^+ = s \wedge b'$ are not related.

Example 3.12. Using the convention of Example 6 we can assert that the element c of the lattice in Figure 31 is simple and invariant and that the element a is simple but it is not invariant. An example of a finite and invariant element which is not simple is furnished by the element 1 of this lattice.

Theorem 3.13. Every simple element of $`P$ is finite.

Proof. Let s be a simple element, $g \leqslant s$ and $g \sim s$. Let $g^+ = s \dot{-} g$. By assumption, $g \sim s = g + g^+$. Hence, by Remark 1, there exist elements e,f such that $g = e + f$, $e \sim g$ and $f \sim g^+$. Since s is simple and $g \geqslant f \sim g^+ \leqslant g^+$, $f = g^+ = 0$. Thus $s = g + g^+ = g$. //

Theorem 3.14. Let s be a simple element of $`P$. Then every element of the set $D(s) = \{x \in P;\ x \lesssim s\}$ is also simple.

Proof. Suppose that $x \sim X \leqslant s$. Let $c \leqslant x$ and $x \wedge c' \geqslant m \sim n \leqslant c$. Since $X \sim x = (x \wedge c') + c$, there exist $\hat{P},\hat{Q}$ such that $X = \hat{P} + \hat{Q}$, $\hat{P} \sim x \wedge c'$ and $\hat{Q} \sim c$. Let $m^+ = (x \wedge c') - m$ and $n^+ = c - n$. From $\hat{P} \sim x \wedge c' = m + m^+$ and $\hat{Q} \sim c = n + n^+$ we conclude that there exist elements p,r,t,u such that $\hat{P} = p + r$, $\hat{Q} = t + u$, $r \sim m$ and $t \sim n$. In summary, we have

$$s \wedge \hat{Q}' \geqslant \hat{P} \geqslant r \sim m \sim n \sim t \leqslant \hat{Q} \leqslant s.$$

Using the fact that s is simple we get $r = t = 0$ and, consequently, $m = n = 0$. //

Theorem 3.15. Let $F(`P)$ denote the set of all finite elements of $`P$ and let $S(`P)$ be the set of all simple elements of $`P$. Then the elements $f_0 = \sup_{`P} F(`P)$ and $s_0 = \sup_{`P} S(`P)$ exist.

In addition, f_0 and s_0 belong to $I(`P)$ and $s_0 \leqslant f_0$.

Proof. By Theorem 9, it suffices to show that the sets $F(`P)$ and $S(`P)$ are hereditary. However, if $f \in F(`P)$ and $g \lesssim f$, then g is finite by Theorem 5.

Suppose that $s \in S(`P)$ and let $h \lesssim s$. Then h is simple by Theorem 14.

Since every simple element is finite, $s_0 \leqslant f_0$. //

Theorem 10 justifies the following definition: Let $a \in P$. The element $|a| = \inf_{`P}\{i \in I(`P);\ a \leqslant i\}$ is called the hull of a.

The reader can find some basic properties of the hull in Exercise VIII;14.

Now we shall show how information gleaned from previous theorems enables us to study the structure of dimension posets. Before doing this, we make a few additional definitions.

A dimension poset $`P$ is said to be of type I, if there is a simple element $t \in P$ such that $|t| = s_0 = 1$. It is of type II, if there are no simple elements in $`P$ and if there exists a finite element g such that $|g| = f_0 = 1$. Finally, it is of type III if all nonzero elements of $`P$ are infinite.

The reader can consult [110] for a proof of the following theorem due to Kalinin:

Theorem 3.16. Every dimension poset $`P$ can be written in the form $`P = `P_1 \oplus `P_2 \oplus `P_3$ where $`P_1$ is of type I, $`P_2$ of type II and $`P_3$ of type III.

Our object now is to obtain an analogue of the Schroeder-Bernstein theorem in a dimension poset $`P$. This can be easily accomplished; we need only one preparatory result:

Lemma 3.17. For every infinite element h of $`P$ there

exists an orthogonal set $\{a_n; n \in \underline{N}\}$ of equivalent elements such that $h \geqslant a_n \neq 0$ for all $n \in \underline{N}$.

Proof. Since $h \notin F(`P)$, there exists an element k such that $k \sim h$ and $k < h$. Set $a_1 = h - k$ and $b_1 = k$ so that $a_1 \neq 0$ and $h = a_1 + b_1$. We repeat this proces inductively. Having found $a_1, a_2, \ldots, a_n$ and $b_1, b_2, \ldots, b_n$ $(n \geqslant 1)$ such that

(i) $h = a_1 + a_2 + \ldots + a_j + b_j$ for every $j = 1,2,\ldots,n$;

(ii) $b_j = a_{j+1} + b_{j+1}$ with $a_{j+1} \sim a_j \neq 0$ and $b_{j+1} \sim b_j$ for every $j = 1,2,\ldots,n-1$,

we prove that it is possible to find a_{n+1} and b_{n+1} satisfying (i) and (ii). Indeed, we have $b_n \sim b_{n-1} = a_n + b_n$. Hence, by Remark 1, there exist elements a_{n+1}, b_{n+1} such that $b_n = a_{n+1} + b_{n+1}$, $a_n \sim a_{n+1}$ and $b_n \sim b_{n+1}$. If $a_n = 0$, then $0 = a_n \sim a_{n-1}$ and so $a_{n-1} = 0$, in contradiction to the induction hypothesis. Thus (ii) is verified also for $j = n$. Moreover,

$$h = a_1 + a_2 + \ldots + a_n + b_n = a_1 + a_2 + \ldots + a_n + a_{n+1} + b_{n+1}$$

and this completes the proof. //

As in Schroeder-Bernstein Theorem we have

Theorem 3.18. If $a \lesssim b$ and $b \lesssim a$ for a,b of a dimension poset $`P$, then $a \sim b$.

Proof. By assumption, $a \sim A \leqslant b$ and $b \sim B \leqslant a$. Let $A^+ = b - A$ so that $B \sim b = A + A^+$. It then follows from Remark 1 that there are elements C,D such that $B = C + D$, $C \sim A$ and $D \sim A^+$. Set $d_1 = D = B \wedge C'$ and $e_1 = B^+ = a \wedge B'$. Since $B \wedge C' \leqslant (a \wedge B')'$, the supremum $d_1 + e_1 = (B \wedge C') + (a \wedge B')$ exists. It is immediate that $s = a \wedge C' \geqslant (B \wedge C') + (a \wedge B') = t'$. Obviously, $C \leqslant (B' \vee C) \wedge (a' \vee B)$ and hence the infimum $C' \wedge [(B' \vee C) \wedge (a' \vee B)]$ exists. Therefore, $s \wedge t = (a \wedge C') \wedge [(B' \vee C) \wedge (a' \vee B)] = a \wedge \{C' \wedge [(B' \vee C) \wedge (a' \vee B)]\}$. Next, we have

$a \wedge B' \leqslant B' \leqslant C'$, $a \wedge B' \leqslant B' \vee C$ and $C \leqslant B' \vee C$. Consequently, $s \wedge t = a \wedge \{[C' \wedge (B' \vee C)] \wedge (a' \vee B)\}$. From $C' \geqslant B'$, $B \leqslant a' \vee B$ and Theorem 2.3 it follows that $s \wedge t = a \wedge [B' \wedge (a' \vee B)] = a \wedge a' = 0$. By orthomodularity, $t' = d_1 + e_1 = a \wedge C' = s$.

Suppose first that a is finite. Then $C \leqslant B \leqslant a$ and $C \sim \sim A \sim a$. This means that $C = a$ and, a fortiori, $b \sim B = a$.

Without loss we may therefore assume that the element a is infinite and that $a > C$.

With the notation already used in the proof of Lemma 17 and with $h = a$, $k = C$ we can write $h - k = a_1 = a \wedge C' = d_1 + e_1$. Since, by Lemma 17, $a_n \sim a_1 = d_1 + e_1$ for every $n \in \underline{N}$, Remark 1 shows that there exist elements b_n, c_n such that $a_n = b_n + c_n$, $b_n \sim d_1$ and $c_n \sim e_1$ with $b_1 = d_1$, $c_1 = e_1$. By (D 2), the supremum $y = \sum\{a_n;\ n \geqslant 1\}$ exists. Let $y^+ = a - y$. Then

$$a = y^+ + y = y^+ + \sum\{a_n;\ n \geqslant 1\} = y^+ + \sum\{b_n + c_n;\ n \geqslant 1\}.$$

Since

$$b_j \leqslant b_j + c_j = a_j \leqslant a_i' = b_i' \wedge c_i' \leqslant c_i'$$

for any $i \neq j$ and since $b_i \leqslant c_i'$, the supremum $\sum b_n + \sum c_n$ exists. We therefore have from (D 2), Lemma I.2 and from $c_n \sim \sim c_{n+1}$ $(n \in \underline{N})$ that

$$a = y^+ + (\sum\{b_n;\ n \geqslant 1\} + \sum\{c_n;\ n \geqslant 1\}) \sim y^+ + (\sum\{b_n;\ n \geqslant 1\} + \\ + \sum\{c_n;\ n \geqslant 2\}) = y^+ + [b_1 + (y - a_1)] = (a - y) + [b_1 + \\ + (y - a_1)] = [(a - y) + (y - a_1)] + b_1.$$

Now take into account that $a_1 \leqslant y \leqslant a$. Let $A_1 = y'$, $A_2 = y \wedge a_1'$ and $C_0 = a$. Then C_0 is compatible with A_1 and A_2, by Remark 2.5 and Theorem 2.8. Since $A_1 \perp A_2$ and $C_0 \wedge A_1 = a \wedge y' \perp \perp C_0 \wedge A_2 = a \wedge (y \wedge a_1') = y \wedge a_1'$, the elements $A_1 \vee A_2$ and

$(C_0 \wedge A_1) \vee (C_0 \wedge A_2)$ exist. Hence, by Theorem 2.13, we have

$$a \wedge [y' \vee (y \wedge a_1')] = C_0 \wedge (A_1 \vee A_2) = (C_0 \wedge A_1) \vee (C_0 \wedge A_2) = \\ = (a \wedge y') \vee (y \wedge a_1').$$

Using Theorem 2.3, we get

$$(a - y) + (y - a_1) = (a \wedge y') \vee (y \wedge a_1') = a \wedge [y' \vee (y \wedge a_1')] =$$

$$= a \wedge [y' \vee (y' \vee a_1)'] = a \wedge a_1' = a \wedge (a \wedge C')' = a \wedge (a' \vee C) = C.$$

Thus $a \sim C + b_1 = C + d_1 = C + (B \wedge C') = B \sim b$ and so $a \sim b$. //

A few illustrative concluding remarks concerning the definition of dimension lattices and posets will be given here. We shall employ descriptions of some typical situations which are useful in helping us to visualize the ideas underlying our purely lattice-theoretic considerations.

Let H be a complex Hilbert space. Recall (cf. III.4) that by an operator on H we mean any linear continuous mapping of H into H. A <u>von Neumann algebra</u> (<u>ring of operators</u>, <u>operator ring</u>) is a nonvoid set A of operators on H which contains cF_1, $F_1 + F_2$, F_1F_2, F_1^* along with F_1 and F_2 for any complex number c, it further contains the identity operator I and it is weakly closed. - This last notion can be defined as follows: Let U be a set of operators on H and let U' be the set consisting of all the operators F on H which commute with all the operators from U, i.e., $F \in U'$ if and only if $FG = GF$ for all $G \in U$. The set U is said to be <u>weakly closed</u> if and only if $U = (U')'$. The postulate that a von Neumann algebra A is weakly closed can be restated as $A = (A')'$.

The investigation of operator rings due to F. J. Murray and J. von Neumann [154] had led to the discovery of a new approach to the dimension theory in the case of a wide class of von Neumann algebras called factors. (A von Neumann algebra A is called a factor if and only if $A' \cap A = \{cI;\ c \in \underline{C}\}$.) Important related results for the general case can be found in the work of I. E. Segal [175] and J. Dixmier [50].

Let F be an operator on A. Then F is said to be a projection in A if and only if it is a projection of the Hilbert space H. If F_1 and F_2 are two projections in A, then we write $F_1 \sim F_2$ if and only if there is an operator F_3 in A such that $F_3^* F_3 = F_1$ and $F_3 F_3^* = F_2$. Let $P(A)$ denote the set of all the projections in A. It can be shown (see [130]) that the relation $\leqslant$ defined on $P(A)$ by $F_1 \leqslant F_2$ if and only if $F_1 = F_1 F_2$ makes $P(A)$ into a complete orthomodular lattice which satisfies the axioms (D 1), (D 2), (B) and (D′) stated above. This shows that the lattice (P(A),) with $\sim$ is a dimension lattice in the sense of Loomis.

Another fundamental concept upon which the origin of a general dimension theory rests is the notion of continuous geometries invented by J. von Neumann (cf. [186], [100], [101] and [94]).

By a continuous geometry (von Neumann lattice) is meant a complete complemented modular lattice which is continuous. - A complete lattice `L is said to be continuous if and only if

$$a \wedge [\bigvee(b_i;\ i \in I)] = \bigvee(a \wedge b_i;\ i \in I)$$

and

$$a \vee [\bigwedge(b_i;\ i \in I)] = \bigwedge(a \vee b_i;\ i \in I)$$

for any $a \in L$ and any chain $(b_i;\ i \in I)$ in `L.

Let a,b be elements of a lattice `L having 0 and 1. Recall that the elements a,b are defined to be perspective (cf.

V.4 and Exercises VII;5 - 6) if and only if there is an element $t \in L$ such that $a \vee t = b \vee t$ and $a \wedge t = b \wedge t = 0$. The relation of perspectivity $\underset{p}{\sim}$ makes any continuous geometry into a dimension lattice in the sense of Loomis (cf. [186]).

The mentioned results opened the possibility of a purely lattice-theoretic approach to a dimension theory in classes of lattices which are more general than the ones of modular lattices. This was accomplished by L. H. Loomis and S. Maeda. In [144] S. Maeda considers a complete lattice and postulates the existence of a binary relation $\perp$ satisfying five axioms not specified here where $a \perp b$ may be interpreted as $a \wedge b = 0$ in a continuous geometry. He then studies some questions of classification and proves the existence of a dimension function.

In her paper [146] on a classification of Boolean algebras D. Maharam defines an abstract measure algebra $(E,\sim)$ as a pair satisfying four postulates. As has been shown by Loomis in his fundamental work [130] a part of her investigation paralles the dimension theory in complete orthomodular lattices.

The Loomis´ ideas were the point of departure for A. Ramsay [167] and for the MacLaren´s characterization of locally finite dimension lattices [133].

The list of known properties of dimension posets in the sense of Šerstnev represented by the theorems derived in this section is far from complete, but we hope that the methods for obtaining them have been worked out clearly enough to illustrate the possibilities of the Šerstnev´s generalization of dimension lattices. Especially the briefly mentioned classification of dimension posets seems to be very promissing.

In this context, we quote von Neumann´s words addressed to

Birkhoff (cf. [28]): "... I do not believe absolutely in Hilbert space any more. After all Hilbert space (as far as quantum-mechanical things are concerned) was obtained by generalizing Euclidean space ... . But if we wish to generalize the lattice of all linear closed subspaces from a Euclidean space to infinitely many dimensions, then one does not obtain Hilbert space, but that configuration, which Murray and I called "case II_1" ... ."

This von Neumann's vision - pertinent to the present-day basic research - of how a more detailed classification concerning dimensional characteristics is needed justifies further studies in this direction and we feel it may have to play an important role in quantum-mechanical interpretations using the framework of orthomodular posets.

4. ORTHOLOGICS

Throughout this final section we have exercised our own prejudices relative to selection and presentation of material. Therefore it is our intention that this be considered an introduction to a broad area rather than a survey of it.

Concerning further information on orthologics we refer to [105], [106] and [64].

We begin by the main definition: A system $`P = (P, \leqslant, 0, 1, \perp)$ is said to be an <u>orthologic</u> if and only if it satisfies the following axioms:

(L 1) $(P, \leqslant)$ is a poset and $0 \leqslant x \leqslant 1$ for all $x \in P$.

(L 2) The binary relation $\perp$ is a symmetric relation defined on P and it is such that $x \perp x$ implies $x = 0$.

(L 3) If $x \perp y$ where $x,y \in P$, then the supremum $x \vee y = \sup_{(P,\leqslant)}\{x,y\}$ exists.

(L 4) If $x,y,z \in P$ are such that $x \perp y$, $x \perp z$ and $y \perp z$, then $x \perp (y \vee z)$.

(L 5) For every $x \in P$ there exists $y \in P$ such that $x \perp y$ and $x \vee y = 1$.

(L 6) The relation $x \leqslant y$ holds if and only if $z \perp y$ implies $z \perp x$ for every $z \in P$.

Remark 4.1. (A) It is immediate that every orthomodular poset is an orthologic, provided we define $a \perp b$ by $a \leqslant b'$.

(B) There exist orthologics which are not orthomodular posets (cf. Exercise VIII;16).

(C) By Proposition 1.6, any logical structure (L,F) can be regarded as an orthomodular poset having a full set of states. In this sense logical structures represent a very special case of orthologics. Notice the following hierarchy of "logical systems":

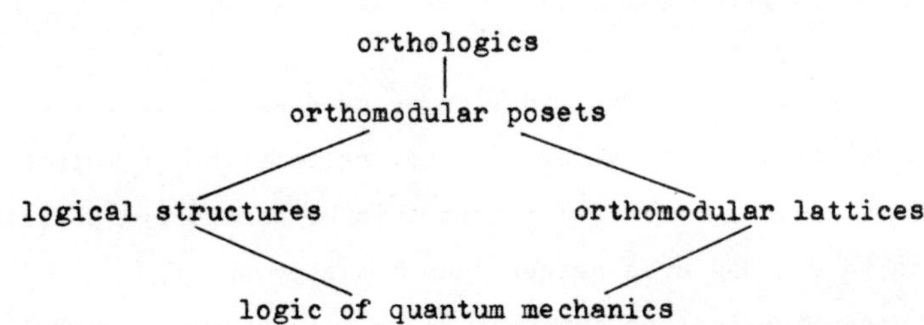

Lemma 4.2. In every orthologic P, $x \perp 0$ for all $x \in P$ and $u \perp 1$ implies $u = 0$.

Proof. 1. Choose $x \in P$. Then, by (L 5), there exists y such that $x \perp y$. Using (L 6), together with $0 \leqslant y$, we get $x \perp 0$.

2. Suppose that $u \perp 1$. It is always true that $u \leqslant 1$. Hence (taking (L 6) into account), $u \perp u$, and, by (L 2), $u = 0$. //

Theorem 4.3. Let $x_1, x_2, \ldots, x_n$ be elements of an orthologic satisfying $x_i \perp x_j$ for every $i \neq j$. Then the supremum $s_j = \sup_{(P, \leqslant)} \{x_1, x_2, \ldots, x_j\}$ exists for every $j = 1,2,\ldots,n$ and $x_{j+1} \perp s_j$ for every $j = 1,2,\ldots,n-1$.

Proof. The proof of this statement will be by induction on j: the case in which $j = 1$ is trivial.

So suppose, inductively, that this statement has been proved when $j = t - 1 \geqslant 0$, and consider now the case in which $j = t$. By induction hypothesis, the supremum $s_{t-1} = \sup \{x_1, x_2, \ldots, x_{t-1}\}$ exists and $x_t \perp s_{t-1}$. From (L 3) we see that also the supremum $s_t = s_{t-1} \vee x_t$ exists and, clearly, $s_t = \sup \{x_1, x_2, \ldots, x_t\}$. The induction hypothesis allows us to assert that $x = x_{t+1} \perp s_{t-1} = y$. By assumption, $x = x_{t+1} \perp x_t = z$ and, by induction hypothesis, $y = s_{t-1} \perp x_t = z$. Hence, by (L 4), $x_{t+1} = x \perp (y \vee z) = s_t$. //

Remark 4.4. Since every orthomodular poset provides an example of an orthologic, an analogous result is valid also for orthomodular posets.

Now we aim to explain a general construction which allows to obtain orthologics from "manuals".

Let E_i $(i \in I \neq \emptyset)$ be nonvoid sets and let $M = \{E_i;\ i \in I\}$. Let H denote the union of the sets E_i. Given $a, b \in H$, we write $a \perp b$ and call the elements a, b orthogonal when there exists E_j, $j \in I$, such that $\{a,b\} \subset E_j$ and when $a \neq b$. A subset D of H is called an event of M if and only if there exists E_k, $k \in I$, such that $D \subset E_k$. The set of all events will be denoted by $E(M)$. A subset $N \subset H$ is said to be orthogonal if and only if $a \perp b$ for all $a \neq b$ of N. Under these conventions we shall say that M is a manual if and only if the following two conditions are satisfied:

(M 1) If $E_i, E_j \in M$ and $E_i \subset E_j$, then $E_i = E_j$.

(M 2) If $E_i, E_j \in M$ and if N is an orthogonal set such that $N \subset E_i \cup E_j$, then there exists $E_k \in M$ with $N \subset E_k$.

For $K \subset H$, we define $K^\perp = \{a \in H; \forall k \in K \ a \perp k\}$. For brevity we write $K^{\perp\perp}$ for $(K^\perp)^\perp$ and $k^\perp$ for $\{k\}^\perp$.

Lemma 4.5. Let $A, B \subset H$. Then

(i) $A \subset B \Rightarrow A^\perp \supset B^\perp$;

(ii) $A \subset A^{\perp\perp}$;

(iii) $A^{\perp\perp\perp} = A^\perp$.

(iv) The following conditions are equivalent:

(1) $A \subset B^\perp$; (3) $A^{\perp\perp} \subset (B^{\perp\perp})^\perp$;

(2) $A^{\perp\perp} \subset B^\perp$; (4) $(A^{\perp\perp})^\perp \supset B^{\perp\perp}$.

Proof. The properties (i) and (ii) follow immediately from the definition of $K^\perp$ and show that here we have a Galois connection (see [29]). This means that (iii) is also valid.

(1) $\Rightarrow$ (2) and (2) $\Rightarrow$ (3). By (i) and (iii), we have that $A^{\perp\perp} \subset B^{\perp\perp\perp} = B^\perp$. (3) $\Rightarrow$ (4). From (i) it is seen that $A^{\perp\perp\perp} \supset$ $\supset (B^{\perp\perp})^{\perp\perp} = (B^{\perp\perp\perp})^\perp = B^{\perp\perp}$. (4) $\Rightarrow$ (1). Using (ii), (i) and (iii), we get $A \subset A^{\perp\perp} = A^{\perp\perp\perp\perp} \subset B^{\perp\perp\perp} = B^\perp$. //

Theorem 4.6. Let Y and Z be subsets of H. Then $(Y \cup Z)^\perp = Y^\perp \cap Z^\perp$.

Proof. Obviously, $c \in Y^\perp \cap Z^\perp$ if and only if $c \in Y^\perp$ and $c \in$ $\in Z^\perp$. This is equivalent to the fact that $c \perp t$ for every $t \in$ $\in Y \cup Z$. Hence the assertion under discussion is equivalent to $c \in (Y \cup Z)^\perp$. //

Let $L(M)$ denote the set $\{K \subset H; \exists G \in E(M) :: K = G^{\perp\perp}\}$. For any $A^{\perp\perp}, B^{\perp\perp} \in L(M)$, we shall write $A^{\perp\perp} \perp B^{\perp\perp}$ if and only if one of the conditions (1)-(4) stated in Lemma 5 is valid.

Theorem 4.7. Let M be a manual. Then $(L(M), \subset, \emptyset, H, \perp)$ is an orthologic.

Proof. (L 1). Use Exercise VIII;15. (L 2). This follows from Lemma 5.

(L 3). Let $X^{\perp\perp}, Y^{\perp\perp} \in L(M)$ be such that $X^{\perp\perp} \perp Y^{\perp\perp}$. Because of $X, Y \in E(M)$, there exist $E_i, E_j \in M$ such that $X \subset E_i$ and $Y \subset E_j$. Hence $X \cup Y \subset E_i \cup E_j$. Since, by Lemma 5, $X \subset Y^{\perp}$, we see that $X \cup Y$ is an orthogonal set. Then, by axiom (M 2), $X \cup Y \in E(M)$ and so $(X \cup Y)^{\perp\perp} \in L(M)$. We shall prove that $(X \cup Y)^{\perp\perp} = \sup\{X^{\perp\perp}, Y^{\perp\perp}\}$. First note that $(X \cup Y)^{\perp\perp}$ contains $X^{\perp\perp}$ and $Y^{\perp\perp}$. Next, suppose that $T^{\perp\perp} \supset X^{\perp\perp}$ and $T^{\perp\perp} \supset Y^{\perp\perp}$. Then $T^{\perp\perp} \supset X$ and $T^{\perp\perp} \supset Y$ so that $T^{\perp\perp} \supset X \cup Y$. Hence, $T^{\perp\perp} = T^{\perp\perp\perp\perp} \supset (X \cup Y)^{\perp\perp}$.

(L 4). Assume that $X^{\perp\perp}, Y^{\perp\perp}, Z^{\perp\perp} \in L(M)$, $X \subset Y^{\perp}$, $X \subset Z^{\perp}$ and $Y \subset Z^{\perp}$. Then $Y \vee Z = (Y \cup Z)^{\perp\perp}$. Furthermore, by Theorem 6, $X \subset Y^{\perp} \cap Z^{\perp} = (Y \cup Z)^{\perp} = ((Y \cup Z)^{\perp\perp})^{\perp} = (Y \vee Z)^{\perp}$. Thus $X \perp Y \vee Z$.

(L 5). Let $X^{\perp\perp} \in L(M)$, i.e., $X \in E(M)$. Then $X \subset E_k$ for an appropriate $E_k \in M$ and, therefore, $Y = E_k \setminus X \subset E_k$. It follows that $Y \in E(M)$. Obviously, $Y \subset X^{\perp}$ and hence $Y^{\perp\perp} \perp X^{\perp\perp}$. Moreover, by Exercise VIII;15, $X \vee Y = (X \cup Y)^{\perp\perp} = (E_k)^{\perp\perp} = H$.

(L 6). Suppose that $X^{\perp\perp} \subset Y^{\perp\perp}$ and $Z^{\perp\perp} \perp Y^{\perp\perp}$. Then, by Lemma 5, $X^{\perp\perp} \subset Y^{\perp\perp} \subset (Z^{\perp\perp})^{\perp}$. Using the same theorem once again we see that $X^{\perp\perp} \perp Z^{\perp\perp}$.

Let $X^{\perp\perp}, Y^{\perp\perp}$ be such that the implication

$$Z^{\perp\perp} \perp Y^{\perp\perp} \Rightarrow Z^{\perp\perp} \perp X^{\perp\perp}$$

is valid for all $Z \in E(M)$. Put $Z = Y^{\perp}$. Clearly, $Y^{\perp} \perp Y^{\perp\perp}$ and so $Y^{\perp} \perp X^{\perp\perp}$. But this means that $X^{\perp\perp} \subset Y^{\perp\perp}$. //

The orthologic $(L(M), \subset, \emptyset, H, \perp)$ will be called the <u>ortho-logic affiliated with the manual</u> M.

Example 4.8. As an illustration of a general guiding principle consider the following experiment (see Figure 59) in which a single electron (e) is confined to move along the x-axis through a small slit (A) in a screen (B). After passing through the slit its y-coordinate q_y and its y-component p_y of momentum can be measured by two measuring devices (C), (D).

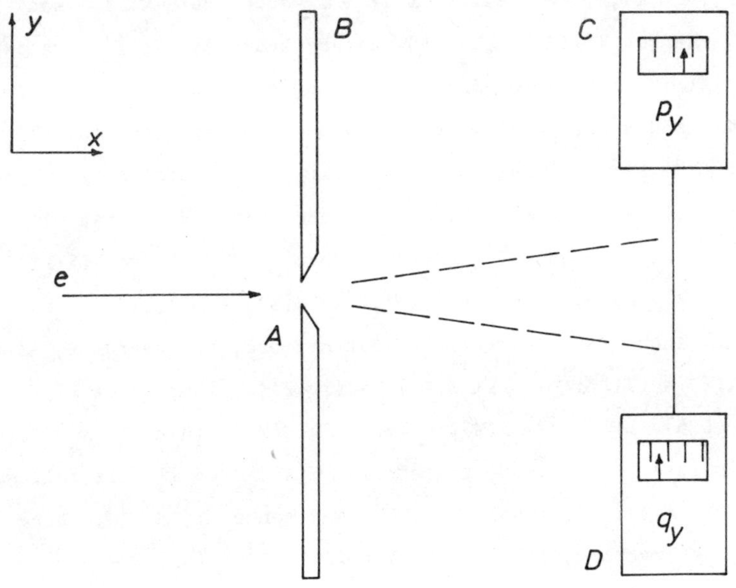

Fig. 59

This experiment can be used to define two physical operations E_1 and E_2 specified below.

Name of the operation:	Description: Measurement at the microlevel establishing that	Symbol of of the outcome:
E_1, E_2	e is not present;	r
E_1	e is present and $q_y > 1$;	a
E_1	e is present and $q_y \in [1/2;1]$;	b
E_1	e is present and $q_y < 1/2$;	c
E_2	e is present and $p_y > 1$;	s
E_2	e is present and $p_y \leqslant 1$.	t

As is customary, identify each operation E_i with its set of outcomes, i.e., write $E_1 = \{a,b,c,r\}$ and $E_2 = \{r,s,t\}$.

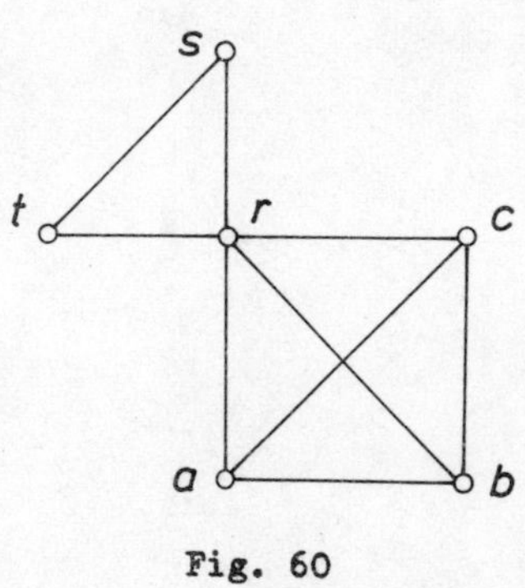

Fig. 60

Let $M = \{E_1, E_2\}$ be a manual consisting of the operations E_1, E_2. The graph of the orthogonality relation $\perp$ defined by M is shown in Figure 60.

Let us now carry out the construction of the orthologic affiliated with the manual $M = \{E_1, E_2\}$.

According to the general procedure, we obtain the elements of the orthologic from the following list of events where we for brevity write abcrst instead of $\{a,b,c,r,s,t\}$ etc.

D	$D^{\perp}$	$D^{\perp\perp}$
∅	H = abcrst	∅
a	bcr	a
b	acr	b
c	abr	c
r	abcst	r
ab	rc	ab
ac	br	ac
ar	bc	ar
bc	ar	bc
br	ac	br
cr	ab	cr
abc	r	abcst

D	$D^{\perp}$	$D^{\perp\perp}$
abr	c	abr
acr	b	acr
bcr	a	bcr
abcr	∅	H
s	rt	s
t	rs	t
rs	t	rs
rt	s	rt
st	r	abcst
rst	∅	H

Hence, as a preliminary remark it may be said that the base set of the orthologic has exactly twenty elements.

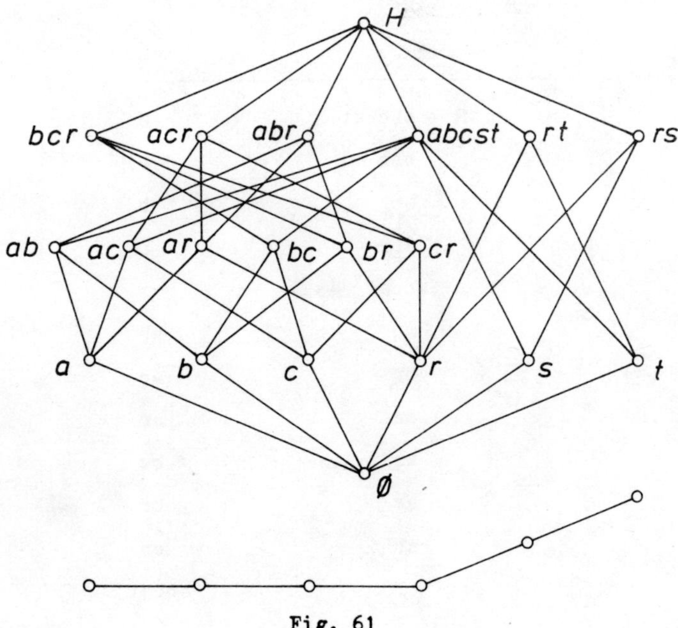

Fig. 61

By means of the $D^{\perp\perp}$'s, the construction of the orthologic is now elementary. The resulting diagram represents an orthomodular lattice which can be obtained as an atomic amalgam of the Boolean algebras $\underline{2}^4$ and $\underline{2}^3$ (see Figure 61 where we below add the corresponding Greechie's diagram and where we - as above - omit the curly brackets).

Some conclusions of the constructed orthologic can be interpreted in this setting physically if we adopt the following definitions (cf. [65] and [66]):

Let $M = \{E_i;\ i \in I\}$ be a manual consisting of nonvoid sets called "operations" and let $H = \bigcup(E_i;\ i \in I)$. By an <u>operational proposition</u> over M is meant any pair (F,G) where F and G are subsets of H. Such an operational proposition (F,G) is said to be <u>testable</u> if and only if there is an operation E of M such that $E \subset F \cup G$.

For example, the operational proposition $(\{a,s\},\{r\})$ is not testable in the considered example.

Notice that the operational propositions of the form $(D^{\perp\perp}, D^{\perp})$ where D is an event are testable.

Indeed, let $D \subset E$. Then $E = D \cup (E \setminus D) \subset D^{\perp\perp} \cup D^{\perp}$.

In this case we write $p(D) = (D^{\perp\perp}, D^{\perp})$. However, since $D^{\perp} = (D^{\perp\perp})^{\perp}$, it is possible to reconstruct $D^{\perp}$ from $D^{\perp\perp}$. Thus $p(D)$ can be identified with $D^{\perp\perp}$. Then we can say that the elements of the constructed orthologic are testable operational propositions and that all the operational propositions which are not testable have been eliminated.

Let G be a simple graph (i.e. undirected and without loops or multiple edges). The set of vertices of the graph G will be denoted by $V(G)$. We shall write $a \sim b$ if and only if the vertices a,b are adjacent. Given a subset W of $V(G)$, we shall use W^* to denote the set $\{v \in V(G); \forall w \in W \ v \sim w\}$, with the convention that we shall write $W^* = w^*$ if $W = \{w\}$. By a clique of the graph G we mean every subset $C \subset V(G)$ which satisfies the following two conditions:

(i) For all $u,v \in C$ either $u = v$ or $u \sim v$.

(ii) If $C \subset D \subset V(G)$ and $C \neq D$, then there are $x,y \in D$ which are not adjacent.

Let $M(G)$ denote the set of all cliques of G.

Theorem 4.9. The set $M(G)$ is a manual.

Proof. We infer from (ii) that the axiom (M 1) holds. Let E_i, E_j be two cliques of G and N an orthogonal subset (in $M(G)$) of $E_i \cup E_j$. Then $a \sim b$ for any two distinct elements $a,b \in N$. It follows from Zorn's lemma that N is contained in a clique of G. //

The orthologic affiliated with the manual $M(G)$ will be denoted by $(L(G), \subset, \emptyset, V(G), \perp)$ and it will be called the orthologic of the graph G.

Since the orthogonality plays a fundamental role in the study of the orthologics, it is not surprizing that one can, beginning with a suitable graph of orthogonality and restricting the attention to the finite case, reconstruct the considered orthologic. The construction proceeds as follows. Let $`P = (P, \leqslant, 0, 1, \perp)$ be an orthologic. Then we define the graph $G(`P)$ of the orthologic $`P$ in the following manner: The vertices are the elements of $V(G(`P)) = P \setminus \{0,1\}$ and two vertices $u,v \in V(G(`P))$ are adjacent if and only if $u \perp v$. The graph constructed in this way is simple and, therefore, we can consider its orthologic $(L(G(`P)), \subset, \emptyset, V(G(`P)), \perp)$.

We recall that two orthologics $(L, \leqslant, 0, 1, \perp)$, $(K, \leqslant, 0, 1, \perp)$ are said to be <u>isomorphic</u> if and only if there exists a bijection f of L onto K such that the following is true: For all $m, n \in L$ the relation $m \perp n$ holds if and only if $f(m) \perp f(n)$.

We are now in a position to state a result in the direction mentioned above.

<u>Theorem 4.10</u>. Let P be a finite set. Then the orthologics $(P, \leqslant, 0, 1, \perp)$ and $(L(G(`P)), \subset, \emptyset, V(G(`P)), \perp)$ are isomorphic.

<u>Proof</u>. Define $f: P \to L(G(`P))$ by $f: p \mapsto p^{\perp\perp}$. Here $0^{\perp} = \{u \in V(G(`P)); u \perp 0\} = V(G(`P))$; $1^{\perp} = \{u \in V(G(`P)); u \perp 1\} = \emptyset$ by Lemma 2. Therefore, $0^{\perp\perp} = V(G(`P))^{\perp} = \emptyset$, $1^{\perp\perp} = \emptyset^{\perp} = V(G(`P))$.

If $p^{\perp\perp} = q^{\perp\perp}$ where $p, q \in P$, then $p^{\perp} = p^{\perp\perp\perp} = q^{\perp\perp\perp} = q^{\perp}$. Thus, if $z \perp q$, then $z \in q^{\perp} = p^{\perp}$ and so $z \perp p$. In view of (L 6) this implies $p \leqslant q$. By symmetry, $q \leqslant p$ so that $p = q$. Hence f is an injection.

Suppose $D^{\perp\perp} \in L(G(`P))$. Then D is an event and, since P is finite, $D = \{d_1, d_2, \ldots, d_n\}$ where $d_i \perp d_j$ for $i \neq j$. By Theorem 3, the supremum $s = \sup_{(P, \leqslant)} \{d_1, d_2, \ldots, d_n\}$ exists. If $c \in D^{\perp}$, then $c \perp d_i$ for all $i = 1, 2, \ldots, n$ by the definition of $D^{\perp}$. From Theorem 3 it follows that $c \perp s$. Conversely, if $d \perp s$, then by (L 6) and $s \geqslant d_i$, we have that $d \perp d_i$. This implies that $d \in D^{\perp}$. Consequently, $D^{\perp} = s^{\perp}$ and it is now clear that $D^{\perp\perp} = s^{\perp\perp}$. Thus f is surjective.

Finally, we have to prove that $x \perp y$ holds if and only if $x^{\perp\perp} \perp y^{\perp\perp}$. However, if $x = 0$, $0^{\perp\perp} = \emptyset$ and the statement is true for every y by Lemma 5. If $x = 1$, then, by Lemma 2, $1 \perp y$ implies that $y = 0$ and the above argument shows that our assertion is valid. Therefore, we may assume that $x, y \in V(G(`P))$. But then, by Lemma 5, $x^{\perp\perp} \perp y^{\perp\perp}$ is equivalent to $\{x\} \subset y^{\perp}$ where

$y^{\perp} = \{c;\ c \perp y \ \&\ c \neq 0, 1\}$. In other words, it is equivalent to $x \perp y$. //

Theorem 4.11. Every finite orthologic is an orthologic of a graph.

Proof. This follows easily from Theorem 10. //

A simple graph G is called a Dacey graph if and only if it has the following property: For every clique C of G and for every pair u,v of $V(G)$ the relation $C \subset u^* \cup v^*$ implies $u \sim v$.

Theorem 4.12. Let $\{e_1, e_2, \ldots, e_n\}$, $n \geq 1$, be a maximal orthogonal subset of an orthomodular poset `P. Suppose that $e_i \neq 0,1$ and let $1 \leq j \leq n$. Then the supremum $s_1 = e_1 \vee e_2 \vee \ldots \vee e_j$ as well as the supremum $s_2 = e_{j+1} \vee \ldots \vee e_n$ exists and $s_2' = s_1$.

Proof. The elements s_1 and s_2 exist by Remark 4. Similarly, the supremum $s_3 = \sup\{e_1, e_2, \ldots, e_n\}$ exists and $s_3 = s_1 \vee s_2$. Suppose first that $s_3 \neq 1$. Then $\{e_1, e_2, \ldots, e_n, s_3'\}$ is an orthogonal subset of $P \setminus \{0,1\}$, in contradiction to the maximality of $\{e_1, e_2, \ldots, e_n\}$. Hence $s_1 \vee s_2 = 1$. If $j = n$, then $s_2 = \sup \emptyset = 0$ and $s_2' = s_1$ as required. Henceforth we assume that $j < n$. We have that $e_{j+1}' \geq e_1, e_2, \ldots, e_j$ so that $e_{j+1}' \geq s_1$. Similarly $e_{j+2}' \geq s_1, \ldots, e_n' \geq s_1$ and, therefore, $s_2' = e_{j+1}' \wedge e_{j+2}' \wedge \ldots \wedge e_n' \geq s_1$. Since, by orthomodularity, $s_2' = s_1 \vee (s_1 \vee s_2)'$ and since $(s_1 \vee s_2)' = 1' = 0$, we get $s_2' = s_1 \vee 0 = s_1$. //

We shall say that an orthologic $(P, \leq, 0, 1, \perp)$ is an orthomodular poset if and only if it is possible to define a unary operation $'$ on P in such a way that $(P, \leq, ', 0, 1)$ is an orthomodular poset and $a \perp b$ (in the orthologic) is true if and only if $a \leq b'$ (in the orthomodular poset).

An easy argument shows that if an orthologic is an orthomodular poset, then the element y of (L 5) is equal to x'. Indeed, we have $x \perp y$ so that $y \leqslant x'$ and, at the same time, $x \vee y = 1$. Hence $x' = x' \wedge (x \vee y) = y$, by orthomodularity.

Theorem 4.13. A finite orthologic $`P = (P, \leqslant, 0, 1, \perp)$ is an orthomodular poset if and only if its graph $G(`P)$ is a Dacey graph.

Proof. 1. Suppose that $`P$ is an orthomodular poset. Let $C = \{e_1, e_2, \ldots, e_n\} \subset P \setminus \{0,1\}$ be a clique in $G(`P)$ and $C \subset \subset u^* \cup v^*$. Here $u^* = (u']$ and $v^* = (v']$. Choose the notation so that $\{e_1, e_2, \ldots, e_j\} \subset (u']$ and $\{e_{j+1}, \ldots, e_n\} \subset (v']$. Let s_1 be as above the supremum $e_1 \vee e_2 \vee \ldots \vee e_j$ and $s_2 = e_{j+1} \vee \ldots \ldots \vee e_n$. Hence $s_1 \leqslant u'$ and $s_2 \leqslant v'$. By Theorem 12, $v \leqslant s_2' = = s_1 \leqslant u'$ and so $u \sim v$ in $G(`P)$.

2. Conversely, suppose that $G(`P)$ is a Dacey graph.

Firstly we shall prove that the element x in (L 5) uniquely determines the element y. Indeed, let v be an element such that $x \perp v$ and $x \vee v = 1$. Let z be an element of P satisfying $z \perp v$. Then $\{x,v\} \subset z^* \cup y^*$. To see that $\{x,v\}$ is a maximal orthogonal set, choose an element c with $c \perp x$ and $c \perp v$. Then, by (L 4), $c \perp (x \vee v)$, i.e. $c \perp 1$. Thus, from Lemma 2, $c = 0$. Since $G(`P)$ is a Dacey graph, $z \perp y$. Consequently, by (L 6), $y \leqslant v$. Symmetrically, $v \leqslant y$ and so $v = y$. From now on, write x' instead of y and consider the unary operation $' : x \mapsto x'$.

We shall now show that $(P, \leqslant, ', 0, 1)$ is an orthocomplemented poset. Since y is uniquely determined, $x'' = x$. From Lemma 2 we see that $0 = 1'$. We wish to show that

(*) $x \leqslant y' \Rightarrow y \perp x$.

In fact, if $x \leq y'$, then $z \perp x$ for every z with $z \perp y'$. Choosing $z = y$ we get $y \perp x$.

However, the converse of (∗) is also true:

(∗∗) $y \perp x \Rightarrow y \leq x'$

To prove (∗∗), suppose that $z \perp x'$ and $y \perp x$. Then $\{x', x\} \subset \subset z^{*} \cup y^{*}$. We shall prove that $\{x', x\}$ is a maximal orthogonal set. If $c \perp x'$ and $c \perp x$, then, by (L 4), $c \perp (x \vee x')$. Hence $c \perp 1$ and so $c = 0$. Since G(`P) is a Dacey graph, $z \perp y$ and it follows from (L 6) that $y \leq x'$.

Assume that $x \leq y$. Then, of course, we have $x \leq (y')'$. Hence, from (∗), $y' \perp x$ and, by (∗∗), $y' \leq x'$.

Combining these results demonstrates that the poset $(P, \leq, ', 0, 1)$ is orthocomplemented.

We conclude this proof by showing that it is even orthomodular.

Suppose that $s \geq t'$ and $t' \vee s' = 1$. Obviously, $\{t', s'\} \subset \subset T^{*} \cup s^{*}$ for every T such that $T \perp t'$. We now assert that $\{t', s'\}$ is a maximal orthogonal set. If $c \perp t'$ and $c \perp s'$, then $c \perp t' \vee s'$ and so $c \perp 1$. Thus $c = 0$. Since G(`P) is a Dacey graph, $T \perp s$. From (L 6) it is seen that $s \leq t'$ and, therefore, $s = t'$. In summary, we have shown that the relations $s \geq t'$ and $s \wedge t = 0$ imply $s = t'$. This proves the theorem. //

EXERCISES

VIII;1 (Varadarajan). Let c and d be elements of an orthomodular poset. Prove that cMd if and only if there exists an element g such that the following conditions are satisfied:

(i) $g \leqslant c$; (ii) $g \leqslant d$; (iii) $d - g \perp c$.

If g has these properties, then $g = c \wedge d$.

VIII;2. Let $`P$ be an orthomodular poset and let $Z(`P) = \{z \in P; \forall p \in P \;\; zMp\}$. Prove that $Z(`P)$ determines a Boolean algebra $(Z(`P), \leqslant, ', 0, 1)$.

VIII;3. An orthocomplemented poset $`P$ is said to be <u>orthogonally complete</u> if and only if the supremum $\bigvee\{a_i;\ i \in I\}$ exists in $`P$ for every orthogonal set $\{a_i;\ i \in I\} \subset P$.

Prove that the Boolean algebra $(Z(`P), \leqslant, ', 0, 1)$ of Exercise VIII;2 is complete in every orthomodular poset $`P$ which is orthogonally complete.

VIII;4. Prove the following assertions concerning an orthogonal subset $\{z_1, z_2, \ldots, z_n\} \subset Z(`P)$ of an orthomodular poset $`P$ such that $z_1 + z_2 + \ldots + z_n = 1$:

(i) Every set $(z_i]$ determines an orthomodular poset $`(z_i] = ((z_i], \leqslant, {}^i, 0, z_i)$ where ${}^i : q \mapsto q' \wedge z_i$.

(ii) $`P = `(z_1] \oplus `(z_2] \oplus \ldots \oplus `(z_n]$.

VIII;5. Let $`P$ be a dimension lattice in the sense of Loomis. Let x and y be elements of P which have a common relative complement in an interval $[0,d]$ where $d \geqslant x \vee y$. Prove that $x \sim y$.

VIII;6. Let $`P$ be a dimension lattice in the sense of Loomis. Let z, Z, a, b be elements of P such that $z + Z = a + b$. Show that $(Z' \vee a') \wedge a \sim (z' \vee b') \wedge z$.

VIII;7. Let x, v and w be elements of a dimension lattice in the sense of Loomis. Show that the relations $x \wedge v = x \wedge w = 0$ and $v \sim w$ imply $x \vee v \sim x \vee w$.

VIII;8. Prove that for any elements z, Z, a, b of a dimension lattice in the sense of Loomis such that $z + Z = a + b$ there exist elements e,f,E,F satisfying $a = e + f$, $b = E + F$, $z \sim e + E$ and $Z \sim f + F$.

VIII;9. Show that in any dimension lattice in the sense of Loomis holds the axiom (D 3).

VIII;10. Let `P be a relatively complemented lattice with 0 and 1. Suppose that $x \wedge y = 0$ for some $x, y \in P$. Prove that there exists a complement z of y such that $x \leqslant z$.

VIII;11. Show that in every dimension lattice in the sense of Loomis holds the axiom (D 4).

VIII;12. Let `P be a finite poset with an equivalence relation $\sim$ which satisfies the axioms (B) and (D 1). Prove that

(i) $(a \leqslant b \,\&\, a \sim b) \Rightarrow a = b$;

(ii) there is no infinite element in `P.

VIII;13. Let $`B_1$ and $`B_2$ be two copies of the Boolean algebra $\underline{2}^3$ with the atoms a,b,c and d,e,f, respectively. Prove that the fourteen-element lattice $`L_{14}$ which arises from $`B_1$ and $`B_2$ when we "paste them together" via the construction of Chapter IV is an example of a dimension poset in the sense of Šerstnev. Show that it is not possible to consider $`L_{14}$ as a dimension lattice in the sense of Loomis.

VIII;14. Prove that the hull $|a|$ of an element a belonging to a dimension poset `P in the sense of Šerstnev has the following properties:

(i) $a \leqslant |a|$;

(ii) if $a \leqslant b$, then $|a| \leqslant |b|$;

(iii) the element a belongs to I(`P) if and only if $a = |a|$;

(iv) $|a| = \sup_{`P}\{x \in P;\ x \lesssim a\}$;

(v) if $a \sim b$, then $|a| = |b|$.

VIII;15. Let $M = \{E_i;\ i \in I\}$ be a manual and $H = \bigcup\{E_i;\ i \in I\}$. Prove that $E_k^{\perp} = \emptyset$, $E_k^{\perp\perp} = H$ and $\emptyset^{\perp\perp} = \emptyset$.

VIII;16. Join new lines to the diagram of the lattice $`L_{14}$ described in Exercise VIII;13 so that, in addition, $a < e'$, $b < f'$, $e < a'$ and $f < b'$ holds in the new ordering. Denote the corresponding new poset by $`P_{14}$.

(i) Show that $`P_{14}$ is not orthomodular.

(ii) Define an orthogonality relation $\perp$ on P_{14} by making $x \perp y$ mean that $x \leq y'$. Prove that $\perp$ makes $`P_{14}$ an ortho-logic.

VIII;17. Every finite orthologic $`P$ is isomorphic to the orthologic affiliated with a manual.

Added in proof. As kindly communicated by T. Katriňák, it is worth pointing out that a definition of a commutator of n elements first appears in the work [38] of G. Bruns and G. Kalmbach and that this paper contains also theorems on the decomposition of an orthomodular lattice on its Boolean and non-Boolean part as well as basic results on the free orthomodular lattice with two generators.

Answers to Exercises

SOLUTIONS TO EXERCISES OF CHAPTER II

1. (a',b), (a',b'), (b,a), (b,a').

2. (i) $\Rightarrow$ (ii). This follows from Theorem 3.1 (iv). (ii) $\Rightarrow$ $\Rightarrow$ (iii). Since $s' \wedge [(s')' \vee (s' \wedge t')] = s' \wedge t'$, De Morgan laws imply (iii). (iii) $\Rightarrow$ (iv). Suppose $s \leqslant t$ and put $r = s' \wedge t$. Then $s \vee r = s \vee (s' \wedge t) = s \vee [s' \wedge (s \vee t)] = s \vee t = t$. (iv) $\Rightarrow$ (v). Apply De Morgan laws. (v) $\Rightarrow$ (i). Let $s \leqslant t'$ and $s \vee t = 1$. Then there exists r such that $s = t' \wedge r \leqslant r$ and $r \geqslant t$. Hence $r \geqslant$ $\geqslant s \vee t = 1$, i.e., $r = 1$. Thus $s = t' \wedge 1 = t'$. The lattice `L is orthomodular by Theorem 3.1 (ii).

3. Suppose (i) is true. Then the lattice is orthomodular and from $s \leqslant t \leqslant u$ it follows that sCt and sCu. By Foulis-Holland Theorem we see that (ii) holds. Hence, (i) implies (ii). This can be proved directly as follows: We always have

$$T' = s \vee (t' \wedge u) \leqslant (s \vee t') \wedge (s \vee u) = (s \vee t') \wedge u = S.$$

In addition,

$$S \wedge T = (s \vee t') \wedge u \wedge [s \vee (t' \wedge u)]' = (s \vee t') \wedge u \wedge s' \wedge (t \vee u').$$

Now $s_1 = u \wedge (t \vee u') \geqslant u \wedge t = t = t_1'$ and $s_1 \wedge t_1 = u \wedge (t \vee u') \wedge$ $\wedge t' = t' \wedge u \wedge (t' \wedge u)' = 0$. Therefore, $s_1 = t_1'$ and so

$$S \wedge T = (s \vee t') \wedge s' \wedge t = (s \vee t') \wedge (s \vee t')' = 0,$$

i.e. $S = T'$.

Conversely, suppose (ii) holds, $s \geqslant t'$ and $s \wedge t = 0$. Since $t' \leqslant t' \leqslant s$, it follows from (ii) that $t' = t' \vee (t \wedge s) = (t' \vee t) \wedge s = s$.

4. (i) $\Leftrightarrow$ (ii). If $a \leqslant b$, then $\overline{\text{med}}\,(a,b,a') = b \wedge (b \vee a') = b$, $\underline{\text{med}}\,(a,b,a') = a \vee (b \wedge a')$. By Theorem 3.1 (iii) these two elements are equal if and only if `L is orthomodular. (iii) $\Rightarrow$ (i). Let (iii) be valid and let $a \leqslant b$. Put $s = a'$, $t = b'$ so that $s \geqslant t$. The considered condition, with $u = a$, yields

$$(b \wedge a') \vee a = [(b \vee a) \wedge a'] \vee (b \wedge a) = (b \vee a) \wedge [a' \vee (b \wedge a)] = b \wedge (a' \vee a) = b.$$

Thus, by Theorem 3.1 (iii), `L is an orthomodular lattice. (i) $\Rightarrow$ (iii). If `L is orthomodular and $s \geqslant t$, then $s \geqslant t \wedge u' = (t' \vee u)'$ and $t' \vee u \geqslant t' \wedge u$. Thus $t' \vee u C s$ and $t' \vee u C t' \wedge u$. Theorem 3.10 implies the equality. (iii) $\Leftrightarrow$ (iv). Obvious.

5. By Theorem 3.10, (i) implies (ii)-(v). (ii) $\Rightarrow$ (i). If (ii) holds and $a \geqslant b = b_1$, $b_2 = a'$, then $a \wedge (b \vee a') = (a \wedge b) \vee (a \wedge a') = b$ and `L is orthomodular. (iii) $\Rightarrow$ (i). If (iii) is true and $b_1 = a' \geqslant b_2$, then $a' \wedge (a \vee b_2) = (a' \wedge a) \vee (a' \wedge b_2) = b_2$ and `L is orthomodular. (iv) $\Rightarrow$ (i). If (iv) is valid and $a \leqslant b_1$, $b_2 = a'$, then $a \vee (b_1 \wedge a') = (a \vee b_1) \wedge (a \vee a') = b_1$ and `L is orthomodular. (v) $\Rightarrow$ (i). From (v), $b_1 = a'$ and $a' \leqslant b_2$ we find that $a' \vee (a \wedge b_2) = (a' \vee a) \wedge (a' \vee b_2) = b_2$. Thus `L is orthomodular.

6. By Theorem 3.7 we know that (i) implies (ii) and (iii). (ii) $\Rightarrow$ (i). If (ii) is true and $a \leqslant b$, then aCb by Theorem 2.3 (iii). Hence $a = a \wedge b = (a \vee b') \wedge b$. Thus `L is orthomodular. (iii) $\Rightarrow$ (i). If (iii) holds and $a \leqslant b$, then $(b \vee a') \wedge a = 1 \wedge a = a = a \wedge b$. Therefore, by assumption, $bCa$, i.e., $b = (b \wedge a) \vee (b \wedge a') = a \vee (a' \wedge b)$ and `L is orthomodular.

7. By Theorems 4.2 and 3.4, (i) implies the conditions (ii)-(v). (iv) ⇒ (i). Suppose (iv) is true. It is immediate that $(a\wedge 0)\vee(a\wedge 0') = 0\vee(a\wedge 1) = a$, i.e., $aC0$. Putting $b_2 = 0$ in (iv), we see that b_1Ca implies aCb_1 for every $a,b_1 \in L$. Hence C is a symmetric relation and, by Theorem 3.4, `L is an orthomodular lattice. Thus (iv) is equivalent to (i). (v) ⇔ (i). This follows similarly. (ii) ⇒ (iii). If aCb_1 and aCb_2, then aCb_1' and aCb_2' by Theorem 2.3. Hence $aCb_1'\vee b_2'$ and, by Theorem 2.3, we find that a commutes with $(b_1'\vee b_2')'$, i.e., $aCb_1\wedge b_2$. (iii) ⇒ (ii). Here we can use a similar argument. However, neither (ii) nor (iii) implies (i). To see this, it suffices to show that (iii) does not imply (i). We shall prove that the condition (iii) is valid in the lattice of Figure 7a which does not satisfy the condition (i). Suppose x,y,z are elements of this lattice which satisfy xCy and xCz, while x does not commute with $y\wedge z$. By Exercise II;1 we have only the following possibilities:

(1) $x = b$ & $y\wedge z = a$; (2) $x = b$ & $y\wedge z = a'$;

(3) $x = a'$ & $y\wedge z = b$; (4) $x = a'$ & $y\wedge z = b'$.

Let us consider in detail the first possibility. From the diagram we conclude that one of the elements y,z is equal to a. But then xCy or xCz is impossible. A similar argument can be applied to the remaining cases. Hence the lattice satisfies (iii) and it is not orthomodular.

8. Set $b = a$ in (BA 3). Then $a = [a\wedge(a\vee c)]\vee(a\wedge a') = a\wedge(a\vee c)$. By Theorem 1.3, `A is an ortholattice. Now put $c = 0$ in (BA 3). As a consequence, we get $a = (a\wedge b)\vee(a\wedge b')$ and so $aCb$ for every $a,b\in A$. Thus, by Corollary 4.6, `A is a Boolean algebra.

9. Let x^* denote the orthocomplement of x in an orthomodular

lattice defined on the considered ~~ortho~~lattice. We shall prove that the orthocomplements x^* are uniquely determined, i.e., $x^* = x'$ for every x. The element c has the following complements: c', b', d', e, a, f', g. Since $c \prec a'$ and a' covers exactly four elements, the orthocomplement of a' must be an element which is covered by exactly four elements and c^* has to be one of these four elements. Hence a is the orthocomplement of a' and $c^* \in \{b', c', d', g\}$. If $c^* = b'$, then $c \vee (c^* \wedge a') = a'$ by orthomodularity. However, $c^* \wedge a' = b' \wedge a' = 0$ and, consequently, $c \vee (c^* \wedge a') = c$, a contradiction. If $c^* = d'$, then we similarly have $c = c \vee (d' \wedge f) = f$, a contradiction. If $c^* = g$, then $c = c \vee (g \wedge f) = f$, a contradiction. Thus $c^* = c'$. Since f is the uniquely determined relative complement of a' in the interval $[c,1]$, we have $f^* = f'$. Similarly, c' has only one relative complement in $[f',1]$, namely e. Hence $e^* = e'$. According to $e \wedge a' = b$, we see that $b^* = e^* \vee (a')^* = e' \vee a = b'$. Since g^* does not belong to $\{a, b, c, d, e, f, a', b', c', d', e', f'\}$, we get $g^* = g'$. Then $d^* = (g \wedge a')^* = g^* \vee (a')^* = g' \vee a = d'$. Note that the considered ortholattice is not orthomodular by Remark 6.7.A.

10. Since a is the uniquely determined complement of a', we have $(a')' = a$. By Theorem 1.2, the De Morgan laws are true in $`L$. Hence $`L$ is an ortholattice. The remainder follows by Theorem 6.13.

11. Denote by x' the uniquely determined complement of x and suppose that $b \leqslant a$. Let r be a relative complement of $a \wedge b'$ in the interval $[0, b']$. Then $r \leqslant b'$, $a \wedge r \leqslant a \wedge b'$ and $a \wedge r \leqslant r$. It follows that $a \wedge r \leqslant r \wedge a \wedge b' = 0$. Furthermore, $a \vee r \geqslant (a \wedge b') \vee r = b'$ and $a \vee r \geqslant b \vee r \geqslant b$. Therefore, $a \vee r \geqslant$

$b' \vee b = 1$, i.e., $a \vee r = 1$. Since `L is uniquely complemented, $r = a'$ and so $a' \leqslant b'$. By Exercise II;10, it is now clear that `L is a Boolean lattice.

12. (i). If $c \leqslant e \leqslant d$ and if e' is a complement of e, then $\bar{e} = d \wedge (c \vee e') = c \vee (d \wedge e')$ is a relative complement of e in $[c,d]$. The statement (ii) is an immediate consequence of (i). The assertion (iii) follows from Exercise II;11.

SOLUTIONS TO EXERCISES OF CHAPTER III

1. If $a = a \dot{\wedge} b = (a \vee b') \wedge b \leqslant b$, then $(b \wedge a') \vee a = b$ by orthomodularity. The implication (ii) $\Rightarrow$ (i) follows dually.

2. Using orthomodularity, we find $(a \dot{\vee} b) \dot{\vee} b = \{[(a \wedge b') \vee b] \wedge \wedge b'\} \vee b = (a \wedge b') \vee b = a \dot{\vee} b$. (ii). This follows from (i) by duality.

3. We have $a \dot{\wedge} b = (a \vee b') \wedge b \leqslant b$. The rest follows dually.

4. If $a \leqslant b$, then $a \dot{\vee} c = (a \wedge c') \vee c \leqslant (b \wedge c') \vee c = b \dot{\vee} c$. A similar argument can be used for the second assertion.

5. Note that $a \dot{\wedge} b = (a \vee b') \wedge b \geqslant a \wedge b \geqslant c$.

6. By orthomodularity, $(a \dot{\vee} b') \dot{\wedge} b = \{[(a \wedge b) \vee b'] \vee b'\} \wedge b = [(a \wedge b) \vee b'] \wedge b = a \wedge b$. The assertion on $a \vee b$ follows by duality.

7. (i). By Exercise III;3, $b \dot{\wedge} a' \leqslant (b \dot{\wedge} a) \vee (b \dot{\wedge} a') = a$. Therefore, $a \geqslant (b \vee a) \wedge a'$ and so $(b \vee a) \wedge a' \leqslant a \wedge a' = 0$. Hence $a = (b \dot{\wedge} a) \vee 0 = b \dot{\wedge} a = (b \vee a') \wedge a$ and it follows that $a \wedge b' = (b \vee a') \wedge a \wedge b' = 0$. Let $s = a'$ and $t = a \vee b$. Then $s \geqslant t'$ and $s \wedge t = b \dot{\wedge} a' = 0$. Since `L is orthomodular, $a' = s = t' = a' \wedge b' \leqslant b'$ which yields $a \geqslant b$. In summary, $s_1 = a \geqslant b = t_1'$ and

$s_1 \wedge t_1 = a \wedge b' = 0$. Thus, by orthomodularity, $a = b$. The necessary condition $a = b$ is evidently also sufficient for the validity of the assertion (i). (ii). Clearly, $b \dot\wedge a' = (b \vee a) \wedge a'$, $b \dot\wedge a = (b \vee a') \wedge a$. Note that a commutes with $b \vee a$. Hence, by Theorem II.4.2, a commutes with $b \dot\wedge a'$. In addition, $b \vee a'$ commutes with $b \vee a$ and a'. Consequently, $b \vee a' C b \dot\wedge a'$. Therefore, again by Theorem II.4.2, $b \dot\wedge a C b \dot\wedge a'$. Then, by Theorem 1.1, $(b \dot\wedge a') \dot\vee (b \dot\wedge a) = (b \dot\wedge a) \dot\vee (b \dot\wedge a')$. Thus the condition of (ii) is identical with the condition of (i). (iii). Here $a = (a \dot\wedge b') \dot\vee (a \dot\wedge b) \geqslant a \dot\wedge b$, by Exercise III;3. Hence (iii) is identical with (ii) of Theorem 1.1.

8. Since $a C b \dot\vee a'$, (ii) is equivalent to (iii) by Theorem 1.1. The rest follows from Exercise III;6 and Theorem 1.1.

9. (i). Write

$$w = a \dot\wedge (a' \dot\vee b) = a \dot\wedge [(a' \wedge b') \vee b] =$$
$$= \{a \vee [(a \vee b) \wedge b']\} \wedge [(a' \wedge b') \vee b].$$

Since $b' C a \vee b$ and $a C a \vee b$, $w = (a \vee b) \wedge (a' \vee b') \wedge [(a' \wedge b') \vee b]$ by Theorem II.3.10. Note that $(a \vee b)' = a' \wedge b'$ and $b \leqslant a \vee b$. By using orthomodularity, we then have $(a \vee b) \wedge [(a' \wedge b') \vee b] = b$. Hence $w = (a \vee b') \wedge b = a \dot\wedge b$ and we see that (i) is always true. (ii). Here we have

$$v = (a' \dot\vee b) \dot\wedge a = \{[(a' \wedge b') \vee b] \vee a'\} \wedge a = [(a' \wedge b') \vee (b \vee a')] \wedge a.$$

However, $a C b \vee a'$ and $a C a' \wedge b'$ and so

$$v = (a' \wedge b' \wedge a) \vee [(b \vee a') \wedge a] = (b \vee a') \wedge a = b \dot\wedge a$$

which means that (ii) is also valid for any a, b.

10. (i) $\Rightarrow$ (ii). Use Exercise III;6. (ii) $\Rightarrow$ (iii) and (iii) $\Rightarrow$ (i). If $a \leqslant b$, then $(a \dot\wedge b') \dot\vee b = \{[(a \vee b) \wedge b'] \wedge b'\} \vee b = b$ and $(b \dot\wedge a') \dot\vee a = \{[(b \vee a) \wedge a'] \wedge a'\} \vee a = (b \wedge a') \vee a$.

11. Let $\underline{c} = \underline{\text{com}}\,(a,b)$, $\bar{c} = \overline{\text{com}}\,(a,b)$, $x = b \vee (a \wedge b') \vee (a' \wedge b') = b \vee \underline{c}$, $y = a \wedge (a' \vee b) \wedge (a' \vee b') = a \wedge \bar{c}$. Note that the elements $\bar{c}, \underline{c}$ commute with a and b. Hence, by Theorem II.3.10, $x \dot{\vee} y = (x \wedge y') \vee y = [(b \vee \underline{c}) \wedge (a \wedge \bar{c})'] \vee (a \wedge \bar{c}) = [(b \vee \underline{c}) \wedge (a' \vee \underline{c})] \vee (a \wedge \bar{c}) = [(b \wedge a') \vee \underline{c}] \vee (a \wedge \bar{c}) = \underline{c} \vee (a \wedge \bar{c}) = (a \vee \underline{c}) \wedge (\underline{c} \vee \bar{c}) = a \vee \underline{c} = a \vee (a' \wedge b) \vee (a' \wedge b')$. Analogously, $y \dot{\wedge} x = [(a \wedge \bar{c}) \vee (b \vee \underline{c})'] \wedge (b \vee \underline{c}) = [(a \wedge \bar{c}) \vee (b' \wedge \bar{c})] \wedge (b \vee \underline{c}) = [(a \vee b') \wedge \bar{c}] \wedge (b \vee \underline{c}) = \bar{c} \wedge (b \vee \underline{c}) = (b \wedge \bar{c}) \vee (\bar{c} \wedge \underline{c}) = b \wedge \bar{c} = b \wedge (a \vee b') \wedge (a' \vee b')$. The interval algebras $`L[y, x \dot{\vee} y]$, $`L[y \dot{\wedge} x, x]$ are orthoisomorphic by Theorem 1.2.

12. First, suppose that $PQ = 0$ and $x \in M$. If $y \in N$, then $\langle x,y \rangle = \langle Px,Qy \rangle$ by Lemma 4.4 (ii). Clearly, $\langle Px,Qy \rangle = \langle x,P^{*}Qy \rangle = \langle x,PQy \rangle = \langle x,o \rangle = 0$ and so $x \in N^{\perp}$.

Conversely, let $M \subset N^{\perp}$, $x \in H$ and $y = Qx$. Then $y \in N = Q(H) \subset M^{\perp}$. By Lemma 4.4 (iii), $PQx = Py = o$.

13. (i). Use Theorem 4.5 (iii) and (ii). (ii). From (i) and from $Q^{\perp} = I - Q$ it follows that $P - PQ = P(I - Q) = (I - Q)P = P - QP$. This is equivalent to $PQ = QP$. (iii). Use (i). (iv). By (i), $P(I - Q) = (I - Q)P$ and so $PQ = QP$.

14. (i). This is immediate from the definition of $\wedge$. (ii). Here $P \wedge Q \leqslant Q \leqslant P^{\perp} \vee Q = (P \wedge Q^{\perp})^{\perp}$ and the result follows from Exercise III;13 (ii). (iii). By (i), $P \wedge Q \leqslant Q$. The assertion is a consequence of Exercise III;13 (i). (iv). Use the fact $P \wedge Q^{\perp} \leqslant Q^{\perp}$ and Exercise III;13 (ii).

15. We have that $B(A + C)x = B(Ax + Cx) = BAx + BCx = ABx + CBx = (A + C)Bx$. In addition, $B(A + C)B(A + C) = BB(A + C)(A + C) = B(A + C)$ and $(B(A + C))^{*} = (A + C)^{*}B^{*} = (A^{*} + C^{*})B^{*} = (A + C)B = B(A + C)$.

16. By Theorem 4.5, $A = AB = BA$ and $C = CD = DC$. Therefore,

$ACBD = CABD = CAD = ACD = AC$ and so $AC \leq BD$ by the same theorem. Similarly, $ACAC = AACC = AC$ and $(AC)^* = C^*A^* = CA = AC$.

17. (i). By Exercise III;14 (i), $P \wedge Q \leq Q$ and $P \wedge Q^{\perp} \leq Q^{\perp}$. Hence, by Exercise III;14 (ii) and Exercise III;16, $(P \wedge Q)(P \wedge Q^{\perp}) \leq QQ^{\perp} = Q(I - Q) = Q - Q^2 = 0$, i.e., $(P \wedge Q)(P \wedge Q^{\perp}) = 0$. (ii). Since, by Exercise III;14 (ii), $(P \wedge Q)(P \wedge Q^{\perp}) = (P \wedge Q^{\perp})(P \wedge Q)$, we can see from Theorem 4.6 that $S = (P \wedge Q) \vee (P \wedge Q^{\perp}) = (P \wedge Q) + (P \wedge Q^{\perp}) - (P \wedge Q)(P \wedge Q^{\perp})$. By (i), this shows that $S = (P \wedge Q) + (P \wedge Q^{\perp})$.

18. First, suppose that $PQ = QP$. By Theorem 4.6, $(P \wedge Q) + (P \wedge Q^{\perp}) = PQ + PQ^{\perp} = PQ + P(I - Q) = PQ + P - PQ = P$.

Conversely, let $P = (P \wedge Q) \vee (P \wedge Q^{\perp})$. By Exercise III;17 (ii), $P = (P \wedge Q) + (P \wedge Q^{\perp})$. Therefore, by Exercise III;14 (iii), $PQ = [(P \wedge Q) + (P \wedge Q^{\perp})]Q = (P \wedge Q)Q + (P \wedge Q^{\perp})Q = Q(P \wedge Q) + Q(P \wedge Q^{\perp}) = Q[(P \wedge Q) + (P \wedge Q^{\perp})] = QP$.

19. Let S,T be two closed subspaces of a Hilbert space such that $S \supset T^{\perp}$ and $S \wedge T = \{o\}$. We shall prove that $S = T^{\perp}$. Let $s \in S$. By Remark 3.12.C and Theorem 3.7, there exist $t \in T$, $w \in T^{\perp}$ such that $s = t + w$. Now $t = s - w$ and $w \in T^{\perp} \subset S$. Hence $t \in S \cap T = S \wedge T = \{o\}$, i.e., $t = o$. Consequently, $s = w \in T^{\perp}$ and so $S \subset T^{\perp}$. Thus $S = T^{\perp}$.

SOLUTIONS TO EXERCISES OF CHAPTER IV

1. The assertion is not true. To see it, it suffices to consider the two-element chain $\underline{2}$ and the three-element chain $\underline{3}$. If we paste together the least elements and the greatest elements of $\underline{2}$ and $\underline{3}$, respectively, we get the corresponding counterexample.

2. Choose $a, b \in S_i^o$ and suppose that $[a] \wedge [b] = [c]$ where $c \in L_j$. Let $m \neq j$. Denote $a_1 = f_{i/m}(a)$, $b_1 = f_{i/m}(b)$. Then $[a_1 \wedge b_1] \leqslant [a_1] = [a]$ and, similarly, $[a_1 \wedge b_1] \leqslant [b]$. Hence $[(a_1 \wedge b_1)^{(m)}] \leqslant [c^{(j)}]$. It follows that there exist $e_o \in S_m^o$, $f_o \in S_j^o$ such that $a_1 \wedge b_1 \leqslant e_o$, $f_{m/j}(e_o) = f_o \leqslant c$. Note that $[e_o] \leqslant [c] \leqslant [a_1]$. Since $\underline{A}$ is convexly pasted, $[c] = [c_1]$ where $c_1 \in S_i^o$. Using $[c_1] \leqslant [a]$, we obtain $c_1 \leqslant a$ and, by symmetry, $c_1 \leqslant b$. Thus $c_1 \leqslant a \wedge b$ (the infimum of $\{a, b\}$ in the lattice $`S_i$) and c_1 is a pasted element. By duality, $c_1 \leqslant$ $\leqslant a \wedge b \leqslant a \vee b \leqslant d_1$ where d_1 is a pasted element of S_i^o. From Lemma 2.3 we conclude that both $a \wedge b$ and $a \vee b$ belong to S_i^o.

3. Suppose $\underline{A}$ is nondistributive. In view of Theorem 2.6 and I.19 it means that $\underline{A}$ contains a diamond $\{[o],[a],[b],[c],[i]\}$ with $[o] < [a] < [i]$.

First assume that $[i]^* \cap [o]^* \neq \emptyset$. By Lemma 2.4, in some S_j there exist elements o_1, a_1, b_1, c_1, i_1 with $[o] = [o_1], [a] = [a_1]$, $\ldots, [i] = [i_1]$. It is clear that these elements form a diamond in $`S_j$, a contradiction.

Next assume that $[i]^* \cap [o]^* = \emptyset$ and write $[a] = [a^{(x)}]$, $[b] = [b^{(y)}]$, $[c] = [c^{(z)}]$. Then, by Lemma 2.2, x, y and z are three different indices of I. However, the same lemma yields $[i] = [i_2^{(s)}] = [i_3^{(t)}] = [i_4^{(u)}]$ with $s \in \{x, y\}$, $t \in \{x, z\}$, $u \in \{y, z\}$. Consequently, at least two indices of s, t, u are distinct and this shows that i is a pasted element, i.e., $[i]^* = I$. Thus, $[i]^* \cap [o]^* = \emptyset$ is contradicted.

4. If $[a] \prec [b]$ and $a < x < b$, then $[a] < [x] < [b]$, a contradiction.

Conversely, suppose that $a \prec b$ and $[a] < [z^{(k)}] < [b]$. Then $[z]^* \cap [a]^* = \emptyset = [z]^* \cap [b]^*$ and there exist a_o, z_o, z_1, b_o

such that $a \leqslant a_o$, $f_{j/k}(a_o) = z_o \leqslant z \leqslant z_1$ and $f_{k/j}(z_1) = b_o \leqslant b$. Since z_o, z_1 are pasted elements, the element z is also pasted. But we have already seen above that this is impossible.

5. Let $[a^{(j)}] \wedge [b^{(k)}] \prec [a^{(j)}]$. We have to prove that $[b] \prec [a] \vee [b]$. Clearly, we can assume that $[a]$ and $[b]$ are incomparable. Let $[d^{(m)}] = [a] \wedge [b]$, $[s^{(n)}] = [a] \vee [b]$. By Lemma 2.2 we may suppose that $m, n \in \{j, k\}$.

Case I: $[a]^{*} \cap [b]^{*} \neq \emptyset$. Then, by Exercise IV;4, $[b] \prec [s]$.

Case II: $[d]^{*} \cap [s]^{*} \neq \emptyset$. By Lemma 2.4, it is apparent that this case implies Case I. Thus, again, $[b] \prec [s]$.

Case III: $[a]^{*} \cap [b]^{*} = \emptyset$ and $[d]^{*} \cap [s]^{*} = \emptyset$. Then $j \neq k$.

Firstly suppose that $m = k$. Since $[d^{(k)}] < [a^{(j)}]$, there are $p_o \in L_k^o$ and $q_o \in L_j^o$ such that $d \leqslant p_o$, $f_{k/j}(p_o) = q_o \leqslant a$. From $[d] \prec [a]$ we conclude that either $[d] = [p_o]$ or $[a] = [q_o]$. Hence we have either Case II or Case I.

Now suppose that $m \neq k$. Then $m = j$ and $n = k$. From $[d^{(j)}] \leqslant [b^{(k)}]$ it follows that there are $d_1 \in L_j^o$ and $d_2 \in L_k^o$ such that $d \leqslant d_1$, $f_{j/k}(d_1) = d_2 \leqslant b$. Let $a_1 = a \vee d_1$. By assumption, Exercise IV;4 and the dual of Lemma 2.2, $[d_1] \prec [a_1] = [a] \vee [d_1] \leqslant [a] \vee [b] = [s]$.

Suppose $[a_1] = [s]$. Then $[d_1] \prec [a_1]$ and $[d_1] \leqslant [b] < [s]$ imply $[d_1] = [b]$. Consequently, $[b]^{*} = I$ and we are in Case I.

If $[a_1] \neq [s]$, then $[d_1] \leqslant [a_1] \wedge [b] < [a_1]$. Hence $[d_1] = [a_1] \wedge [b] \prec [a_1]$. By Case II, $[b] \prec [a_1] \vee [b] = [a] \vee [d_1] \vee [b] = [s]$.

6. (1). The set of the conditions (i),(ii),(iv) from the definition of an orthocomplemented lattice is self-dual. If $z \leqslant a$ and $z \leqslant a'$, then $a' \leqslant z'$ and $a \leqslant z'$. Hence, by (iii), $z' = 1$. Thus $z = 1' = 0$ and we see that the dual of the condition (iii)

is also valid. (2) (A). We shall show that $(s \vee t)'$ has the properties of the infimum of the set $\{s', t'\}$. Since $s \leqslant s \vee t$, $s' \geqslant$ $\geqslant (s \vee t)'$ and, similarly, $t' \geqslant (s \vee t)'$. If $d \leqslant s'$ and $d \leqslant$ $\leqslant t'$, then $d' \geqslant s$ and $d' \geqslant t$. Consequently, $d' \geqslant s \vee t$. This implies that $d \leqslant (s \vee t)'$. (B). This assertion follows from (1) and (A) by duality.

7. (a). If $Y' = x \geqslant y' = X$, then the element $X \vee Y = y' \vee x'$ there exists by (i). By Exercise IV;6 we see that $X' \wedge Y' = x \wedge y$. (b). This is a consequence of (a).

8. The existence of all the least upper bounds and the greatest lower bounds follows easily by (i) and Exercise IV;7. (1) $\Rightarrow$ (2). Let $S = t' \leqslant s = T$. By (1), $s = T = S \vee (S \vee T')' = t' \vee (t' \vee s')'$. In view of Exercise IV;6, $t' \vee (t' \vee s')' = t' \vee (t \wedge s) = t' \vee 0 = t'$. (2) $\Rightarrow$ (3). Clearly, $S = (s \vee t') \wedge u \geqslant s \vee (t' \wedge u) = T'$. Let $s_1 =$ $= u \wedge (t \vee u') \geqslant t = t_1'$. By Exercise IV;6, $s_1 \wedge t_1 = u \wedge (t \vee u') \wedge$ $\wedge t' = (t \vee u') \wedge (t \vee u')' = 0$. Hence, by (2), $s_1 = t_1'$. Therefore,

$$S \wedge T = (s \vee t') \wedge u \wedge s' \wedge (t \vee u') = (s \vee t') \wedge s' \wedge t =$$
$$= (s \vee t') \wedge (s \vee t')' = 0,$$

i.e., $S = T'$. (3) $\Rightarrow$ (4). Put $s = t = S$, $u = T$ and assume that $S \leqslant T$. By (3), $S \vee (S' \wedge T) = (S \vee S') \wedge T = 1 \wedge T = T$. (4) $\Rightarrow$ (1). If $s \leqslant t$, then $t = s \vee (s' \wedge t) = s \vee (s \vee t')'$ by (4) and Exercise IV;6. Since (3) is self-dual, the remainder follows by duality.

9. If N is a subset of the set M which has an even number of elements, then the set $N' = M \setminus N$ has also an even number of elements. Evidently, the conditions (i)-(iv) of the definition of an orthocomplemented poset are satisfied. If $X \perp Y$, then $X \subset Y'$. Moreover, the set $X \cup Y$ has an even number of elements and it has the properties of the supremum of $\{X, Y\}$. If $X \subset Y$, then $Y' \subset X'$ so that $X \vee Y' = X \cup Y'$. Hence, if $X \subset Y$, we have

$$X \vee (X \vee Y')' = X \cup (X \vee Y')' = X \cup (X \cup Y')' = X \cup (X' \cap Y) =$$
$$= (X \cup X') \cap (X \cup Y) = M \cap Y = Y.$$

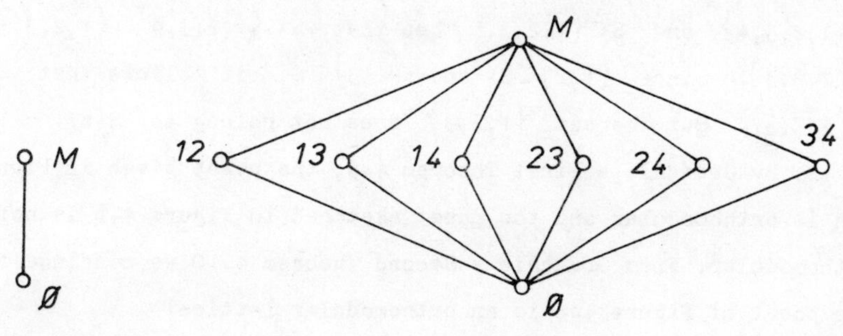

Fig. 62a

Fig. 62b

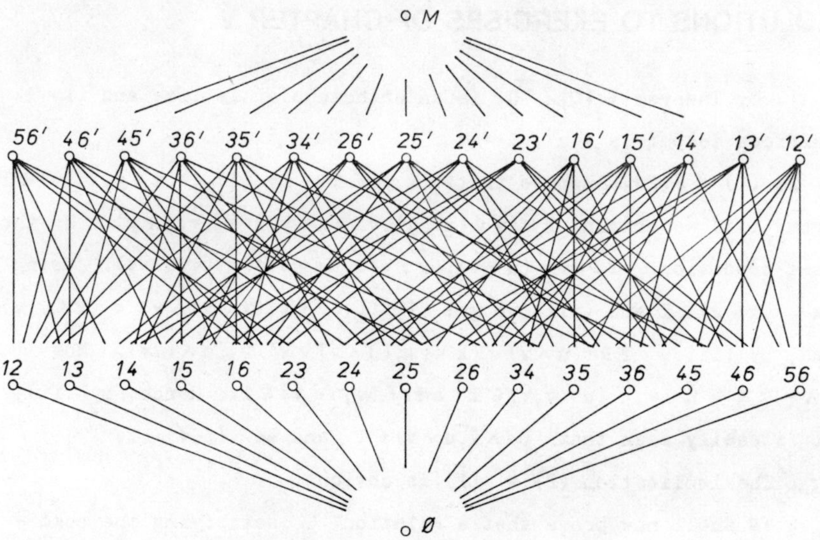

Fig. 62c

10. The diagrams of the posets $\mathcal{S}(M)$ are shown in Figures 62 a–c. Here we write 12 instead of $\{1,2\}$ etc.

11. Use the results of Exercise IV;10 (c): Denote by 1,2,3,4, 5 five distinct elements of M. Suppose the existence of the supremum S of $\{\{1,2\},\{1,3\}\}$ in the poset $`S(M)$. Then $S \subset \subset\{1,2,3,4\}$ and $S \subset \{1,2,3,5\}$ so that $S \subset \{1,2,3,4\} \cap \{1,2,3,5\} = = \{1,2,3\}$. Since $\{1,2\} \subset S$ and $\{1,3\} \subset S$, it follows that $S = = \{1,2,3\}$. But the set $\{1,2,3\}$ does not belong to $S(M)$.

12. By Greechie's First Theorem 4.9, the poset given in Figure 44a is orthomodular and the poset sketched in Figure 44b is not orthomodular. From Greechie's Second Theorem 4.10 we conclude that the poset of Figure 44c is an orthomodular lattice.

SOLUTIONS TO EXERCISES OF CHAPTER V

1. By Theorem 1.10, $`G$ is an orthomodular lattice and it suffices to put $s = u'$, $t = v'$.

2. (1) $\Rightarrow$ (2). Suppose that $(x,y) \in T$. By (i), $(x \wedge y, x \vee y) \in T$. From (iii) we obtain $T \ni (v \vee (x \wedge y), v \vee (x \vee y)) = (v, x \vee y)$. By the same condition, $T \ni (v \wedge (u \vee v), (x \vee y) \wedge (u \vee v)) = (v, u \vee v)$. Using $(x \wedge y, x \vee y) \in T$ and (iii) we get $T \ni (x \wedge y \wedge v, (x \vee y) \wedge v) = (x \wedge y, v)$ and, by (iii), $T \ni ((u \wedge v) \vee (x \wedge y), (u \wedge v) \vee v) = (u \wedge v, v)$. Now $u \wedge v \leqslant v \leqslant u \vee v$, $(u \wedge v, v) \in T$ and $(v, u \vee v) \in T$. Hence, by (ii), it is easily seen that $(u \wedge v, u \vee v) \in T$ and so $(u,v) \in T$.

The implication (2) $\Rightarrow$ (1) is obvious.

We shall now prove that a relation T satisfying the conditions (i)-(iii) of Theorem 3.3 is a congruence relation. If $(x,y) \in T$ and $(y,z) \in T$ (see Figure 63), then $(x \wedge y, y) \in T$ by (2). Therefore, by (iii) of Theorem 3.3, $(x \wedge y \wedge z, y \wedge z) \in T$. From (2) we infer that $(y \wedge z, z) \in T$. In view of (ii) this implies

$(x\wedge y\wedge z, z)\in T$. It follows dually that $(z, x\vee y\vee z)\in T$. Thus by (ii), $(x\wedge y\wedge z, x\vee y\vee z)\in T$. Using (2) and the relations $x\wedge y\wedge z\leqslant x$, $z\leqslant x\vee y\vee z$, we see that $(x,z)\in T$. If $(x,y)\in T$ and $z\in L$, then $(x\wedge y, x\vee y)\in T$ by (i). Hence, by (iii), $(x\wedge y\wedge z, (x\vee y)\wedge z)\in T$. However, $x\wedge y\wedge z\leqslant x\wedge z\leqslant (x\vee y)\wedge z$, $x\wedge y\wedge z\leqslant y\wedge z\leqslant (x\vee y)\wedge z$

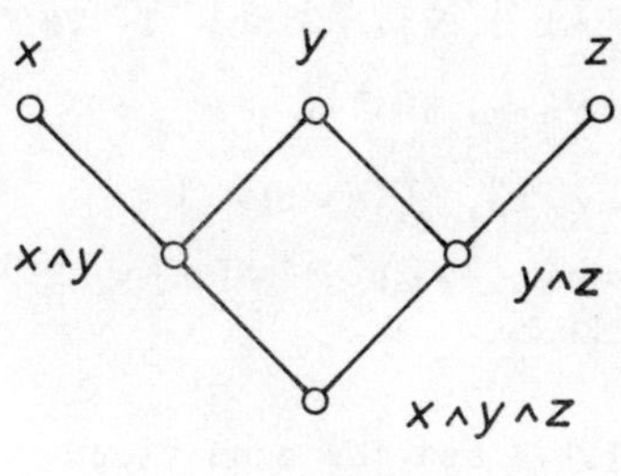

Fig. 63

and (2) imply that $(x\wedge z, y\wedge z)\in T$. The conclusion $(x\vee z, y\vee z)\in T$ follows dually. Clearly, any congruence relation T satisfies the conditions (i)-(iii).

3. Every p-ideal satisfies (3) and (4). Conversely, suppose that (3) and (4) are true. If $i\in I$ and $k\leqslant i$, then $k = i\wedge k = i\mathbin{\dot\wedge} k\in I$.

4. Let I be a p-ideal and $i,j\in I$. Since $i\mathbin{\dot\vee} j = (i\wedge j')\vee j\leqslant i\vee j$ and since I is an ideal, $i\mathbin{\dot\vee} j\in I$. Hence $p({}^{\backprime}L)\subset R({}^{\backprime}L)$. If J is a right ideal and $i\in J$, $j\in J$, then $i\mathbin{\dot\wedge} j'\in J$ by (6). According to $j\in J$, we have $(i\mathbin{\dot\wedge} j')\mathbin{\dot\vee} j\in J$ by (5). From Exercise III;6 we conclude that $i\vee j = (i\mathbin{\dot\wedge} j')\mathbin{\dot\vee} j$, i.e., $i\vee j\in I$. Analogously to the solution of Exercise V;3 we see that $k\leqslant i\in J$ implies $k\in J$. Thus $R({}^{\backprime}L)\subset I({}^{\backprime}L)$. We know that every right ideal satisfies (3). Therefore, by Exercise V;3, $R({}^{\backprime}L)\subset p({}^{\backprime}L)$ and, consequently, $R({}^{\backprime}L) = p({}^{\backprime}L)$. If I is an ideal of the lattice ${}^{\backprime}L$, then $i\wedge j'\in I$ for every $i,j\in I$ so that $i\mathbin{\dot\vee} j = (i\wedge j')\vee j\in I$. If $i\in I$ and $k\in L$, then $k\mathbin{\dot\wedge} i = (k\vee i')\wedge i\leqslant i$. Thus $k\mathbin{\dot\wedge} i\in I$ and so $I({}^{\backprime}L)\subset L({}^{\backprime}L)$.

5. By Remark 3.9.A, we have exactly four congruence relations in this lattice. From Theorem 4.10 we conclude that there are four

p-ideals, namely $I_1 = \{0\}$, $I_2 = (c']$, $I_3 = (c]$, $I_4 = L$.

6. By Theorem II.1.3, $(a \vee b) \vee c = (c' \wedge b')' \vee a$. Since T is a congruence relation of the algebra $(L, \vee, \wedge, ', 0, 1)$,

$$[(a \vee b) \vee c] = [a \vee b] \vee [c] = ([a] \vee [b]) \vee [c], \quad [(a \vee b) \vee c] =$$
$$= [(c' \wedge b')' \vee a] = [(c' \wedge b')'] \vee [a] = ([c'] \wedge [b'])' \vee [a] =$$
$$= ([c]' \wedge [b]')' \vee [a].$$

The postulates (b 2), (b 3) of Theorem II.1.3 and the condition od Theorem II.5.1 can be verified similarly.

Observe that the validity of the assertion stated in the exercise follows also immediately from the fact that orthomodular lattices form an equational class of algebras.

7. The assertion follows from the definition of an upper commutator.

8. We shall first find the set K of Theorem 5.4. By Theorem III.2.11, $\mathrm{com}\,(a,b) = 0$ whenever aCb. Hence $0 \in K$ and we can confine our attention only to the elements x,y which do not commute. Now, the elements e, d, e', d' do not commute with a and b. To determine K, it therefore, by Exercise V;7, suffices to observe that $\mathrm{com}\,(a,d) = c' = \mathrm{com}\,(a,e) = \mathrm{com}\,(b,d) = \mathrm{com}\,(b,e)$. Combining Theorems 5.4 and 5.6 we get $L' = (c']$.

9. By Lemma 5.2, $\overline{\mathrm{com}}_{[0,b]}(c,d) = \overline{\mathrm{com}}_{[0,1]}(c,d) = \mathrm{com}\,(c,d)$ so that the sets K and M defined in Theorems 5.4 and 5.6 are equal.

10. Write $\bar{c} = \overline{\mathrm{com}}\,(x,y)$ and $\underline{c} = \bar{c}'$. (i) $\Rightarrow$ (ii). Since xCy, $x \wedge \bar{c} = y \wedge \bar{c} = 0$ by Theorem III.2.11. (ii) $\Rightarrow$ (iii). By definition and by Theorem II.3.10, $x \wedge \bar{c} = [(x \wedge \bar{c}) \wedge (y \wedge \bar{c})] \vee$ $\vee [(x \wedge \bar{c}) \wedge (y \wedge \bar{c})'] = (x \wedge \bar{c}) \wedge (y' \vee \underline{c}) = x \wedge [(y' \wedge \bar{c}) \vee (\bar{c} \wedge \underline{c})] =$ $= x \wedge y' \wedge \bar{c} = 0$. Hence, again by Theorem II.3.10, $\underline{c} = \underline{c} \vee (x \wedge \bar{c}) =$ $= x \vee \underline{c}$. (iii) $\Rightarrow$ (i). Clearly, $x \leqslant \underline{c}$ and so $1 = x' \vee x \leqslant x' \vee \underline{c} =$

$= x' \vee (x \wedge y) \vee (x \wedge y')$. Put $s = (x \wedge y) \vee (x \wedge y')$ and $t = x'$ so that $s \vee t = 1$. From $s \leqslant t'$ we obtain $s = t'$ by orthomodularity.

SOLUTIONS TO EXERCISES OF CHAPTER VI

1. (a). $`L' = (c']$. (b). $F_2' = (\text{com}\,(x,y)]$.

2. In this lattice we have

$$e/0 \nearrow 1/b \searrow d/a \nearrow 1/c \searrow b/a \nearrow 1/d \searrow c/a \nearrow 1/b \searrow d/a.$$

Consequently, the quotients $1/b$, $1/c$, $1/d$, b/a, c/a and d/a belong to the projective allelomorph $A(`L; \approx)$. Thus the set $A(`L; \approx)$ consists of these six quotients and of all the quotients of the form x/x where x runs through L.

3. Here $e/0 \nearrow 1/b \searrow d/a \searrow_w d/0 \nearrow 1/e$. Hence $A(`L) = A(`L; \approx) \cup \cup \{e/0\}$ (see the solution of Exercise VI;2).

4. We have that $r/p \nearrow^w q/p \approx_w q/r$ and so $r/p \approx_w q/r$. This shows that $r/p \in A(`L)$.

5. (1) Let $`L$ be a modular lattice and let $b/a \nearrow^w d/c$. Suppose $a \leqslant u < v \leqslant b$. Then there exist $x, y \in [c,d]$ such that

(*) $v/u \nearrow y/x$.

Indeed, put $x = u \vee c$ and $y = v \vee c$. Then $c \leqslant x \leqslant b \vee c \leqslant d$ and, similarly, $y \in [c,d]$. By modularity,

$$u \leqslant v \wedge x = v \wedge (u \vee c) = u \vee (v \wedge c) \leqslant u \vee (b \wedge c) = u \vee a = u.$$

Moreover, $v \vee x = v \vee u \vee c = v \vee c = y$. Thus (*) is true.

(2) Let $b/a \approx_w d/c$ and $a \leqslant u < v \leqslant b$. Then there exist x, $y \in [c,d]$ such that $v/u \approx y/x$.

In fact, this follows from (1) and its dual by induction.

(3) Set $b/a = q/p$, $d/c = q/r$, $v/u = r/p$ and use (2). Then $r/p \approx y/x$, $r \leqslant x \leqslant y \leqslant q$ and so $r/p \in A(`L; \approx)$.

6. The possibilities for the elements p,q,r in the lattice of Figure 54 indicates Table 2. From the table we conclude that the mentioned implication holds. In the lattice of Figure 3 we have $i/a \searrow b/o \nearrow i/c$ and so $i/a \approx_w i/c$. However, c/a does not belong to the projective allelomorph.

p:	a	a	a
r:	b	c	d
q:	1	1	1

Table 2

7. All the elements of this lattice are $\hat{C}$-accessible.

8. The only $\hat{B}$-accessible elements are $0, c'$.

9. (i). Consider I.25.B. (ii). By (i) and by the proof of Theorem VI.1.10 where we show that $L/\hat{C}$ is distributive, $`L/Q$ is isomorphic to a sublattice of a distributive lattice and, therefore, it is distributive. (iii). Suppose to the contrary that $b/a \in A(`L)$ and that (a,b) does not belong to Q. Then $[b]/[a] \in A(`L/Q)$ and $[a] \neq [b]$ in the distributive lattice $`L/Q$. This contradicts Theorem 1.3. (iv). This follows from the definition of $\hat{C}$ and from (iii).

10. Certain technicalities which arise from the fact that in general $(h,k)^{P(a)}$ depends on the choice of the element a are needed: (i). It is easily seen that

$$(h,k)^{P(B,c)} = (h,k)^{P(b \vee B,c)} \wedge (B,c),$$

$$(h,k)^{P(b,c)} = (h,k)^{P(b \vee B,c)} \wedge (b,c).$$

Let $(h,k)^{P(b \vee B,c)} = (m,n)$. Then $(x,y) = (m,n) \wedge (B,c)$ and $(u,v) = (m,n) \wedge (b,c)$. Hence $y = n \wedge c = v$.

(ii). Note that we may assume $h_1 \geqslant h$, since otherwise we replace h_1 by $h_1 \vee h$. (Indeed, $(h_1 \vee h,k)^{P(b,c)} = [(h_1,k) \vee \vee (h,k)]^{P(b,c)} = (h_1,k)^{P(b,c)} \wedge (h,k)^{P(b,c)} = (s,t) \wedge (u,v) = (s \wedge u, t \wedge v)$. Our assertion applied to $h \leq h_1 \vee h$ and $h_1 \leq h_1 \vee h$ and the symmetry of the assumptions then give $t = t \wedge v = v$.) By Theorem II.6.2 (i), a relative complement of the element (h,k) in the interval $[(0,0),(h_1,k)]$ can be expressed in the form $(h,k)^{P(b,c)} \wedge (h_1,k)$. Therefore,

$$(h_1,k) = (h,k) \vee [(h,k)^{P(b,c)} \wedge (h_1,k)],$$

i.e.,

$$(h_1,k)^{P(b,c)} = (h,k)^{P(b,c)} \wedge [(h,k) \vee (h_1,k)^{P(b,c)}].$$

It follows that $(s,t) = (u,v) \wedge [(h,k) \vee (s,t)]$ and this gives $t = v \wedge (k \vee t)$. However, (s,t) is a relative complement of (h_1,k) in $[(0,0),(b,c)]$ and so $k \vee t = c$. Since $v \leq c$, $t = v \wedge c = v$.

(iii). Write

$$(h,k)^{P(b,c)} = (u,v), \quad (h_1,k)^{P(B,c)} = (w,z),$$

$$(h_1,k)^{P(b,c)} = (e,f)$$

where $(h,k) \leq (b,c)$, $(h_1,k) \leq (B,c)$, $(h_1,k) \leq (b,c)$. We have to show that $v = z$. However, by (i), $z = f$ and by (ii), $v = f$. Thus $k^{R(c)}$ is well defined. A similar argument proves that also $h^{Q(b)}$ is well defined.

(iv). Obviously, `H is a lattice with the least element which we denote by 0.

First, note that $(0,0)^{P(0,0)} \leq (0,0)$ by Lemma V.1.3 (i). Hence $(0,0) = (0^{Q(0)}, 0^{R(0)})$ and so $0^{Q(0)} = 0$, $0^{R(0)} = 0$.

Further, let $b \in H$. We shall prove that $([0,b], \vee, \wedge, {}^{Q(b)}, 0, b)$ is an orthomodular lattice. To do this, apply Theorem II. 5.2. To verify (OM 1), let $x,y \in [0,b]$. We wish to show that

$x = x \vee (y \wedge y^{Q(b)})$. To prove this statement, observe that the condition (OM 1) applied to $`G[(0,0),(b,0)]$ implies that

$$(x,0) = (x,0) \vee [(y,0) \wedge (y,0)^{P(b,0)}] = (x,0) \vee [(y,0) \wedge$$
$$\wedge (y^{Q(b)}, 0^{R(0)})] = (x \vee (y \wedge y^{Q(b)}), 0).$$

This establishes the statement. An analogous argument can be used to verify the conditions (OM 2) and (OM 3).

It remains to prove that $x^{Q(m)} = x^{Q(n)} \wedge m$ whenever $x \leqslant m \leqslant n$ are elements of H. By (G 2) of Section V.1, it is evident that

$$(x^{Q(m)}, 0) = (x^{Q(m)}, 0^{R(0)}) = (x,0)^{P(m,0)} = (x,0)^{P(n,0)} \wedge$$
$$\wedge (m,0) = (x^{Q(n)}, 0) \wedge (m,0) = (x^{Q(n)} \wedge m, 0).$$

Therefore, the second condition (G 2) in the definition of a generalized orthomodular lattice is also true for `H.

Thus `H is a generalized orthomodular lattice. In the same way one proves that `K is a generalized orthomodular lattice.

11. Evidently, $G' = \{0\}$ if and only if $\overline{\mathrm{com}}_{[0,b]}(c,d) = 0$ for every $b \in G$ and every $c,d \in [0,b]$. By Theorem III.2.11 and Theorem II.4.5 this holds if and only if every interval $[0,b]$, $b \in G$, is a base set of a distributive lattice. But this is equivalent to the distributivity of $(G, \vee, \wedge)$.

12. First, suppose `G is solvable. Then $`G^{(n)} = \langle 0 \rangle \in \underline{D}$ and, therefore, it is solvable in the class of distributive lattices.

Conversely, let `G be solvable in the class $\underline{D}$. Then there exists $n \in \underline{N}_0$ such that $`G^{(n)} \in \underline{D}$ and so $`G^{(n+1)} = \langle 0 \rangle$, by Exercise VI;11.

13. If `G is distributive, then `G′ - as a sublattice - is also distributive.

We shall prove that if `G is nondistributive, then `G′

is also nondistributive. By Theorem III.2.11 and Theorem II.4.5, there exists $b \in G$ such that $[0,b]$ contains at least two elements c,d with $k = \mathrm{com}_{[0,b]}(c,d) \neq 0$. By Corollary III.2.4, $(k] \subset G'$ contains a sublattice isomorphic to $`M_{O2}$. Consequently, G' is nondistributive.

Another proof of the assertion stated in Exercise VI;13 can be found in [147; Thm 9, p. 36] (see Exercise VI;12) and in [11].

14. If $`G$ is modular, then its sublattice $`G'$ is also modular.

We shall show that if $`G$ is nonmodular, then $`G'$ is also nonmodular. By Theorem II.6.8, there exists u such that $`G$ contains as a sublattice a pentagon $\{0,a,b,c,u\}$ with $0 < a < b < u$. We shall write $x'$ to denote the orthocomplement $x^{P(u)}$ of $x \in [0,u]$ in the orthomodular lattice $`G[0,u]$. We have

$$g = \mathrm{com}_{[0,u]}(a,c) = (a \vee c) \wedge (a \vee c') \wedge (a' \vee c) \wedge (a' \vee c') = $$
$$= u \wedge (a \vee c') \wedge (a' \vee c) \wedge (a \wedge c)' = (a \vee c') \wedge (a' \vee c)$$

and, similarly, $h = \mathrm{com}_{[0,u]}(b,c) = (b \vee c') \wedge (b' \vee c)$. Note that $b' \leqslant a' \vee c$ and that $a \leqslant b \vee c'$. By Theorem II.4.4 it follows that (i) $a' \vee c$ as well as (ii) $b \vee c'$ and (iii) $d = (a' \vee c) \wedge \wedge (b \vee c')$ commute with every element of the subalgebra $`S = = \langle a,b,c \rangle$ of $`G[0,u]$ generated by a,b,c. Clearly, $g = (a \vee \vee c') \wedge d$ and $h = (b' \vee c) \wedge d$. Hence, by (iii) and Theorem II. 3.10, $g \vee h = [(a \vee c') \vee (b' \vee c)] \wedge d = d \in G'$. Now we shall transform the original pentagon into a smaller one: According to (iii) and Theorem II.3.10, $f: x \mapsto x \wedge d$ defines a lattice-homomorphism of $`S$ into $`G[0,d]$. We claim that $f(a) \neq f(b)$. Suppose to the contrary that $a \wedge (a' \vee c) = a \wedge d = b \wedge d = b \wedge (a' \vee c)$. Then, by orthomodularity,

$$a = (a \wedge c') \vee [a \wedge (a' \vee c)] = (a \wedge c') \vee [b \wedge (a' \vee c)] = b,$$

a contradiction. Hence the elements $0 = f(0)$, $f(a)$, $f(b)$, $f(c)$, $d = f(u)$ form a pentagon in $`G'$. Thus $`G'$ is nonmodular.

Note. Using the more general approach explained in Chapter VII, we easily find that $\overline{\mathrm{com}}\,(a,b,c) = (b \vee c') \wedge (a' \vee c) = d$ and so we immediately have $d \in G'$ (see Corollary VII.2.11).

SOLUTIONS TO EXERCISES OF CHAPTER VII

1. Here $C(`L) = \{0,1,c,c'\}$.

2. The first upper commutator is the meet of all the elements of the form

$$u = (a_1^{e_1} \vee a_2^{e_2} \vee \dots \vee a_n^{e_n} \vee a_{n+1}) \wedge (a_1^{e_1} \vee a_2^{e_2} \vee \dots \vee a_n^{e_n} \vee a_{n+1}')$$

where $e_i \in \{-1,1\}$. By Remark II.4.3, the element a_{n+1} commutes with the elements $a_1^{e_1} \vee a_2^{e_2} \vee \dots \vee a_n^{e_n}$. Hence, by Theorem II.3.10,

$$u = (a_1^{e_1} \vee a_2^{e_2} \vee \dots \vee a_n^{e_n}) \wedge (a_{n+1} \vee a_{n+1}') = a_1^{e_1} \vee a_2^{e_2} \vee \dots \vee a_n^{e_n}.$$

Observe that these elements form the second upper commutator.

3. If $(u,x) \in \hat{B}$ where $u = (x \wedge y) \vee (x \wedge y')$, then $(u,x) \in \hat{C}$ by Corollary 1.9. Hence, $(u,x) \in \hat{B} \cap \hat{C} = \Delta$ and this yields $u = x$, i.e., xCy.

If xCy, then $(u,x) = (x,x) \in \Delta \subset \hat{B}$.

4. Let x,y be elements with the indicated property. By Theorem 1.4, $\overline{\mathrm{com}}_{[x \wedge y, x \vee y]}(x,y) = x \vee y \neq x \wedge y$ and so $\overline{\mathrm{com}}\,(x,y) \neq$ $\neq 0$. In view of Theorem III.2.11 we see that x,y do not commute.

The elements x,y of the free orthomodular lattice $`F_2$ do not commute while there is no sublattice of $`F_2$ having the mentioned property.

5. If c is a common complement for a,b, put $t = c$ in the

definition of perspective elements (see Remark 1.15).

Suppose t is an element such that $a \vee t = b \vee t$ and $a \wedge t = b \wedge t = 0$. Let r be a relative complement of the element $a \vee t$ in the interval $[t,1]$ (see Figure 64 which shows one of the possible configurations).

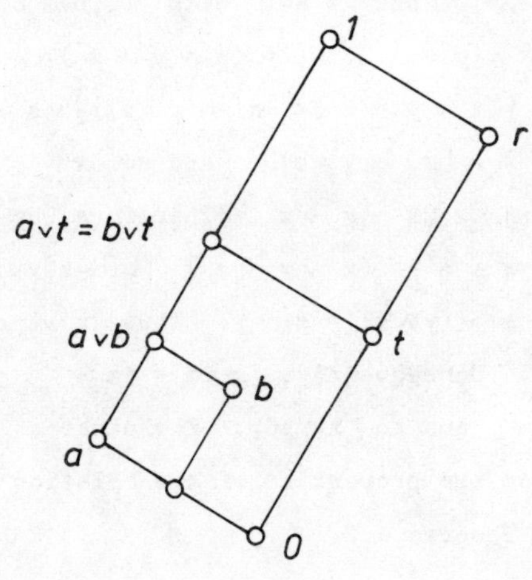

Fig. 64

It is straightforward that $a \vee r \geqslant a \vee t$ and so $a \vee r \geqslant r \vee (a \vee t) = 1$, i.e., $1 = a \vee r$. On the other hand, $a \wedge r \leqslant (a \vee t) \wedge r = t$ and, therefore, $a \wedge r \leqslant a \wedge t = 0$. Thus $a \wedge r = 0$ and it is clear that r is a complement of a and b.

6. By Remark 1.15.A, it suffices to show that $a \underset{p}{\sim} b$ implies $a \underset{sp}{\sim} b$. Using the notation of the definition of $a \underset{p}{\sim} b$, putting $s = (a \vee b) \wedge t$ and applying modularity, we get $a \vee s = a \vee [(a \vee b) \wedge t] = (a \vee b) \wedge (a \vee t)$. Since $a \vee b \leqslant a \vee t$, $a \vee s = a \vee b$. In addition, $a \wedge s \leqslant a \wedge t = 0$, i.e., $a \wedge s = 0$. By symmetry it follows that s is a common relative complement for the elements a,b in the interval $[0,a \vee b]$. Thus $a \underset{sp}{\sim} b$.

7. If `L is distributive, then $a_i C a_j$ for every i,j. By Exercise VII;2 and by Theorem III.2.11,

$$\mathrm{com}\,(a_1,\dots,a_n) = \mathrm{com}\,(a_1,\dots,a_{n-1}) = \dots = \mathrm{com}\,(a_1,a_2) = 0.$$

Conversely, if the indicated condition is true, then it suffices to use Theorems III.2.11 and II.4.5.

8. Similarly as in the proof of Corollary 2.13, (i) $\Rightarrow$ (ii).

Put $a = x$, $b = y$, $x_1 = x$, $x_2 = y$ in the free orthomodular lattice $`F_2$. Then (ii) holds, while (i) is not valid.

9. If $(a,b) \in \hat{B}$, then $(a \wedge b, a \vee b) \in \hat{B}$. Hence, by Corollary 1.9, $((a \wedge b) \vee (a \wedge b'), a) \in \hat{B} \cap \hat{C} = \Delta$. Thus (i) implies (ii). The converse does not hold: Choose $a = 0$ and $b \neq 0$ with $(0,b) \in \hat{C}$.

10. Since $x = [x \wedge (y \wedge z)] \vee [x \wedge (y \wedge z)']$, $x' = (x' \vee y' \vee z') \wedge \wedge [x' \vee (y \wedge z)]$. Hence $x' \vee z' = \{(x' \vee y' \vee z') \wedge [x' \vee (y \wedge z)]\} \vee z'$. However, $x' \vee y' \vee z' \geqslant x \wedge (y' \vee z') = [x' \vee (y \wedge z)]'$ and so $x' \vee \vee (y \wedge z) C x' \vee y' \vee z'$. In addition, $z' C x' \vee y' \vee z'$. Therefore, by Foulis-Holland Theorem II.3.10, $x' \vee z' = (x' \vee y' \vee z') \wedge [x' \vee z' \vee \vee (y \wedge z)]$. From II.3.10 we get $z' \vee (y \wedge z) = z' \vee y$. Thus $x' \vee \vee z' = (x' \vee z' \vee y') \wedge (x' \vee z' \vee y)$. Consequently, $x \wedge z = (x' \vee \vee z')' = [(x \wedge z) \wedge y] \vee [(x \wedge z) \wedge y']$ and so $x \wedge z C y$. The other assertions follow immediately from the properties of the relation C, similarly as in the proof of Theorem 4.2.

11. (a). We shall prove that the element $d = (\bigvee a_i) \wedge c$ has the properties of the supremum of the set $\{a_i \wedge c;\ i \in I\}$. First note that $\bigvee a_i \geqslant a_j$ for every $j \in I$. Hence $d \geqslant a_j \wedge c$. Let $t \in L$ be an element such that $a_i \wedge c \leqslant t$ for every $i \in I$. It has to be shown that $d \leqslant t$. From $a_i \wedge c \leqslant t$ we conclude that $a_i \wedge c \leqslant t \wedge a_i$. Now $c C c'$ and $c C a_i$. Therefore, by Theorem II.3.10,

$$c' \vee (t \wedge a_i) \geqslant c' \vee (a_i \wedge c) = c' \vee a_i \geqslant a_i .$$

This means that $c' \vee (t \wedge a_i) \geqslant \bigvee a_i$, i.e., $d = (\bigvee a_i) \wedge c \leqslant [c' \vee \vee (t \wedge a_i)] \wedge c$. Since $a_i \wedge c \leqslant t$, $t C a_i \wedge c$. Using $a_i C c$ and Theorem 4.2, we see that $t \wedge a_i C c$. This together with $c C c'$ and Theorem II.3.10 implies that $d \leqslant [c' \vee (t \wedge a_i)] \wedge c = (t \wedge a_i) \wedge \wedge c \leqslant t$.

(b). Since $c C a_i$, we have $c' C a_i$. Hence, by (a),

$$[(\bigvee a_i)\wedge c]\vee[(\bigvee a_i)\wedge c'] = \bigvee(a_i\wedge c)\vee\bigvee(a_i\wedge c') =$$
$$= \bigvee[(a_i\wedge c)\vee(a_i\wedge c')] = \bigvee a_i,$$

that is, $\bigvee a_i Cc$. The remaining assertions follow by duality.

12. (i). Note that $c, c', a\vee b'$ and $a\vee b$ belong to the center of the subalgebra $\langle a,b,c\rangle$ generated by the elements a,b,c. Hence Foulis-Holland Theorem II.3.10 and Theorem III.1.1 yield

$$L = a\dot{\wedge}(b\dot{\vee}c) = a\dot{\wedge}(b\vee c) = [a\vee(b\vee c)']\wedge(b\vee c) =$$
$$= [a\vee(b'\wedge c')]\wedge(b\vee c) = (a\vee b')\wedge(a\vee c')\wedge(b\vee c) =$$
$$= (a\vee b')\wedge\{[(a\vee c')\wedge b]\vee[(a\vee c')\wedge c]\} = (a\vee b')\wedge[(a\wedge b)\vee$$
$$\vee(b\wedge c')\vee(a\wedge c)] = (a\wedge b)\vee[(a\vee b')\wedge b\wedge c']\vee(a\wedge c).$$

On the other hand,

$$R = (a\dot{\wedge}b)\dot{\vee}(a\dot{\wedge}c) = (a\dot{\wedge}b)\dot{\vee}(a\wedge c) = \{[(a\vee b')\wedge b]\wedge(a\wedge c)'\}\vee$$
$$\vee(a\wedge c) = [(a\vee b')\wedge b\wedge(a'\vee c')]\vee(a\wedge c) = (a\vee b')\wedge$$
$$\wedge\{[b\wedge(a'\vee c')]\vee(a\wedge c)\} = (a\vee b')\wedge\{[b\wedge(a'\vee c')]\vee a\}\wedge$$
$$\wedge\{[b\wedge(a'\vee c')]\vee c\} = (a\vee b')\wedge[(b\wedge a')\vee(b\wedge c')\vee a]\wedge(b\vee c) =$$
$$= ([(a\vee b')\wedge(b\wedge a')]\vee\{(a\vee b')\wedge[(b\wedge c')\vee a]\})\wedge(b\vee c) =$$
$$= (a\vee b')\wedge[a\vee(b\wedge c')]\wedge(b\vee c) = (a\vee b')\wedge(a\vee b)\wedge(a\vee c')\wedge$$
$$\wedge(b\vee c) = (a\vee b')\wedge(a\vee b)\wedge\{[(a\vee c')\wedge b]\vee[(a\vee c')\wedge c]\} =$$
$$= (a\vee b')\wedge(a\vee b)\wedge[(a\wedge b)\vee(b\wedge c')\vee(a\wedge c)] =$$
$$= (a\wedge b)\vee[(a\vee b')\wedge b\wedge c']\vee(a\wedge c) = L.$$

(iii). Observe that $cC(b\vee a')\wedge a$ and $aC(b\vee a')\wedge a$. Therefore, $c\wedge a$ commutes with $b\dot{\wedge}a$. From Theorem III.1.1 we conclude that $c\dot{\wedge}a$ commutes with $b\dot{\wedge}a$. By Theorem III.1.1 and Theorem 7.3,

$$(b\dot{\vee}c)\dot{\wedge}a = (c\dot{\vee}b)\dot{\wedge}a = (c\dot{\wedge}a)\dot{\vee}(b\dot{\wedge}a) = (b\dot{\wedge}a)\dot{\vee}(c\dot{\wedge}a).$$

The statements (ii) and (iv) follow by duality.

13. Case I: aCb and aCc. Then the first equality holds by Theorem 7.1. Note that $bCb\wedge c$ and $aCb\wedge c$. Consequently, $a\wedge bCb\wedge c$. Applying Theorem 7.1, we get the second equality.

Case II: bCa and bCc. It follows from Theorem 7.6 that (i) is true if and only if $cCa \wedge b$. Because of Theorem 4.2, this is equivalent to $c \wedge bCa \wedge b$. The statement (ii) is a consequence of Theorem 7.3.

Case III: cCa and cCb. Then (i) and (ii) hold by Exercise VII;12. Moreover, in this case we have $a \wedge bCc$ and $a \wedge bCb$. Thus $a \wedge bCb \wedge c$.

14. If x commutes with y, then $x = (x \wedge y) \vee (x \wedge y')$ by definition. From Theorem II.3.6 we can see that $x = (x \vee y) \wedge (x \vee y')$ and so $I_1 = \{x\}$. A similar argument using Theorems II.2.3, II.3.4 and II.3.5 shows that $I_2 = \{y\}$, $I_3 = \{y'\}$, $I_4 = \{x'\}$.

The converse follows from Theorem 1.7.

15. Every orthomodular lattice satisfies the indicated implication by Theorem 4.2.

Suppose an ortholattice is not orthomodular. By Theorem II.5.4, it contains a benzene ring $\{0,s,t,s',t',1\}$. Putting $a = s$, $b = t$, $c = 1$, one obtains $b \wedge c = t$. Since sCt and tC1, we have $aCb \wedge c$ and bCc. However, $(t \wedge s) \vee (t \wedge s') = s \vee 0 = s \neq t$ and, consequently, $b = t$ does not commute with $a \wedge c = s$.

16. Put $a = (x' \vee y') \wedge [x \vee (x' \wedge y)]$ and $b = (x' \vee y') \wedge [y \vee (x \wedge y')]$. By Remark III.2.2, $x' \vee y'$, $x' \wedge y$ and $x \wedge y'$ belong to the center of the subalgebra $\langle x,y \rangle$. Hence, by Theorem VII.4.2 and Theorem II.3.10,

$$aCb \Leftrightarrow (x' \vee y') \wedge aCy \vee (x \wedge y') \Leftrightarrow aCy \vee (x \wedge y') \Leftrightarrow$$
$$\Leftrightarrow (x \wedge y') \vee \{(x' \vee y') \wedge [x \vee (x' \wedge y)]\}Cy \Leftrightarrow (x' \vee y') \wedge [x \vee (x' \wedge y)]Cy \Leftrightarrow (x' \vee y') \wedge [x \vee (x' \wedge y)]Cy' \Leftrightarrow$$
$$\Leftrightarrow [x \vee (x' \wedge y)]Cy' \wedge (x' \vee y') \Leftrightarrow x \vee (x' \wedge y)Cy' \Leftrightarrow$$
$$\Leftrightarrow x \vee (x' \wedge y)Cy \Leftrightarrow xCy \vee (x' \wedge y) \Leftrightarrow xCy.$$

17. Let $a \in L$ and let s be standard. By assumption, $a \wedge (s \vee a') = (a \wedge s) \vee (a \wedge a') = a \wedge s$. Hence, by Theorem II.3.7, sCa.

Conversely, if $s \in C({}^{\backprime}L)$, then use Theorem II.3.10.

18. Use Corollary VII.1.9 and Duality Principle.

SOLUTIONS TO EXERCISES OF CHAPTER VIII

1. First, suppose that cMd. By Theorem 2.12, cCd and, therefore, the element $g = c \wedge d$ exists. Now cMd implies $c'Md$ and so $d - g = d \wedge (c' \vee d') \leqslant c'$ by Theorem 2.12. Thus $d - g \perp c$.

Conversely, let g be an element satisfying (i)-(iii). Then $c = g \vee (c \wedge g')$, by (i) and by orthomodularity. Similarly, $g \vee (d \wedge g') = d$. Since, by (iii), $d \wedge g' \leqslant c' \leqslant c' \vee g$, the elements g, $c \wedge g'$, $d \wedge g'$ are mutually orthogonal. Thus cMd and, consequently, the element $c \wedge d$ exists. Moreover, $t' = g \leqslant c \wedge d = s$. We have $s \wedge t = (c \wedge d) \wedge g' \leqslant c \leqslant (d - g)' = (d \wedge g')'$. But $s \wedge t \leqslant d \wedge g'$. Hence $s \wedge t = 0$ and, by orthomodularity, $c \wedge d = s = t' = g$.

2. By Remark 2.5, 0 and 1 belong to $Z({}^{\backprime}P)$. If $z \in Z({}^{\backprime}P)$, then $z' \in Z({}^{\backprime}P)$ by Theorem 2.8. Suppose that $c, d \in Z({}^{\backprime}P)$ and $a \in P$. Because of Theorem 2.8, the element $c \wedge d$ exists. Now, referring to the same theorem and to $d \in Z({}^{\backprime}P)$, we conclude that also $g = (a \wedge c) \wedge d$ exists. Since $g \leqslant a = C$ and $g \leqslant c \wedge d = D$, the first two conditions of Exercise VIII;1 are verified. Thus in order to prove that $c \wedge d \in Z({}^{\backprime}P)$ we need only show that $w = D - g = D \wedge g' \leqslant C' = a'$. However,

$$w = (c \wedge d) \wedge [(a \wedge c) \wedge d]' = (c \wedge d) \wedge [(a' \vee c') \vee d'] = \\ = c \wedge \{d \wedge [(a' \vee c') \vee d']\}.$$

From $dMa' \vee c'$ and Theorem 2.12 it follows that $d \wedge [(a' \vee c') \vee \vee d'] = (a' \vee c') \wedge d$. Similarly, $c \wedge (a' \vee c') = c \wedge a'$. Hence

$$w = c \wedge [d \wedge (a' \vee c')] = d \wedge [c \wedge (a' \vee c')] = d \wedge (c \wedge a') \leqslant a'.$$

Next, if $c, d \in Z(`P)$, then $c', d' \in Z(`P)$ and so $c' \wedge d' \in Z(`P)$. Therefore, $c \vee d = (c' \wedge d')' \in Z(`P)$. Finally, from Theorem 2.13 we can see that $Z(`P)$ determines a Boolean algebra.

3. Let $\emptyset \neq H \subset Z(`P)$. We shall prove that $\sup_{(P, \leqslant)} H$ exists. Let M denote the set of all the orthogonal sets $\underline{a} = \{a_i;\ i \in \in I\} \subset P$ such that there exists a subset $P(\underline{a})$ of H with

$$s(\underline{a}) = \sup_{(P, \leqslant)} \{a_i;\ i \in I\} = \sup_{(P, \leqslant)} P(\underline{a}).$$

Given $\underline{a} \in M$, let $Q(\underline{a})$ denote the union of all the sets $P(\underline{a})$. Then $s(\underline{a}) = \sup_{(P, \leqslant)} Q(\underline{a})$. Note that $M \neq \emptyset$. Indeed, if $h \in H$, then $\{h, 0\} \in M$. It is easy to see that the conditions for Zorn's lemma are satisfied in $(M, \subset)$. It follows that there exists a maximal orthogonal subset $\underline{c} = \{c_k;\ k \in K\} \in M$. Suppose $s_0 = = \sup_{(P, \leqslant)} Q(\underline{c})$ is not the supremum of H. Then there is $h_0 \in H$ such that $s_0 \geqslant h_0$ does not hold. Since $h_0 \in Z(`P)$, $h_0 M s_0$. By Theorem 2.12, $h_0 = (h_0 \wedge s_0) \vee (h_0 \wedge s_0')$. Now $\underline{d} = \underline{c} \cup \{h_0 \wedge s_0'\} \in M$. In fact, $x \leqslant s_0 \leqslant (h_0 \wedge s_0')'$ for all $x \in \underline{c}$. If $h_0 \wedge s_0' \in \underline{c}$, $h \wedge \wedge s_0' = 0$ and so $h_0 = h_0 \wedge s_0 \leqslant s_0$, a contradiction. Note that $t = \sup_{(P, \leqslant)} (\underline{c} \cup \{h_0 \wedge s_0'\}) = [\sup_{(P, \leqslant)} \underline{c}] \vee (h_0 \wedge s_0') = s_0 \vee (h_0 \wedge s_0')$. We have $s_0 M h_0$ and $s_0 M s_0'$. Clearly, the infimum $(s_0 \vee h_0) \wedge (s_0 \vee \vee s_0')$ exists. Hence, by Theorem 2.13 and Remark 2.15,

$$t = s_0 \vee h_0 = [\sup_{(P, \leqslant)} Q(\underline{c})] \vee h_0 = \sup_{(P, \leqslant)} (Q(\underline{c}) \cup \{h_0\}).$$

In summary, $\underline{c} \subset \underline{d}$ and $\underline{c} \neq \underline{d}$, contrary to the choice of $\underline{c}$. Thus $H = Q(\underline{c})$ and $\sup H = \sup_{(P, \leqslant)} \underline{c}$. It remains to use I.13.

4. (i). If $q \leqslant z_i$, then $(q^i)^i = (q' \wedge z_i)' \wedge z_i = (q \vee z_i') \wedge z_i =$

$= q \wedge z_i = q$ by Theorem 2.3. Similarly, if $a \leqslant z_i$, then $a \vee a^i =$ $= a \vee (a' \wedge z_i) = z_i$. The remaining axioms follow easily.

(ii). Without loss of generality we may assume that $n = 2$. Then $1 = z_1 \vee z_2$ and $p = p \wedge 1 = p \wedge (z_1 \vee z_2)$ for every $p \in P$. Since $p \wedge z_1 \leqslant z_1 \leqslant z_2' \leqslant (p \wedge z_2)'$, the supremum $s = (p \wedge z_1) \vee$ $\vee (p \wedge z_2)$ exists and, by Theorem 2.13, $s = p$. If $p = q_1 \vee q_2$ where $q_1 \leqslant z_1$ and $q_2 \leqslant z_2$, then $z_1 \wedge p = z_1 \wedge (q_1 \vee q_2)$ exists. Also the element $(z_1 \wedge q_1) \vee (z_1 \wedge q_2)$ exists, since $z_1 \wedge q_2 \leqslant$ $\leqslant z_1 \wedge z_2 \leqslant z_1 \wedge z_1' = 0$. Hence, by Theorem 2.13, $z_1 \wedge p = z_1 \wedge$ $\wedge q_1 = q_1$. Similarly, $q_2 = z_2 \wedge p$.

5. Let c be a common relative complement for x and y and let d^+ be the relative orthocomplement of d in $[c,1]$. Then d^+ is a common complement for x and y. In fact, let $z \in \{x, y\}$. Then $z \vee d^+ \geqslant z \vee c = d$ and so $z \vee d^+ \geqslant d \vee d^+ = 1$. Since $z \wedge d^+ \leqslant d \wedge d^+ = c$, $z \wedge d^+ \leqslant z \wedge c = 0$. Now it is sufficient to use (D').

6. It suffices to prove that the element $(Z \wedge a) \vee a'$ is a common complement for the elements $(Z' \vee a') \wedge a$, $(z' \vee b') \wedge z$. To this end write $s = z + Z$ and note that $Z \leqslant s \wedge z'$. Since $Z^+ =$ $= (s \wedge z') - Z = s \wedge z' \wedge Z' = s \wedge s' = 0$, $s \wedge z' = Z$. Similarly, $a = s \wedge b'$. Write

$$h = [(z' \vee b') \wedge z] \vee [(Z \wedge a) \vee a'] = \{[(z' \vee b') \wedge z] \vee a'\} \vee (Z \wedge a).$$

It is evident that $zCz' \vee b'$ and $a \leqslant b' \leqslant z' \vee b'$. We therefore have $h = [(z' \vee b' \vee a') \wedge (z \vee a')] \vee (Z \wedge a)$. Furthermore, because $a' \geqslant b$, we find that $b' \vee a' = 1$. Hence

$$h = (z \vee a') \vee (Z \wedge a) = (z \vee a') \vee [(s \wedge z') \wedge a] = (z \vee a') \vee (z' \wedge a) = 1.$$

Evidently, $(Z \wedge a) \vee a' = [(s \wedge z') \wedge a] \vee a' = (z' \wedge a) \vee a'$ and so

$$d = [(z' \vee b') \wedge z] \wedge [(Z \wedge a) \vee a'] = [(z' \vee b') \wedge z] \wedge [(z' \wedge a) \vee a'] =$$

$$= (z' \vee b') \wedge \{z \wedge [(z' \wedge a) \vee a']\}.$$

It is readily shown that $zCz' \wedge a$ and $z' \wedge aCa'$. Consequently, $d = (z' \vee b') \wedge z \wedge a'$. But $a = s \wedge b'$ and it follows that $z \wedge a' = z \wedge (s' \vee b) = (z \wedge s') \vee (z \wedge b)$, since $s'Cz$ and $s'Cb$. Here $z \wedge s' = z \wedge z' \wedge Z' = 0$. Thus $d = (z' \vee b') \wedge z \wedge b = 0$ and the conclusion follows from (D$'$).

7. Let $e = (x \vee v) - x$ and $f = (x \vee w) - x$. Obviously, by orthomodularity, $e \vee x = [(x \vee v) \wedge x'] \vee x = x \vee v$. Since $e \wedge x = 0$, it is clear that x is a common relative complement for v and e. By Exercise VIII;5, $v \sim e$. Similarly, $w \sim f$ and, therefore, $e \sim f$. From (D 2) we find that $x \vee v = x + e \sim x + f = x \vee w$.

8. Let $e = (a' \vee Z') \wedge a$, $f = a \wedge Z$, $E = b \wedge z$ and $F = (b' \vee z') \wedge b$. Evidently, $a = e + f$ and $b = E + F$. All that has to be shown is that $z \sim e + E$, as the remainder follows immediately from symmetry. Set $x = b \wedge z = E$, $v = (Z' \vee a') \wedge a = e$, $w = (z' \vee b') \wedge z$. Then $x \wedge v = (Z' \vee a') \wedge a \wedge b \wedge z \leqslant a \wedge b \leqslant b' \wedge b = 0$ and, in a similar manner, $x \wedge w = (z' \vee b') \wedge z \wedge b \wedge z = (z \wedge b)' \wedge (z \wedge b) = 0$. By Exercise VIII;6, $v \sim w$. Combining this with Exercise VIII;7 we can deduce easily $e + E = x + v \sim x + w$. From orthomodularity we then have $x + w = (b \wedge z) \vee [(b' \vee z') \wedge z] = z$.

9. Suppose we have $x = a + b \sim c + d$. By (B), there exist elements z, Z such that $x = a + b = z + Z$, $z \sim c$ and $Z \sim d$. Then by Exercise VIII;8 we see that there exist elements e, f, E, F satisfying $c \sim z \sim e + E$, $d \sim Z \sim f + F$, $a = e + f$ and $b = E + F$.

10. Let z be a relative complement of the element $x \vee y$ in the interval $[x, 1]$. Then $x \leqslant z$ and $z \vee y \geqslant x \vee y$ so that $z \vee y \geqslant (x \vee y) \vee z = 1$. At the same time, $z \wedge y \leqslant z \wedge (x \vee y) = x$

and, consequently, $z \wedge y \leqslant x \wedge y = 0$.

11. Let $a, b \in P$ be elements which are not orthogonal. Suppose $(a \vee b') \wedge b = 0$ and set $s = a \vee b'$, $t = b$. Then $s \geqslant t'$ and $s \wedge t = 0$. By orthomodularity, $a \leqslant a \vee b' = s = t' = b'$, a contradiction. Thus $x = (a \vee b') \wedge b \neq 0$. Let $y = a'$. It is clear that $x \wedge y = (a \vee b') \wedge b \wedge a' = (a \vee b') \wedge (a \vee b')' = 0$. By Exercise VIII;10, there exists a complement z of the element y such that $x \leqslant z$. Since a is also a complement of $y = a'$, we can conclude from (D′) that $z \sim a$. Let x^+ be the relative orthocomplement of the element x in the interval $[0, z]$. It follows that $a \sim z = x + x^+$. Hence, by the axiom (B), there exist elements $c, D \in P$ such that $a = c + D$, $c \sim x$ and $D \sim x^+$. It is evident that $b \geqslant (a \vee b') \wedge b = x \sim c \leqslant a$.

12. (i). Suppose $a < b$. Then $a \sim b = a + a^+$ where $a^+ = b - a$. Hence, by (B), there exist elements z, Z such that $a = z + Z$, $z \sim a$ and $Z \sim a^+$. If $Z = 0$, then $a^+ \sim 0$ and, by (D 1), $a^+ = 0$. It follows that $b = a + a^+ = a$, a contradiction. If $z = a$, then $Z \leqslant z + Z = a = z$ and $Z \leqslant z'$ so that $Z = 0$ which is not possible. Thus $z < a < b$ and $z \sim a$. Clearly, it is always possible to repeat this proces and this can only happen when there is an infinite decreasing sequence of elements in `P. This is a contradiction, because we have assumed that `P is finite.

(ii). The assertion follows from (i) by the definition of an infinite element.

13. The lattice $`L_{14}$ is shown in Figure 65. It is a dimension poset in the sense of Šerstnev under the equivalence relation defined by its classes $\{0\}$, $\{a,b,c,d,e,f\}$, $\{a',b',c',d',e',f'\}$, $\{1\}$.

Suppose an equivalence relation $\sim$ defined on L_{14} makes $`L_{14}$ into a dimension lattice in the sense of Loomis. Note that d is a common complement for a and b'. Hence, by (D´), we have $a \sim b'$, which contradicts Exercise VIII;12.

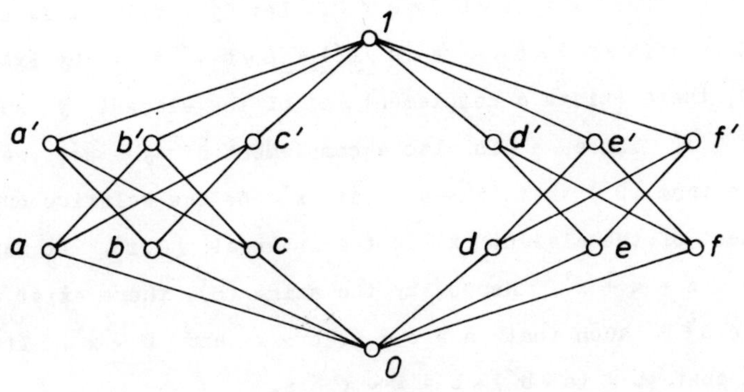

Fig. 65

14. (i)-(iii). Obvious. (iv). Let $D(a) = \{x \in P;\ x \lesssim a\}$. We shall prove that it is a hereditary set. Suppose that $y \sim Y \leqslant \leqslant x \sim X \leqslant a$. Let $Y^+ = x - Y$ so that $X \sim x = Y + Y^+$. By Remark 3.1, there exist elements e, f such that $e \sim Y$, $f \sim Y^+$ and $X = e + f$. Since $y \sim Y \sim e$ and $e \leqslant X \leqslant a$, $y \lesssim a$, i.e., $y \in D(a)$. From Theorem 3.9 we conclude that the supremum $s = = \sup D(a)$ exists and that $s \in I(`P)$. Evidently, $a \in D(a)$ and so $a \leqslant s$. Thus, by the definition of $|a|$, $|a| \leqslant s$. Now, let $x \in D(a)$. Then $x \sim X \leqslant a \leqslant |a|$ and, therefore, $x \lesssim |a|$. Since $|a| \in I(`P)$, Theorem 3.7 yields $x \leqslant |a|$. Hence, $|a|$ is an upper bound of the set $D(a)$ and it follows that $s \leqslant |a|$. In summary, $s = |a|$.

(v). Let $D(b) = \{y \in P;\ y \lesssim b\}$ and $x \in D(a)$. Then $x \sim X \leqslant a$. Let $X^+ = a - X$. From $b \sim a = X + X^+$ and Remark 3.1 we infer that there exist elements e,f such that $X \sim e$, $X^+ \sim f$ and $b = e + f$. Using $x \sim X \sim e \leqslant b$, we get $x \lesssim b$. Consequently, $D(a) \subset D(b)$ and, by symmetry, $D(a) = D(b)$. Thus, by (iv), $|a| = \sup D(a) = \sup D(b) = |b|$.

15. Suppose $E_k^\perp \neq \emptyset$ and choose $y \in E_k^\perp$. If $u \in E_k$, then $u \perp y$. Hence there exists E_m such that $\{u,y\} \subset E_m$. The set $N = E_k \cup \cup \{y\}$ is orthogonal and $N \subset E_k \cup E_m$. We therefore have from (M 2) that there exists E_s with $E_k \subset E_k \cup \{y\} \subset E_s$ and it follows from (M 1) that $y \in E_k = E_s$. Consequently, $y \in E_k^\perp \cap E_k = \emptyset$, a contradiction. Thus $E_k^\perp = \emptyset$ and $E_k^{\perp\perp} = H$. Finally, $\emptyset^{\perp\perp} = (E_k^\perp)^{\perp\perp} = E_k^\perp = \emptyset$.

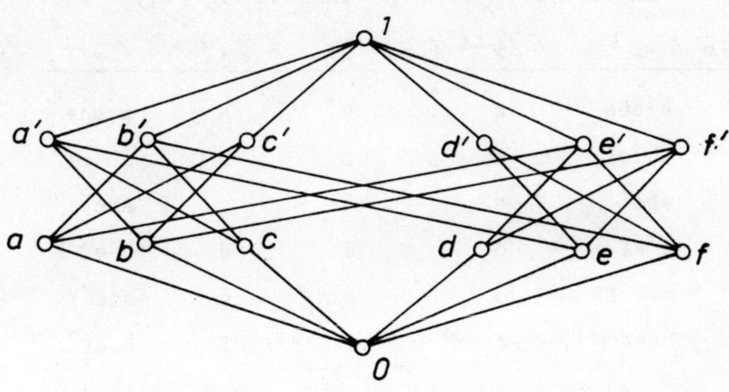

Fig. 66

16. (i). The poset `P_{14} is shown in Figure 66. We have that $s = f' > b = t'$ and $s \wedge t = f' \wedge b' = 0$. Thus it is not an orthomodular poset.

(ii). Let $G_{12} = G(`P_{14})$ be the graph of `$P_{14}$ indicated in Figure 67. It is not difficult to show that `P_{14} is isomorphic to the orthologic of the graph G_{12}. The sets $E_1 = \{a,a'\}$, $E_2 = \{a,b,c\}$, $E_3 = \{a,e\}$, $E_4 = \{b,b'\}$, $E_5 = \{b,f\}$, $E_6 = \{c,c'\}$,

$E_7 = \{d,e,f\}$, $E_8 = \{d,d'\}$, $E_9 = \{e,e'\}$, $E_{10} = \{f,f'\}$ represent the cliques and we can consider them as the elements of the corresponding manual M. The events of M are the sets E_i, the sets $F_1 = \{a,b\}$, $F_2 = \{a,c\}$, $F_3 = \{b,c\}$, $G_1 = \{d,e\}$, $G_2 = \{d,f\}$, $G_3 = \{e,f\}$, all the one-element subsets of the E_i's and the empty set $\emptyset$. By Exercise VIII;15, $E_i^{\perp\perp} = H = \bigcup M$ and $\emptyset^{\perp\perp} = \emptyset$. The results for $\{x\}^{\perp\perp}$ are tabulated below. (For brevity we write x instead of $\{x\}$, xy instead of $\{x,y\}$ etc.)

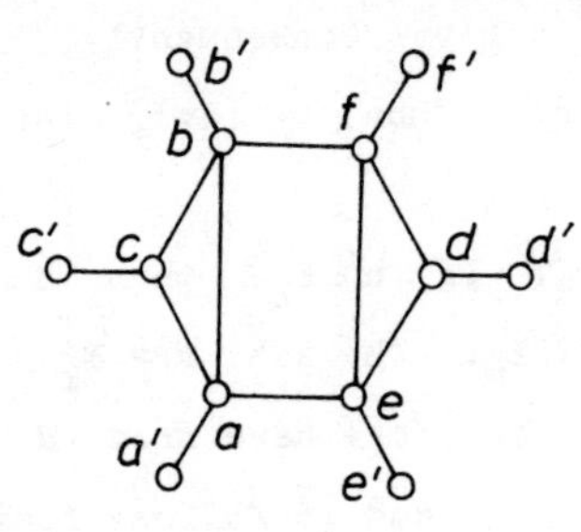

Fig. 67

x	$x^\perp$	$x^{\perp\perp}$	x	$x^\perp$	$x^{\perp\perp}$
a	a'bce	a	a'	a	a'bce
b	ab'cf	b	b'	b	ab'cf
c	abc'	c	c'	c	abc'
d	d'ef	d	d'	d	d'ef
e	ade'f	e	e'	e	ade'f
f	bdef	f	f'	f	bdef'

We also have $F_i^{\perp\perp} = p^\perp$ where $p \in \{a,b,c\} \setminus F_i$ and $G_i^{\perp\perp} = q^\perp$ where $q \in \{d,e,f\} \setminus G_i$. Now it is easy to construct the orthologic of the graph G_{12} and to prove directly that it is isomorphic to the orthologic $`P_{14}$.

17. By definition, the orthologic $(L(G(`P)), \subset, \emptyset, V(G(`P)), \perp)$ is affiliated with the manual $M(G(`P))$. Hence it suffices to use Theorem 4.10.

References

1. J. Abbott: Trends in Lattice Theory. Van Nostrand, New York, 1970.
2. N. J. Achieser, I. M. Glasmann: Theorie der linearen Operatoren im Hilbert-Raum. Akademie-Verlag, Berlin, 1954.
3. D. H. Adams: A note on a paper by P. D. Finch, J. Austral. Math. Soc. 11(1970), 63-66.
4. K. Baker: Equational classes of modular lattices, Pacific J. Math. 28(1969), 9-15.
5. E. G. Beltrametti, G. Cassinelli: The Logic of Quantum Mechanics. Addison-Wesley, Reading, 1981.
6. E. G. Beltrametti, G. Cassinelli: Quantum mechanics and p-adic numbers, Found. Phys. 2(1972), 1-7.
7. M. K. Bennett: States on orthomodular lattices, J. Natur. Sci. and Math. 8(1968), 47-52.
8. M. K. Bennett: A finite orthomodular lattice which does not admit a full set of states, SIAM Review 12(1970), 267-271.
9. L. Beran: Treillis sous-modulaires. Séminaire Dubreil-Pisot: Algèbre et théorie des nombres, 22e année, 1968/69 (exposé du 12 mai).
10. L. Beran: Amalgames des ensembles ordonnés et des treillis. Summer session on the theory of ordered sets and general algebra held at Cikháj 1969, University Press, Brno, 1969 (Lecture held on the 1st September 1969).
11. L. Beran: An approach to solvability in orthomodular lattices,

Acta Univ. Carolinae - Math. et Phys. 13(1972), 41-42.

12. L. Beran: On a construction of amalgamation, Acta Univ. Carolinae - Math. et Phys. 14(1973), 31-39.

13. L. Beran: Modularity in generalized orthomodular lattices, Comment. Math. Univ. Carolinae 15(1974), 189-193.

14. L. Beran: Groups and Lattices. SNTL, Praha, 1975 (Czech).

15. L. Beran: Reflection and coreflection in generalized orthomodular lattices, Acta Univ. Carolinae - Math. et Phys. 16(1975), 57-61.

16. L. Beran: On solvability of generalized orthomodular lattices, Pacific J. Math. 57(1975), 331-337.

17. L. Beran: Three identities for ortholattices, Notre Dame J. Formal Logic 17(1976), 251-252.

18. L. Beran: Ordered Sets. Mladá Fronta, Praha, 1978 (Czech); (Russian translation), Nauka, Moscow, 1981.

19. L. Beran: Über gewisse Sätze vom Foulis-Holland-Typ in Boole´schen Zwerchverbänden, J. reine angew. Math. 297(1978), 214-220.

20. L. Beran: On finitely generated orthomodular lattices, Math. Nachr. 88(1979), 129-139.

21. L. Beran: Formulas for orthomodular lattices, Studia Sci. Math. Hungar. 11(1979), 451-455.

22. L. Beran: Some applications of Boolean skew-lattices, Studia Sci. Math. Hungar. 14(1979), 183-188.

23. L. Beran: Central and exchange properties of orthomodular lattices, Math. Nachr. 97(1980), 247-251.

24. L. Beran: Extension of a theorem of Gudder and Schelp to polynomials of orthomodular lattices, Proc. Amer. Math. Soc. 81(1981), 518-520.

25. S. K. Berberian: Introduction to Hilbert Space. Oxford Univ. Press, New York, 1961.

26. J. H. Bevis, C. K. Martin: A note on matrices over orthomodular lattice, The Symposium on Orthomodular Lattices (Univ. of Mass. Press, 1969).

27. G. Birkhoff: Lattice Theory. 2nd edition. Amer. Math. Soc. Colloq. Publ., New York, 1948.

28. G. Birkhoff: Lattices in applied mathematics. Proc. Symp. Pure Math. Vol. II. Amer. Math. Soc., Providence, 1961.

29. G. Birkhoff: Lattice Theory. 3rd edition. Amer. Math. Soc. Colloq. Publ. vol. XXV. Providence, 1967.

30. G. Birkhoff, T. C. Bartee: Modern Applied Algebra. McGraw-Hill, New York, 1970.

31. G. Birkhoff, J. von Neumann: The logic of quantum mechanics, Ann. of Math. 37(1936), 823-843.

32. A. Böhm: The Rigged Hilbert Space and Quantum Mechanics. Lecture Notes in Physics 78(1978). Springer, Berlin, 1978.

33. C. C. Brown: On the finite measures on the closed subspaces of a Hilbert space, Proc. Amer. Math. Soc. 19(1968), 470-472.

34. J. Brown, J. Greechie: Reductions and level products of orthomodular posets, Notices Amer. Math. Soc. 21(1974), A-45.

35. G. Bruns: Free ortholattices, Canad. J. Math. 28(1976), 977-985.

36. G. Bruns, G. Kalmbach: Varieties of orthomodular lattices, Canad. J. Math. 23(1971), 802-810.

37. G. Bruns, G. Kalmbach: Varieties of orthomodular lattices. II, Canad. J. Math. 24(1972), 328-337.

38. G. Bruns, G. Kalmbach: Some remarks on free orthomodular lattices. Proc. Univ. of Houston, Lattice Theory Conf. Houston, 1973, 397-403.

39. G. Bruns, R. Greechie: Some finiteness conditions for orthomodular lattices, Canad. J. Math. 34(1982), 535-549.

40. G. Bruns, R. Greechie: Orthomodular lattices which can be covered by finitely many blocks, Canad. J. Math. 34(1982), 696-699.

41. J. Bub: The Interpretation of Quantum Mechanics. Reidel, Dordrecht, 1974.

42. L. A. Cammack: A new characterization of orthomodular partially ordered sets, Matematički Vesnik 12(27)(1975), 319-328.

43. D. E. Catlin: Irreducibility conditions on orthomodular lattices, J. Natur. Sci. and Math. 8(1968), 82-87.

44. G. Cattaneo, A. Manià: Abstract orthogonality and orthocomplementation, Proc. Cambr. Phil. Soc. 76(1974), 115-132.

45. E. Christensen: Measures on projections and physical states, Comm. Math. Phys. 86(1982), 529-538.

46. P. Crawley, R. P. Dilworth: Algebraic Theory of Lattices. Prentice-Hall, Englewood Cliffs, 1973.

47. J. O. Dacey: Orthomodular spaces and additive measurement, caribbean J. Sci. Math. 1(1969), 51-67.

48. A. S. Davydov: Quantum Mechanics. Pergamon Press, Oxford, 1965.

49. R. P. Dilworth: The structure of relatively complemented lattices, Ann. of Math. 51(1950), 348-359.

50. J. Dixmier: Applications $\natural$ dans les anneaux d´opérateurs, Compositio Math. 10(1952), 1-55.

51. J. Dixmier: Les algèbres d´opérateurs dans l´espace Hilbertien. Gauthier-Villars, Paris, 1957.

52. A. Dvurečenskij: Signed states on a logic, Math. Slovaca 28 (1978), 33-40.

53. A. Dvurečenskij, S. Pulmanová: On the sum of observables in

a logic, Math. Slovaca 30(1980), 393-399.

54. Gy. Fáy: Transitivity of implication in orthomodular lattices, Acta Sci. Math. (Szeged) 28(1967), 267-270.

55. P. D. Finch: Sasaki projections on orthocomplemented posets, Bull. Austral. Math. Soc. 1(1969), 319-324.

56. P. D. Finch: A transposition principle in orthomodular lattices, Bull. London Math. Soc. 2(1970), 49-52.

57. P. D. Finch: On orthomodular posets, J. Austral. Math. Soc. 11(1970), 57-62.

58. P. D. Finch: Orthogonality relations and orthomodularity, Bull. Austral. Math. Soc. 2(1970), 125-128.

59. P. D. Finch: Quantum logics as an implication algebra, Bull. Austral. Math. Soc. 2(1970), 101-106.

60. J. Flachsmeyer: Note on orthocomplemented posets, Proceedings of the Conference Topology and Measure III, Part 1,2 (Vittel/ Hiddensee, 1980), pp. 65-73, Wissensch. Beitr., Ernst-Moritz-Arndt. Univ., Greifswald, 1982.

61. J. Flachsmeyer: Note on orthocomplemented posets. II, Rend. Circ. Math. Palermo (2), Suppl. No. 2, 31(1982), 67-74.

62. D. J. Foulis: Conditions for the modularity of an orthomodular lattice, Pacific J. Math. 11(1961), 889-895.

63. D. J. Foulis: A note on orthomodular lattices, Portugal. Math. 21(1962), 65-72.

64. D. J. Foulis, C. H. Randall: An approach to empirical logic, Amer. Math. Monthly 77(1970), 363-374.

65. D. J. Foulis, C. H. Randall: Operational statistics. I. Basic concepts, J. Math. Phys. 13(1972), 1667-1675.

66. D. J. Foulis, C. H. Randall: Operational statistics. II. Manuals of operations and their logics, J. Math. Phys. 14 (1973), 1472-1480.

67. G. R. Gerelle: Representation of finite orthomodular posets, Notices Amer. Math. Soc. 22(1975), 720-06-9.

68. A. M. Gleason: Measures on the closed subspaces of a Hilbert space, J. Math. Mech. 6(1957), 885-893.

69. R. Godowski: Varieties of orthomodular lattices with a strongly full set of states, Demonstratio Math. 14(1981), 725-733.

70. R. Godowski: States on orthomodular lattices, Demonstratio Math. 15(1982), 817-822.

71. K. Gottfried: Quantum Mechanics. Vol. I. Benjamin, New York, 1966.

72. G. Grätzer, E. T. Schmidt: Ideals and congruence relations in lattices, Acta Math. Acad. Sci. Hungar. 9(1958),137-175.

73. G. Grätzer: Universal Algebra. Van Nostrand, Princeton, 1968.

74. G. Grätzer: Lattice Theory (First Concepts and Distributive Lattices). Freeman, San Francisco, 1971.

75. G. Grätzer: Lattice Theory (General Theory). Akademie-Verlag, Berlin, 1978.

76. R. J. Greechie: A class of orthomodular nonmodular lattices, Notices Amer. Math. Soc. 11(1964), 219.

77. R. J. Greechie: On the structure of finite orthomodular lattices, Notices Amer. Math. Soc. 13(1966), 66T-123.

78. R. J. Greechie: Hyper-irreducibility in an orthomodular lattice, J. Natur. Sci. and Math. 8(1968), 109-111.

79. R. J. Greechie: On the structure of orthomodular lattices satisfying the chain condition, Journ. Comb. Theory 4(1968), 210-218.

80. R. J. Greechie: A particular non-atomic orthomodular poset, Comm. Math. Phys. 14(1969), 326-328.

81. R. J. Greechie: An orthomodular poset with a full set of states not embeddable in Hilbert space, Caribbean J. Sci. Math. 1(1969), 15-26.

82. R. J. Greechie: Orthomodular lattices admitting no states, Journ. Comb. Theory 10(1971), 119-132.

83. R. J. Greechie: Any complete atomic orthomodular lattice with countably many atoms is a sublattice of one generated by three elements, J. Natur. Sci. and Math. 17(1977), 33-41.

84. R. J. Greechie: On generating distributive sublattices of orthomodular lattices, Proc. Amer. Math. Soc. 67(1977), 17-22.

85. R. J. Greechie: Finite groups as automorphism groups of orthocomplemented projective planes, J. Austral. Math. Soc. 25 (1978), 19-24.

86. S. Gudder: Axiomatic quantum mechanics and generalized probability theory; in Probabilistic Methods in Applied Mathematics, Vol. II, ed. A. Bharucha-Reid (Academic Press, New York, 1970), 53-129.

87. S. P. Gudder, R. H. Schelp: Coordinatization of orthocomplemented and orthomodular posets, Proc. Amer. Math. Soc. (1970), 229-237.

88. S. P. Gudder: Representations of groups as automorphisms on orthomodular lattices and posets, Canad. J. Math. 23(1971), 659-673.

89. S. P. Gudder: Partial algebraic structures associated with orthomodular posets, Pacific J. Math. 41(1972), 717-730.

90. J. Gunson: On the algebraic structure of quantum mechanics, Comm. Math. Phys. 6(1957), 262-285.

91. P. R. Halmos: Measure Theory. Van Nostrand, Princeton, 1950.

92. P. R. Halmos: Introduction to Hilbert Space. Chelsea Publ., New York, 1957.

93. P. R. Halmos: Lectures on Boolean algebras. Van Nostrand, Princeton, 1963.

94. I. Halperin: Dimensionality in reducible geometries, Ann. of Math. 40(1939), 581-599.

95. L. Herman, E. L. Marsden, R. Piziak: Implication connectives in orthomodular lattices. Notre Dame J. Formal Logic 16 (1975), 306-338.

96. S. S. Holland, Jr.: A Radon-Nikodym theorem in dimension lattices, Trans. Amer. Math. Soc. 108(1963), 66-87.

97. S. S. Holland, Jr.: Distributivity and perspectivity in orthomodular lattices, Trans. Amer. Math. Soc. 112(1964), 330-343.

98. S. S. Holland, Jr.: Isomorphisms between interval sublattices of an orthomodular lattice, Hiroshima Math. J. 3(1973), 227-241.

99. L. Iturrioz: A simple proof of a characterization of complete orthocomplemented lattices, Bull. London Math. Soc. 14(1982), 542-544.

100. T. Iwamura: On continuous geometries I, Japan J. Math. 19 (1944), 57-71.

101. T. Iwamura: On continuous geometries II, J. Math. Soc. Japan 2(1950), 148-164.

102. M. F. Janowitz: A note on generalized orthomodular lattices, J. Natur. Sci. and Math. 8(1968), 89-94.

103. J. M. Jauch: Foundations of Quantum Mechanics. Addison-Wesley, Reading, 1968.

104. J. M. Jauch, C. Piron: On the structure of quantal proposition systems, Helv. Phys. Acta 42(1969), 842-848.

105. B. Jeffcott: The center of an orthologic, J. Symbolic Logic 37(1972), 641-645.

106. B. Jeffcott: Decomposable orthologics, Notre Dame J. Formal Logic 16(1975), 329-338.

107. J. Ježek, V. Slavík: Primitive lattices, Czechoslovak Math. J. 29(104)(1979), 595-634.

108. T. F. Jordan: Linear Operators for Quantum Mechanics. Wiley, New York, 1969.

109. S. Kakutani, G. W. Mackey: Ring and lattice characterizations of complex Hilbert space, Bull. Amer. Math. Soc. 52(1946), 727-733.

110. V. V. Kalinin: Orthomodular dimension posets, Algebra i logika 15(1976), 535-537 (Russian).

111. G. Kalmbach: Orthomodular Lattices. Academic Press, London, 1983.

112. G. Kalmbach: Orthomodular lattices do not satisfy any special lattice equation, Arch. Math. (Basel) 28(1977), 7-8.

113. G. Kalmbach: Orthomodular logic. Z. Math. Logik Grundlagen Math. 20(1974), 395-406.

114. I. Kaplansky: Any orthocomplemented complete modular lattice is a continuous geometry, Ann. of Math. 61(1955), 524-541.

115. T. Katriňák: Notes on Stone lattices, Mat. Časopis Sloven. Akad. Vied 17(1967), 20-37 (Russian).

116. T. Katriňák: Eine Charakterisierung der fast schwach modularen Verbände, Math. Z. 114(1970), 49-58.

117. F. Katrnoška: On the representation of orthocomplemented posets, Comment. Math. Univ. Carolinae 23(1982), 489-498.

118. J. Klukowski: On Boolean orthomodular posets, Demonstratio Math. 8(1975), 5-14.

119. J. Klukowski: On the representation of Boolean orthomodular partially ordered sets, Demonstration Math. 8(1975), 405-423.

120. A. N. Kolmogorov: Foundations of the Theory of Probability. 2nd ed. Chelsea Publ., New York, 1956.

121. D. Krausser: On orthomodular amalgamations of Boolean algebras, Arch. Math. (Basel) 39(1982), 92-96.

122. H. Kröger: Zwerch-Assoziativität und verbandsähnliche Algebren, Bayer. Akad. Wiss. Math.-Natur. Kl. S.-B. 1973, 23-48.

123. H. Kröger: Das Assoziativgesetz als Kommutativitätsaxiom in Boole´schen Zwerchverbänden, J. reine angew. Math. 285 (1976), 53-58.

124. H. Kröger: One-sided ideals in lattices, in Bericht Nr. 7601, Institut für Informatik und Praktische Math., Universität Kiel, 1976.

125. H. Kröger: Absorptionsaxiome in verallgemeinerten Verbänden, Proceedings of the Klagenfurt Conference, 1978, 164-176.

126. H. Kröger: Assoziativähnliche Gesetze, Publ. Inst. Math. 23 (37)(1978), 129-135.

127. H. Kröger: Ein Assoziativitätskriterium vom Foulis-Holland-Typ, J. reine angew. Math. 298(1979), 196-198.

128. A. G. Kurosh: Lectures on General Algebra. GIFML, Moscow, 1962 (Russian).

129. S. Kyuno: An inductive algorithm to construct finite lattices, Math. Comp. 33(1979), 409-421.

130. L. H. Loomis: The lattice theoretic background of the dimension theory of operator algebras, Mem. Amer. Math. Soc. No. 18(1955), 1-36.

131. G. W. Mackey: Quantum mechanics and Hilbert space, Amer. Math. Monthly 64(1957), 45-57.

132. G. W. Mackey: Mathematical Foundations of Quantum Mechanics. Benjamin, New York, 1963.

133. M. D. MacLaren: Nearly modular orthocomplemented lattices, Trans. Amer. Math. Soc. 114(1965), 401-416.

134. M. D. MacLaren: Notes on axioms for quantum mechanics. Argonne National Laboratory Report, ANL-7065, 1965.

135. M. J. Mączyński: Hilbert space formalism of quantum mechanics without the Hilbert space axiom, Reports on math. phys. 3 (1972), 209-219.

136. M. J. Mączyński: On a functional representation of the lattice of projections on a Hilbert space, Studia Math. 47(1973), 253-259.

137. M. J. Mączyński: Functional properties of quantum logics, Internat. J. Theoret. Phys. 11(1974), 149-156.

138. M. J. Mączyński: On a lattice characterization of Hilbert spaces. Colloq. Math. 31(1974), 243-248.

139. M. J. Mączyński: Orthomodularity and lattice characterization of Hilbert spaces. Bull. Acad. Polon. Sci. Sér. Sci. Math. Astronom. Phys. 24(1976), 481-484.

140. M. J. Mączyński, T. Traczyk: A characterization of orthomodular partially ordered sets admitting a full set of states, Bull. Acad. Polon. Sci. Sér. Sci. Math. Astronom. Phys. 21 (1973), 3-8.

141. M. J. Mączyński, T. Traczyk: Some representations of orthomodular and similar posets, Acta Fac. Rerum Natur. Univ. Comenian, Special Number 1975, 25-28.

142. F. Maeda: Kontinuerliche Geometrien. Springer, Berlin, 1950.

143. F. Maeda, S. Maeda: Theory of Symmetric Lattices. Springer, Berlin, 1970.

144. S. Maeda: Dimension functions on certain general lattices, J. Sci. Hiroshima Univ., Ser. A 19(1955), 211-237.

145. S. Maeda: On conditions for the orthomodularity, Proc. Japan Acad. 42(1966), 247-251.

146. D. Maharam: The representation of abstract measure functions, Trans. Amer. Math. Soc. 65(1949), 279-330.

147. E. L. Marsden: The commutator and solvability in a generalized orthomodular lattice, Pacific J. Math. 33(1970), 357-361.

148. E. L. Marsden: Distribution in orthomodular lattices, Notices Amer. Math. Soc. 20(1973), A-51.

149. E. Merzbacher: Quantum Mechanics. Wiley, New York, 1961.

150. A. Messiah: Quantum Mechanics, Vols. I and II. North-Holland, Amsterdam, 1961, 1963.

151. P. D. Meyer: An orthomodular poset which does not admit a normed orthovaluation. Bull. Austral. Math. Soc. 3(1970), 163-170.

152. P. J. G. Meyer: On the structure of orthomodular posets, Discrete Math. 9(1974), 119-146.

153. M. K. Mukherjee: A note on the characterization of orthogonality and compatibility of elements of a quantum logic, Portugal. Math. 38(1979), 107-112.

154. F. J. Murray, J. von Neumann: On rings of operators, Ann. of Math. 37(1936), 116-229.

155. M. Nakamura: The permutability in a certain orthocomplemented lattice. Kōdai Math. Sem. Rep. 9(1957), 158-160.

156. H. Nakano, S. Romberger: Cluster lattices, Bull. Acad. Sci. Polon. Sér. Sci. Math. Astronom. Phys. 19(1971), 5-7.

157. E. A. Nordgren: The lattice of operator ranges of a von Neumann algebra, Indiana Univ. Math. J. 32(1983), 63-68.

158. W. Poguntke: Finitely generated ortholattices: The commutator

and some applications. Technische Hochschule Darmstadt, preprint Nr. 564 (1980).

159. W. Poguntke: On finitely generated simple complemented lattices, Canad. Math. Bull. 24(1981), 69-72.

160. J. C. T. Pool: Simultaneous observability and the logic of quantum mechanics. University of Iowa, Ph. D., 1963.

161. J. C. T. Pool: Baer ✱-semigroups and the logic of quantum mechanics, Comm. Math. Phys. 9(1968), 118-141.

162. J. C. T. Pool: Semimodularity and the logic of quantum mechanics. Comm. Math. Phys. 9(1968), 212-228.

163. E. Prugovečki: Quantum Mechanics in Hilbert Space. Academic Press, New York, 1970.

164. E. Prugovečki: Geometrization of quantum mechanics and the new interpretation of the scalar product in Hilbert space, Phys. Rev. Lett. 49(1982), 1065-1068.

165. S. Pulmannová: Superpositions of states and a representation theorem, Ann. Inst. Henri Poincaré, 32(1980), 351-360.

166. S. Pulmannová: Compatibility and partial compatibility in quantum logics, Ann. Inst. Henri Poincaré, 34(1981), 391-403.

167. A. Ramsay: Dimension theory in complete orthocomplemented weakly modular lattices. Trans. Amer. Math. Soc. 116(1965), 9-31.

168. C. H. Randall: A mathematical foundation for empirical science - with special reference to quantum theory. Ph. D. Thesis, Rensselaer Polytechnic Institute, New York, 1966.

169. J. Sallantin: Informations pures sur un système de propositions, C. R. Acad. Sc. Paris 275(1972), A-65 - A-68.

170. L. Ya. Savel´ev: Measures on ortholattices, Dokl. Akad. Nauk SSSR 264(1982), 1091-1094 (Russian).

171. E. Scheibe: The Logical Analysis of Quantum Mechanics, Pergamon Press, Oxford, 1973.

172. G. C. Schrag: Every finite group is the automorphism group of some finite orthomodular lattice, Proc. Amer. Math. Soc. 55(1976), 243-249.

173. E. A. Schreiner: Modular pairs in orthomodular lattices, Pacific J. Math. 19(1966), 519-528.

174. D. Schweigert: Affine complete ortholattices, Proc. Amer. Math. Soc. 67(1977), 198-200.

175. I. E. Segal: A non-commutative extension of abstract integration, Ann. of Math. 57(1953), 401-457.

176. A. N. Šerstnev: On Boolean logics, Učenye Zapiski Kazanskogo Universiteta 128(1968), 48-62.

177. J. Šimon: Opérations dérivées des treillis orthomodulaires. I, Acta Univ. Carolinae - Math. et Phys. 22(1981), 7-14.

178. J. Šimon: Opérations dérivés des treillis orthomodulaires. II, Acta Univ. Carolinae - Math. et Phys. 23(1982), 29-36.

179. G. Szász: Introduction to Lattice Theory. Academic Press, New York, 1963.

180. O. Tamaschke: Submoduläre Verbände, Math. Z. 74(1960), 186-190.

181. G. L. Trigg: Quantum Mechanics. Van Nostrand, Princeton, 1964.

182. D. M. Topping: Asymptoticity and semimodularity in projection lattices, Pacific J. Math. 20(1967), 317-325.

183. V. S. Varadarajan: Probability in physics and a theorem on simultaneous observability. Comm. Pure Appl. Math. 15(1962), 189-217.

184. V. S. Varadarajan: Geometry of Quantum Theory. Vols. 1 and 2. Van Nostrand, Princeton, 1968, 1970.

185. J. von Neumann: Mathematische Grundlagen der Quantenmechanik. Dover Publ., New York, 1943.

186. J. von Neumann: Continuous Geometry. Princeton University Press, Princeton, 1960.

187. J. von Neumann: Collected Works of John von Neumann. Vols. I and IV. Pergamon Press, Oxford, 1961.

188. R. Wille: Primitive subsets of lattices, Algebra Universalis 2(1972), 95-98.

189. N. Zierler: On the lattice of closed subspaces of Hilbert space, Pacific J. Math. 19(1966), 583-586.

190. N. Zierler, M. Schlessinger: Boolean embeddings of orthomodular sets and quantum logic, Duke Math. J. 32(1965), 251-262.

Subject Index